Springer Water

The book series Springer Water comprises a broad portfolio of multi- and interdisciplinary scientific books, aiming at researchers, students, and everyone interested in water-related science. The series includes peer-reviewed monographs, edited volumes, textbooks, and conference proceedings. Its volumes combine all kinds of water-related research areas, such as: the movement, distribution and quality of freshwater; water resources; the quality and pollution of water and its influence on health; the water industry including drinking water, wastewater, and desalination services and technologies; water history; as well as water management and the governmental, political, developmental, and ethical aspects of water.

Nurudeen Abiola Oladoja ·
Emmanuel Iyayi Unuabonah
Editors

Strategic Management of Wastewater from Intensive Rural Industries

Springer

Editors
Nurudeen Abiola Oladoja
Department of Chemical Sciences
Adekunle Ajasin University
Akungba-Akoko, Nigeria

Emmanuel Iyayi Unuabonah
African Centre of Excellence for Water
and Environment Research and Department
of Chemical Sciences
Redeemer's University Ede
Akoda, Nigeria

ISSN 2364-6934 ISSN 2364-8198 (electronic)
Springer Water
ISBN 978-3-031-90313-7 ISBN 978-3-031-90314-4 (eBook)
https://doi.org/10.1007/978-3-031-90314-4

This Springer imprint is published by the registered company Springer Nature Switzerland AG
The registered company address is: Gewerbestrasse 11, 6330 Cham, Switzerland

Contents

Impacts of Rural Intensive Industries Wastewater on Ecological Systems

Y. I. Bulu and N. A. Oladoja

Abstract Rural industries play a vital role in the economic development of a country, offering employment opportunities and providing a means for income generation while reducing migration to cities. However, these industries often lack proper wastewater management facilities, resulting in a significant consumption of large volumes of water and subsequent direct release of untreated wastewater into freshwater bodies or the soil, thereby releasing toxic substances into the environment with its attendant costs. The wastewater released is often laden with heavy metals, nutrients, and organic compounds with high Biochemical Oxygen Demand (BOD) and Chemical Oxygen Demand (COD). This review therefore examines the impact of releasing such wastewater into freshwater on the ecological structure and function of the environment without prior treatment.

1 Introduction

A rural industry in the context of this article refers to the small-scale economic activities and businesses that revolve around natural product extraction, primarily located in non-urban areas. These industries typically involve the production, processing, or manufacturing of goods and services in rural regions and they cover a wide range of sectors, including agriculture, forestry, mining, fishing, agribusiness, etc. These activities are often carried out far away from densely populated urban centres. Rural industries are essential for the economic development of rural areas, providing employment

Y. I. Bulu (✉)
Department of Plant Science and Biotechnology, Adekunle Ajasin University, Akungba-Akoko, Nigeria
e-mail: yetunde.bulu@aaua.edu.ng

N. A. Oladoja
Hydrochemistry Research Laboratory, Department of Chemical Sciences, Adekunle Ajasin University, Akungba-Akoko, Nigeria
e-mail: bioladoja@yahoo.com

N. A. Oladoja and E. I. Unuabonah (eds.), *Strategic Management of Wastewater from Intensive Rural Industries*, Springer Water,
https://doi.org/10.1007/978-3-031-90314-4_1

opportunities, income generation, and contributing to the overall well-being of rural communities as well as reducing the rate of migration to urban centres.

Rural industries have simple production techniques and can vary significantly from one region to another based on local conditions and resources. They include textile and dye, brick making, meat processing, aquaculture etc. The environmental cost of rural industries is enormous because they often lack basic technology for waste management. They use and generate large volumes of water, which are usually discharged indiscriminately into local water bodies (Wang et al., 2008; Wheeler et al., 2000) or directly into the surrounding soil. The proclivity to generate large volumes of wastewater hinges on the use of substandard equipment and the absence of hazardous wastewater treatment facilities (Shen et al., 2005; Wang et al., 2008). Several reports have also pointed out that the lack of proper regulation and enforcement of environmental measures enables numerous small rural businesses to operate without adequate waste management facilities (Lin, 1997; Swanson et al., 2001).

Water pollution and consumption by rural industries are related to the type of industrial activity. However, the major water polluters include food processing, paper and pulp milling, chemical manufacturing, metal casting, and brick making industries. These industries use large volumes of water and produce large quantities of wastewater that contain high concentrations of nutrients (nitrogen, phosphorus), organic compounds (phenolic compounds, carboxylic acids etc.) heavy metals (lead, cadmium, mercury, etc.), and other pollutants that may alter the quality of water and useful services provided by the environment to man. Wastewater generated by rural industries proves highly challenging to monitor and regulate because they are developed and are situated in a decentralized manner with many operating without the awareness or endorsement of local environmental authorities.

2 Wastewater from Selected Rural Industries

2.1 Food Processing

The Food and Agricultural Organization (FAO) of the United Nations estimated that the global water withdrawal grew 1.7 times faster than the population over the last century. This intensified sustainability concerns over the rising demands for water in agriculture, industry, and households. The United Nations Department of Economic and Social Affairs projected that the global population will reach 8.4–8.6 billion by the year 2030. It has been estimated that livestock industries consume 8% of the global water supply (Schlink et al., 2010), and the freshwater requirement for food and livestock production is projected to increase by 65% within the next 30 years (Venugopal & Sasidharan, 2021). The processing of food results in the release of voluminous amounts of wastewater, which are rich in nutrients, organic carbon, nitrogenous organics, inorganics, suspended and dissolved solids, and high biochemical and chemical oxygen demands.

2.1.1 Fisheries and Aquaculture Industries

The seafood industry comprises of both capture fisheries and aquaculture that provide consumer needs. The marine seafood includes finfish (i.e., pollock, tuna, herring, mackerel, whiting, and others), crustaceans (e.g., shrimp, krill, crab, and lobster), bivalves (e.g., mussels, oysters, clams, and scallops), cephalopods (e.g., squid and cuttlefish). The major items of freshwater aquaculture include carp, tilapia and shellfish species. The industry processes about 78% of the total harvest into diverse products such as chilled, frozen, smoked, dried, fermented or marinated or other items, which are traded in diverse markets (Venugopal, 2006). Declining wild fish stocks and continually increasing seafood demand have largely contributed to the prevalence of aquaculture farming, which has become a major pillar in food security around the world, with a global rise of 527% between 1990 and 2018 (Choudhury et al., 2022). Currently, land-based recirculating aquaculture systems, though highly intensive, are the most sustainable fish farming technique.

It is estimated that about 8% of seafood production is wasted annually. According to Coppola et al. (2021) about two-thirds of the total amount of fish harvested is discarded as waste, creating huge economic and environmental concerns. The amount of water used for aquaculture is abundant, thereby generating massive amounts of wastewater (Ghaly et al., 2005), which contains a variety of pollutants that are high in organic matter in the form of total suspended solids (TSS), fats, oils, and grease (FOG).

Wastewater from fish farms is often rich in organic materials, including unconsumed feed, faecal matter, shellfish drop-off, and other materials (Venugopal & Sasidharan, 2021). Uneaten feed pellets and faecal droppings lead to the buildup of suspended and dissolved substances in the water (Akinwole et al., 2016). The decomposition of these wastes increases the activity of aerobic bacteria, thereby causing an increase in the Chemical Oxygen Demand (COD) and Biological Oxygen Demand (BOD), and subsequently depletes the dissolved oxygen concentration in the water.

The excessive feed pellets and fish faecal droppings also contribute to nutrient accumulation in the wastewater. Since fishery wastes, unlike municipal wastes, are very rich in protein, rapid putrefaction occurs. Fish feeds contain high quantities of nitrogen and phosphorus, but less than half are retained within the fish's body (Piedrahita, 2003). Therefore, a large portion is deposited in the wastewater, which is eventually discharged into the environment. Other sources of nutrients in the wastewater are blood, bone, manure, detergents, and sanitizing compounds. Thus, the excess nutrients are available in the form of nitrogenous compounds (ammonia, nitrite and nitrate, proteins), phosphorus (almost solely as phosphates), and dissolved organic carbon, hormones, and biomasses (Cardoso et al., 2022; Dauda et al., 2019). The proteins encompass myofibrillar proteins, collagen, gelatin, enzymes, soluble peptides and volatile amines, which remain in soluble, colloidal or particulate forms. The major compositions of dissolved organic carbon in aquaculture wastewater are humic-like substances, carbohydrates, protein-like substances, low molecular weight aldehydes, fulvic acids, phenols and organic peroxides (Mook et al., 2012; Mostofa et al., 2005).

Moreover, antibiotics such as oxytetracycline, chloramphenicol and oxolinic acid are commonly used in aquaculture against bacterial infection. Other chemicals are also used for disinfection and as antifouling agents. Aside from the discharge of wastewater during fish harvesting and processing (cleaning and packaging), aquaculture systems often discharge wastewater regularly into the environment during routine operations and maintenance of water quality parameters (temperature, dissolved oxygen content, alkalinity etc.) and cleaning of equipment and facilities. Unforeseen circumstances, such as equipment failure and weather events could also lead to release of wastewater into the environment.

2.1.2 Meat Processing Industries

The global demand for meat has continued to rise at an unprecedented rate. This rise is driven by factors such as increasing incomes, evolving consumer preferences in emerging economies, and cost efficiencies in meat production and pricing (Devine, 2003). This places meat processing at the forefront of rural economic activities, bridging the gap between local agricultural production and the global demand for high-quality meat products. In many tropical countries, livestock farming forms a substantial part of smallholder farming systems. In such regions, meat processing assumes a vital role within rural industries, serving as a cornerstone for local communities and contributing significantly to the agricultural economy.

The rural meat-processing industry involves establishments primarily dedicated to slaughtering various animal species, including cattle, hogs, sheep, lambs, and calves. This is done to obtain meat for sale or for various on-site purposes. Meat processing involves a sequence of activities, starting from animal slaughter, meat cutting, and meticulous inspection to ensure its safety for consumption, packaging, and further processing into products like sausages or lunchmeats. These products are then distributed to stores and made available to customers. In recent times, meat production has shifted from urban areas, near the consumers of meat products, to rural areas, nearer to the cattle and poultry producers (MacDonald & Ollinger, 2000). This relocation was mainly prompted by factors that include lower land and labour costs, less stringent environmental restrictions, and easier access to animals and lower labour costs in rural areas (Van Alfen, 2014).

Meat-processing industry has a considerable impact on water resources, consuming a substantial portion of the total freshwater used by the food and beverage industry (24%) and the agricultural sector (up to 29%). Consequently, the industry generates substantial volumes of wastewater from the slaughterhouse operations. This wastewater necessitates extensive treatment to ensure safe and sustainable discharge into the environment (Bustillo-Lecompte & Mehrvar, 2015). Most developing countries release the effluent from abattoirs untreated directly onto land or into waterbodies (Fearon et al., 2014; Olawuyi et al., 2020) (Fig. 1).

For hygienic purposes, the use of large amounts of water in slaughtering and cleaning caused the generation of large volumes of wastewater. The infiltration of this wastewater into groundwater is the main cause of apprehension, specifically due

Fig. 1 Butchers washing carcasses of slaughtered animals in a river (Olawuyi et al., 2020)

to the blood therein, which is the major dissolved pollutant in the wastewater. For example, the blood from a single slaughtered cow is posited to be equivalent to the daily effluent load of the total sewage produced by 50 people (Aniebo et al., 2009). In Gaza City where the daily capacity of the slaughterhouse is 190 heads of calves and 300 heads of sheep and goats, the cleaning process consumes approximately $100m^3$/day of water, with a daily wastewater generation of 66 m^3 (Husam & Nassar, 2019).

Slaughterhouse wastewater (SWW) has a large variation in pH, chemical oxygen demand (COD), biochemical oxygen demand (BOD_5), TSS, FOG and other parameters (Farzadkia et al., 2016; Kupusovic et al., 2007; Malollari et al., 2019; Valta et al., 2014). It is also characterized by an unpleasant odour, along with pathogenic and non-pathogenic microorganisms, detergents, and cleaning disinfectants (Bustillo-Lecompte & Mehrvar, 2015). The variations in the characteristics of slaughterhouse wastewater are caused by several factors such as production capacity, type, and weight of the animal, methods of transportation, animal receiving and holding, processing technology, amount of carcass, washing temperature, cleaning and sanitizing procedure, and labourers behaviour (Budiyono et al., 2011).

The wastewater is equally rich in nitrates (from blood), phosphate (Bustillo-Lecompte et al., 2014; Farzadkia et al., 2016; Kupusovic et al., 2007; Malollari et al., 2019), salts, heavy metals, bacteria, viruses and other microbes. The general characteristics of slaughterhouse wastewater (Bustillo-Lecompte & Mehrvar, 2015) are pH, 4.90–8.10; TSS, 270–6400 mg/L; TOC, 70–1200 mg/L, COD, 500–15,900 mg/L; BOD, 150–4635 mg/L, total phosphorus, 25–200 mg/L and total nitrogen, 50–841 mg/L. For example, effluent from Gaza slaughterhouse was characterized as follows; pH (7.1 ± 0.1), electrical conductivity (3300 ± 500 μs/m), dissolved oxygen (1.95 ± 0.30), BOD (2350 ± 100), Total Suspended Solid (TSS) (3500 ± 120), COD (4502.5 ± 170), Ammonia nitrogen (30 ± 7), and total Kjeldahl Nitrogen (154 $\pm$

12) and Orthophosphate (9 ± 4) mg/L (Husam & Nassar, 2019). The BOD and COD values from chicken slaughterhouse wastewater ranged between 1341–1821 ± 242.7 mg/L, and 3154.19–7719.3 ± 2282.69 mg/L respectively (Yaakob et al., 2018). Lee et al. (2010) affirmed that in Korea, raw piggery wastewater contains 11,533 mg/L of BOD, and 21,894 mg/L of COD.

Antibiotics, such as beta-lactams, macrolides, quinolones, aminoglycosides, tetracyclines, sulphonamides etc., are used in livestock production for both therapeutic and as growth promoters. These have been reported to increase the incidences of antibiotic residues, antimicrobial-resistant bacteria and antibiotic resistance genes (ARG) in abattoir wastewater (Foyle et al., 2023; Van Boeckel et al., 2015). The repeated use of low doses of antibiotics in animals, particularly for growth promotion and prevention, is linked to the rise and spread of antibiotic-resistant bacteria (ARBs) (Van Boeckel et al., 2015; You & Silbergeld, 2014). A strong connection is established between the prevalence of ARBs in animals and humans, as well as the levels of antibiotic use in animals and the occurrence of ARBs in both animal and human populations (Foyle et al., 2023; Van Boeckel et al., 2015).

The bacteria group known as the ESKAPE pathogens (*Enterococcus faecium, Staphylococcus aureus, Klebsiella pneumoniae, Acinetobacter baumannii, Pseudomonas aeruginosa,* Enterobacter species) and Extended-Spectrum Beta-Lactamase (ESBL)-producing *Escherichia coli* have been identified as pathogens of great concern because of their potentials to cause life-threatening infections and exhibition of high levels of resistance against multiple classes of commonly used antibiotics (Foyle et al., 2023; Mulani et al., 2019). Savin et al. (2020) reported that ESKAPE bacteria and ESBL-producing E. coli occurred in water samples from different processing stages of the wastewater treatment plants in two independent German pig slaughterhouses. Meanwhile, frequently detected antibiotic residues included doxycycline (7.5%), chlortetracycline (4.5%), sulfadiazine (3.0%), tetracycline (1.5%), and trimethoprim (1.5%). The antimicrobial residues detected were suggested to originate from the muscle, fat tissue, and intestines of the slaughtered animals, or be excreted as unmetabolized residue from faeces and urine (Savin et al., 2020).

Wastewater from slaughterhouses is prone to pathogens from the digestive tracts of slaughtered animals, washing and rinsing of intestinal contents and cleaning of used equipment. According to Mittal (2004), abattoir wastewater contains several million cfu/100 mL of total coliform, faecal coliform, and Streptococcus groups of bacteria. The presence of some non-pathogenic microbes suggests the potential presence of enteric pathogens such as *Salmonella* species and *Campylobacter jejuni* and of gastrointestinal parasites such as *Ascaris* sp., *Giardia lamblia, Cryptosporidium parvum,* and enteric viruses such as human adenoviruses, norovirus, rotavirus, and hepatitis A virus (HAV) etc. (Osuolale & Okoh, 2017). Campylobacter is usually a nonpathogenic component of the gastrointestinal microbiota in livestock such as cattle, pigs, sheep and poultry (the main reservoir of *C. jejuni*) (Skarp et al., 2016; Tang et al., 2020). Non-pathogenic or commensal *Escherichia coli* strains naturally inhabit the digestive systems of animals and are excreted in faeces (Jang et al., 2017; Ramos et al., 2020). The presence of *E. coli* in waters has traditionally served as an indicator of recent faecal contamination. However recent findings indicated their ability to survive and reproduce in non-enteric, natural environments (Jang et al., 2017). Salmonella is commonly found in the intestines of various warm-blooded animals, including healthy cattle, sheep, swine, chickens, ducks, and turkeys.

2.1.3 Dairy Industries

The production of livestock for meat and milk in developing countries is primarily one of the main agricultural strategies for raising living standards. Dairying, an integral component of livestock production, is widely practised worldwide and involves a multitude of farms, varying in size and orientation, from small-scale subsistence to market-oriented farms. Over time, dairy farming has evolved from being a mere family operation to a significant occupation for rural women that has now transformed into a specialized and well-established industry marked by mechanization across its various aspects.

Dairy farming involves the breeding, raising, and utilization of dairy animals, primarily cows, for the production of milk and its derived dairy products. Although cow milk is predominantly meant for human consumption, other essential dairy animals include buffalo, goats, sheep and camel (Chung et al., 2022). The dairy industry covers operations that transform raw milk into a range of products, which include pasteurized and sour milk, yoghurt, various types of cheese (hard, soft, and cottage cheese), cream, butter, ice cream, milk and whey powders, lactose, condensed milk, and a variety of desserts.

However, the dairy industry poses environmental challenges due to the generation of highly polluted wastewater, laden with milk constituents, such as casein, inorganic salts, nitrogen, phosphorus, lactose, and sanitizing agents used during cleaning. On average, the dairy industry produces 6–10 L of wastewater per litre of processed milk (Kolhe & Powar, 2011), with approximately 2% of the total milk processed being sent into drains (Munavalli & Saler, 2009). This wastewater is characterized by elevated levels of BOD, total suspended solids (TSS), fat, oil, and grease (FOG) (Kushwaha et al., 2011; Shivsharan et al., 2013). It also contains dissolved solids, suspended solids, chlorides, sulphates, and various compounds like albumin, casein, lactose, and milk fat, which are highly biodegradable and decompose rapidly, resulting in unpleasant odours. The decomposition of dairy waste further depletes dissolved oxygen levels in receiving streams, leading to anaerobic conditions and potential ecological harm (Ahmad et al., 2019; Raghunath et al., 2016).

Typically, the appearance of dairy wastewater is pale white, with an alkaline pH value range, but becomes acidic during the fermentation process, where the lactose in the milk is converted to lactic acid (Shete & Shinkar, 2013). The casein therein undergoes precipitation and decomposition, leading to the formation of a highly odorous black sludge (Kolhe et al., 2009). The suspended solids fraction of the wastewater is attributed to the gelatinous milk and curd fines, while the total suspended solids (TSS) are ascribed to the presence of organic and inorganic matters (Garcha et al., 2016; Shete & Shinkar, 2013).

2.2 Textile and Dye Industries

In terms of its output or production and employment, the textile industry is one of the oldest and largest industries in the world (Verma et al., 2012). The fundamental strength of this industry flows from its strong production base of a wide range of fibres/yarns from natural fibres like cotton, jute, silk, and wool to synthetic/man-made fibres like polyester, viscose, nylon, and acrylic (Khan & Malik, 2014). In Nigeria, studies showed that the textile industry had been a major employer of labour in the manufacturing sector and contributed immensely to the socio-economic and cultural development of the rural communities. This is because all raw materials were initially locally sourced. For instance, the cotton is planted and locally spun, and the dye used in dyeing the yarn is extracted from plants fetched from the farms.

The most important chemical constituent used in textile industries is dye. Historical records of the use of natural dyes extracted from plants and certain insects date back to 3500 BC. Adire, known as "tie and dye" in Yoruba culture, denotes an indigo-dyed textile traditionally crafted in the southwestern region of Nigeria by Yoruba women. This textile artistry encompasses a diverse array of resist-dyeing methodologies. The Adire industry plays a pivotal role in the socioeconomic landscape of southwestern Nigeria, particularly empowering women by providing substantial employment opportunities and income. Historically, the creation of Adire entailed the utilization of fermented Indigo plant leaves to create intricate patterns on locally spun and woven fabrics. Indigo dye is characterized by its unique blue hue and is derived from organic compounds. Traditionally, indigo was sourced from the leaves of some selected plants in the Indigofera genus, mainly the *Indigofera tinctoria*, primarily in Asia. In the southwestern Nigerian context, Yoruba women have historically relied on the aerial parts of the *Philenoptera cyanescens*, colloquially known as 'elu,' as their indigo dye source for Adire production, a practice dating back at least to the eleventh century.

Synthesis of natural dyes was a long and tedious process which also gave a limited and dull range of colours and low colour fastness when exposed to washing and sunlight. Therefore, contemporary practices have largely replaced natural indigo with synthetic indigo, along with other synthetic dyes and chemical additives. Currently, synthetic dyes such as acid, azo, basic, and mordant dyes, which offer advantages of cost-efficiency, enhanced brightness, superior colourfastness, and ease of application to textile materials are now favoured over naturally occurring dyes. However, it is crucial to note that many synthetic dyes contain chemical components such as heavy metals (Yaseen & Scholz, 2019) such as mercury, lead, chromium, copper, etc. toluene, and benzene (Tara et al., 2020; Zare et al., 2018). Most of these synthetic dyes exhibit non-biodegradable properties, rendering them environmentally persistent, resulting in long-term environmental impacts.

Textile dyes are highly persistent and are difficult to degrade because of the existence of complex aromatic structures in which electrons are delocalized, molecules with conjugated double bonds that are photocatalytic, and chemically stable (Maddhinni et al., 2006). Among the dyes, azo dyes are the largest group of dyes commonly used and frequently discharged into the environment (Zhao & Hardin, 2007). These dyes are characterized by the presence of one or more azo groups (–N = N–) linked to the –OH or $-NH_2$ type auxochrome groups, namely the monoazo, the diazo and the triazo. The presence of heterocyclic rings, which give its intense and bright shade also confer on it resistant to biodegradability (Sharma et al., 2021).

2.3 *Pulp and Paper Industry*

The pulp and paper industry is one of the largest industries with a rich history, dating back to approximately 105 AD in China. Paper and paper products (tissue and board) are typically made from cellulose pulp derived from wood, and lignocellulosic materials and are used for writing, printing, and packaging. While wood and agricultural waste are the primary sources for paper production, recycled waste paper is used for newsprint-grade paper production. Paper production involves several key steps: pulping (separating fibres from lignin), bleaching (reducing pulp colour by chemically oxidizing residual lignin), and finishing (papermaking) (Mboowa, 2021). At certain stages of the process, dyes, coating materials or preservatives may be introduced to achieve the desired quality of the paper or related products (Bajpai & Bajpai, 2018).

The industry has a significant environmental impact with substantial water consumption and the generation of large volumes of wastewater. It has been estimated that about 60,000–100,000 gallons of water are used for the production of a tonne of paper, with the corresponding discharge of 47,000 and 80,000 gallons of wastewater, containing chlorinated organic compounds (e.g., chloroform, chlorate, resin acids, chlorinated hydrocarbons, phenols, catechols, guaiacols, furans, dioxins, syringes, vanillin etc.) to water and soil (Badar & Farooqi, 2012). Therefore, the pulp and paper industry is rated as one of the largest industrial polluters of water, air and land (Sharma et al., 2022; Verma et al., 2012). The persistent dark brown colour of pulp and paper wastewater results from the presence of lignin and its derivatives (e.g., chlorolignin) during pulp bleaching, which involve the use of chlorine-based chemicals (Prasongsuk et al., 2009). The paper and pulp industry effluent are characterized by high pH, BOD, COD, chlorinated compounds measured as adsorbable organic halides (AOX), total dissolved solids, suspended solids mainly fibers, fatty acids, tannins, resin acids, lignin and its derivatives, chloride, phosphate, sulfate, nitrate, and several toxic mutagenic pollutants along with various heavy metals (Hossain & Ismail, 2015; Singh et al., 2021, 2022).

Moreover, the effluent contains various toxic chemicals and volatile organic compounds, such as terpenes, alcohols, phenols, methanol, acetone, chloroform, methyl ethyl ketone, surfactants, dyes and pigments, acids and alkaline solutions (Sharma et al., 2022). Consequently, the paper mill effluents pose several negative effects on receiving soil and aquatic ecosystems (flora and fauna). According to Yadav and Chandra (2018), the effluent from paper mill were characterized by an alkaline nature, heavy metals and several organic pollutants (2-methyl-4-keto-2-pentan-2-ol; 3,7-dioxa-2,8-disilanonane,2,2,8,8-tetramethyl5-[(trimethylsilyl)oxy]; 2-methyl-4-keto-2-trimethylsiloxypentane; 4-ethyl-2-methoxyphenol, ethyl-2-octynoate; cis-9-hexadecenoic acid and octadecenoic acid).

3 Impact of Wastewater Discharge from Rural Industries on the Environment

3.1 Water

Wastewater from rural industries usually contain high levels of organic compounds characterized by high concentrations of biochemical oxygen demand (BOD), chemical oxygen demand (COD), total organic carbon (TOC), and total suspended solids (TSS) and dissolved solids (TDS) (Table 1) mainly from urine, excrement, flesh, blood, tallow and mucosa, fat, grease, proteins (FOG) and plant residues. In addition to these are detergents and disinfectants used for cleaning activities (Bustillo-Lecompte & Mehrvar, 2015), soluble microbial products (SMPs) (i.e., collection of organic compounds released by cells into solution from substrate metabolism, biomass growth and biomass decay) (Soh et al., 2020), and trace harmful chemicals and recalcitrant organic chemicals that are resistant to biodegradation (Bustillo-Lecompte et al., 2014).

Elevated TSS levels increase water turbidity, reduce light penetration in receiving water bodies, and potentially affect aquatic life. Increased TSS can lead to flocculation and sinking of phytoplankton, reduced light transmission through the water, decreased photosynthesis with subsequent reduction in phytoplankton density, growth rates and productivity (Boyd & Boyd, 2020; Chapman et al., 2017). Ingestion of sediment particles with no nutritional value by zooplankton results in starvation (Chapman et al., 2017). *Acartia tonsa*, copepod zooplankton eaten by fishes, ingestion of phytoplankton decreased at increased sediment concentration ($t = -2.895$, $p = 0.004$) with a further decrease in egg production, resulting in a lower survival rate (Sew et al., 2018). The exposure of juvenile *Barbonymus schwanenfeldii* and *Oreochromis niloticus* fishes to 1000 mg/L of TSS significantly reduced (p-value $<$ 0.05) their survival rate to less than 70% when compared with the control ($\approx$ 89.2%) (Nyanti et al., 2018). The fish's low survival was suggested to be associated with respiratory failure caused by clogging and physical damage of the gills at the high TSS concentration (Nyanti et al., 2018).

Organic matter in wastewater acts as a substrate and provides nutrients to microbial growth. The biomass produces soluble microbial products and extracellular polymeric substances that are released during cell lysis, diffuse through the cell membrane, or excreted from some other purposes (Laspidou & Rittmann, 2002; Sampaio et al., 2021). These cellular products are ubiquitous in biological treatment and constitute the major COD of wastewater. While the organic matter may

Table 1 Composition of wastewater from selected rural industries

Parameters	Slaughterhouse (Bustillo-Lecompte & Mehrvar, 2015)	Dairy industry (Shivsharan et al., 2013)	Dairy industry (Srivastava et al., 2016)	Textile dye (Kehinde & Aziz, 2014)	Pulp and paper (Sharma et al., 2022)
TOC (mg/L)	70–1200				
BOD5 (mg/L)	150–4635	1200–1800	720–830	80–6000	7136–9364
COD (mg/L)	500–15,900	1900–2700	1320–1670	150–30,000	26,324–56,872
TN (mg/L)	50–841				174–249
TSS (mg/L)	270–6400	500–740	195–321	15–8000	124–248
TDS (mg/L)		900–1350	1580–1835	2900–3100	1354–2465
TS			1420–2156		
DO			0.9–1.5		
pH	4.90–8.10	7.2–8.8	8.6–9.6	6.95–11.8	8.03–8.07
TP (mg/L)	25–200		10–100		
Orto-PO4 (mg/L)	20–100				
K(mg/L)	0.01–100				21.01–34.17
Color (mg/L Pt scale)	175–400				

TOC—total organic carbon, BODs—biological oxygen demand, total nitrogen, TSS—total suspended solids, total dissolved solids, TS—total solids, DO—dissolved oxygen, TP—total phosphorus

influence the colour and odourodour of water, the consumption of oxygen during its breakdown by microorganisms causes hypoxia/anoxia, leading to such problems as reproduction, developmental abnormalities, and even death, in fish and other benthic organisms. This could also trigger biodiversity losses.

Suspended solids found in wastewater include protozoa, algae, bacterial floc and single cells, microbial waste products, and other miscellaneous debris. Dissolved organic matter (smaller than 0.45 μm) are typically cell fragments and macromolecules such as polysaccharides, proteins, lipids, nucleic acids, and vitamins. All proteins contain carbon, hydrogen, and oxygen, as well as a high and constant proportion of nitrogen (about 16%). Foul odours are produced during protein decomposition (Tchobanoglous & Burton, 1991). Proteins are degraded to produce free amino acids, which are further degraded to form volatile sulfur compounds (VSCs), the main odour-causing chemicals, such as hydrogen sulfide (H_2S), methanethiol (methyl mercaptan), dimethyl sulfide (DMS), dimethyl disulfide (DMDS), and dimethyl trisulfide (DMTS). For example, the degradation of methionine forms methanethiol, which is further oxidized to form DMDS (Higgins et al., 2006, 2008).

During the dyeing process in textile industries, a lot of the dye is lost and released into wastewater due to the inadequate process of dyeing and improper absorption of the dyes into the dyeing stuff. The wastewater generated contains chemicals that are toxic, carcinogenic, mutagenic or teratogenic to various organisms (Sarayu & Sandhya, 2012). The wastewater released is characterized by strong colour, high pH, high chemical oxygen demand (COD), and low biodegradability (Kehinde & Aziz, 2014; Yaseen & Scholz, 2019). The main pollutants include organic and inorganic chemicals such as sizing agents, surfactants, inhibitors, chlorine compounds, salts, dyeing waste, phosphate as well as dissolved, suspended, and total solids.

According to Kehinde and Aziz (2014), the composition and characteristics of textile dye depends on the country, mill, fabric types and the process involved such as the chemical applied and equipment used. Aside from reduction in water clarity, colour alters the thermal condition in the water as well as dissolved oxygen concentration, decreases light penetration (Ganapuram et al., 2015; Houser, 2006), and thus impairing the photosynthetic activities of aquatic flora, thereby affecting the food sources of aquatic organisms (Ahmaruzzaman & Gupta, 2011; Sharma et al., 2021). This in turn affects primary production (Kritzberg et al., 2020) with a consequent influence on the aquatic food webs, prey-predator interactions and ecosystems services of aquatic ecosystems, e.g., fish production, drinking water quality, and recreation services (Kritzberg et al., 2020).

According to Sandhya (2010), dyes are persistent environmental pollutants that can biomagnify within the food chains causing organisms at higher trophic levels to show higher levels of contamination than their prey. Contamination by dye inhibits the development of microalgae (Gita et al., 2018; Hu & Wu, 2001; Moorthy et al., 2021; Sharma et al., 2021) and causes a disturbance in the trophic transfer of energy and nutrients in aquatic ecosystems. The effect of Optilan Red on parameters, such as specific growth rate (SGR), protein, chlorophyll, carotenoid, and elemental composition of the microalgae *Chlorella vulgaris* was monitored for 96 h. Results showed that specific growth rate, protein, pigments and elemental composition such as carbon and sulfur of the microalgae were reduced significantly when the dye concentration increased. Maximum percentage reduction was observed for total carotenoid and chlorophyll at 50 ppm (Gita et al., 2018).

The growth of *Anabaena* sp.,nitrogen-fixing cyanobacteria, was inhibited by RP2B azo dye, with the degree of inhibition increasing from 50 to 81% with an increasing concentration of RP2B (5–50 mg/l) (Hu & Wu, 2001). This indicated that the azo dye is non-biodegradable as *Anabaena* could not utilize it as a nitrogen source. A study by Hernández-Zamora et al. (2016) showed that Congo red significantly inhibited the population growth of the microalgae Chlorella vulgaris after 96 h exposure. The dye, at concentrations of 3.11 mg/L (LC_{10}) and 13.58 the g/L (LC_{50}) affected females' population of *Ceriodaphnia dubia* with a 100% mortality by day 6 and 3, respectively. Although, some organisms exposed to 3.11 mg/L of the Congo red lived up to 6 days, yet they did not reach sexual maturity (Hernández-Zamora et al., 2016).

Sulfur dioxide, a preservative used in meat processing in some abattoirs, raises sulfur levels in the liquid waste and contributes to the obnoxious odour of the abattoir environment. The high FOG consists of the most objectionable components of the effluents, the decomposition of which generates unpleasant odours. FOG concentrations in the abattoir wastewater cause severe problems, due to their insolubility and slow rate of degradation, forming scums and coating surfaces which impede the exchange of air and water, thereby increasing the oxygen demand.

The release of wastewater into water bodies impacts water quality since they contain highly biodegradable organic compounds that tend to reduce DO, which can make aquatic life perish due to hypoxic and, in extreme cases, anoxic conditions. Eutrophication, an excessive growth of algae and macrophytes in the aquatic bodies, results from the introduction of macronutrients such as nitrogen (N) and phosphorus (P) from the wastewater. Subsequent algae and aquatic plants accumulation, biomass growth and decomposition (Bulu & Oladoja, 2023; Bulu & Saliu, 2021) results in depletion of DO, as bacteria decompose the dead aquatic flora biomass, increasing the pH and creating an anaerobic condition, with considerable impact on aquatic life (Chen et al., 2018; Cui et al., 2021).

Enrichment of water bodies with excess nutrients (N and P) leads to a shift in the composition of species from oligotrophic to eutrophic species. According to Sand-Jensen et al. (2017), five decades of eutrophication resulted in a decline in macrophyte richness from 36 species to 12, causing the disappearance of deep-rooted mosses, charophytes, and short angiosperms while tall angiosperms survived and pollution-resistant macroalgae thrived. Generally, eutrophic lakes are characterized by the presence of phytoplankton blue-green algae (cyanobacteria), some of which produce toxic compounds. The presence of toxic phytoplankton-producing toxins (e.g., microcystin) can have deleterious effects on the entire ecosystem, causing the death of large numbers of macroalgae, invertebrates, and fish, reducing the edibility of freshwater and sea foods (Anderson & Garrison, 1997). Certain species of diatom such as the genus *Pseudo-nitzschia* produce domoic acid (DA), a neurotoxin responsible for amnesic shellfish poisoning (Pulido, 2008; Sauvey et al., 2023),

Aside from the production of toxins, many fish kills result from hypoxic (low oxygen) or anoxic (no oxygen) conditions created by blooms, which depletes oxygen from the surrounding water by putrefying bacterial following the death of the alga mass. Due to high nutrient concentrations (maximum NO_3^-–N 103.1, NO_2^-–N 26.0, NH_4^+–N 104.0, and PO_4^{3-}–P 99.0 μmol/L) in the lower section of the Tapi estuary, there has been a persistent deficiency in dissolved oxygen, ranging from 2.0 to 5.0 mg O_2/L. This has created an anoxic/hypoxic condition that is responsible for fish kills in the estuary (Ram et al., 2014). In the southeast of India, a fish kill incident at Puducherry's Chunnambar backwater was attributed to high levels of ammonia (61.4 μM) and phosphate (6.2 μM) (Mishra et al., 2022). These levels indicated eutrophic conditions in the water, coupled with low dissolved oxygen levels (1.62 mg/L at station 1 and 2.4 mg/L at station 5), as well as high organic matter and bacterial loads in both the water and sediment (Mishra et al., 2022). Furthermore, the presence of algae blooms e.g., Pseudo-nitzschia cells, which occurred in the backwater and on the gills of the deceased fish indicated that the fish may have experienced suffocation due to gill clogging or irritation, ultimately leading to asphyxiation (Mishra et al., 2022).

In the pulp and paper industry, the pulp-washing process results in the deposition of materials such as cellulose, lignin, phenols, fatty acids, tannins, and resins. These materials released from the pulp mix together to form an alkaline, viscous, dark black-coloured waste known as black liquor (Chandra & Abhishek, 2011; Sharma & Singh, 2021), which may also impact water colour. Ratia et al. (2012) extensively documented the impacts of the pulp and paperboard industry in aquatic body. Their findings revealed that, despite advancements in technology over two decades leading to improved water quality recovery in a watercourse receiving effluents from a paper and mill industry, observable effects remained. Specifically, the increase in insect species richness was noted, however, morphological abnormalities persisted in the tracheal gills of caddis larvae, particularly Trichoptera, Hydropsychidae. This persistence suggests the continued presence of a mixture of harmful organic compounds, such as resin acids, alkylated PAHs, and chlorinated organic compounds historically released from the mills and potentially more recently via resuspension from the water-sediment (Ratia et al., 2012).

The low molecular weight (LMW) compounds e.g., chlorophenol, and other organochlorinated and bleaching compounds from pulp and paper due to their hydrophobicity nature and capability to penetrate cell membranes have been reported to bioaccumulate in the body fat of animals. The chlorinated organic compounds e.g., dioxins and furans, aside been listed as group 1 human carcinogens by the International Agency for Research on Cancer, can cause genetic changes in exposed organisms (Kanan & Samara, 2018; Mudhoo et al., 2013). This has resulted in a growing concern about the potential adverse effects of genotoxicants on aquatic biota and public health through the contamination of drinking water supplies, recreational water or edible organic species.

Effluents from textile and dye, pulp etc., industries are laden with heavy metals such as nickel, cobalt, copper, and chromium. Wastewater from food processing industries contains heavy metals, as certain metals are added as additives in animal feeds. Heavy metals found in abattoir liquid waste can affect the aquatic ecosystem in various ways. Commonly reported heavy metals include Cu, Zn, Fe, Mn, Cr, Ni, Cd, Hg, and Pb (Nandomah & Tetteh, 2023). The disturbances caused by the heavy metals in the aquatic ecosystem include, the death of aquatic organisms, acute/chronic toxicity, habitat destruction due to sedimentation, and alteration of water chemistry. These metallic cations can be assimilated by fish gills, oral inlet and dermal intake and get accumulated in their tissues at the release of such wastewater into water bodies. For example, the assessment of the time-dependent accumulation of cadmium in different tissues of the benthic fish, *Solea senegalensis*, showed an increase in cadmium concentration for all the tissues over time, in the following order: muscle (0.003 μg/g) < gills (0.34 μg/g) < liver (0.76 μg/g) < intestine (2.1 μg/g) (Pavlaki et al., 2021). The fish intestine showed the highest cadmium uptake, followed by the liver, where depuration was not detected, suggesting the presence of a storage compartment for this metal in the liver (Pavlaki et al., 2021).

The toxicity of heavy metals from rural industry wastewater varies due to the presence of other inorganic non-metallic compounds. For instance, the toxic effects of organometallic pollutants from paper mill effluent on stinging catfish (*Heteropneustes fossilis*) were observed as notable changes in the secondary lamellae, primary lamellae, mucous cell, epithelial cell, and chloride cell, while in sludge worm (*Tubifex tubifex*), reduction in haemoglobin content, whitening of rear body part, disintegration of body part was observed (Singh et al., 2021).

Heavy metals cause DNA damage and instigate genotoxicity in aquatic animals such as fishes, resulting in aberrant gene expression, DNA strand breakage, and improper DNA-protein interaction. Genotoxicity stimulates the biosynthesis of reactive oxygen species (ROS), which causes oxidative modifications of DNA. It has been noted that the exposure of plants to heavy metals may inhibit plant growth and biosynthetic pathways or the induction of reactive oxygen species (ROS) (Bulu & Oladoja, 2023). Arán et al. (2017) reported a 30% decrease in the synthesis of photosynthesis pigment in the leaves of *Limnobium laevigatum* (Humb. & Bonpl. ex Willd.) Heine after 21 days exposure to higher concentration of a mixture of Pb (1000 mg/L), Cr (1000 mg/L), Ni (500 mg/L), and Zn (100 mg/L). Das et al. (2014) reported a decrease in the root, shoot biomass, and leaf necrosis at 15 mg/L when *P. stratiotes* was exposed to Cd for 21 days. Therefore, at higher heavy metal concentrations, impairment of plant cell function occurs and sometimes death of plant tissues (Bulu & Oladoja, 2023).

3.2 Soil

Following the large volume of water used by the rural industries, generated wastewater is sometimes indiscriminately disposed on soil with various associated risks to soil quality. Such polluted water is mostly laden with high concentrations of salt, heavy metals, nutrients, toxic organic compounds, and organic matter. Out of the mineral resources required for plant growth, inorganic N and P are the two principal nutrients that limit the growth of plants (Ågren et al., 2012; Jiang et al., 2019) with N considered the most limiting nutrient for most terrestrial ecosystems, while phosphorus is most limiting in the aquatic environment (Vitousek & Howarth, 1991). Nitrogen is removed from soil by plant uptake, leaching as nitrate and released into the atmosphere by denitrifying organisms. Much of the nitrogen from animal wastes is present as urea or ammonium salts and the latter can be adsorbed onto negatively charged sites in soil. In aerobic conditions, the nitrification process converts ammonium to nitrate, while gaseous loss (denitrification) occurs in anaerobic conditions, such as in waterlogged soils or wetland sediments. However, the contribution of denitrification to nitrogen removal depends on the concentration, and probably the type of COD present in the irrigation water.

Excessive nitrogen uptake by plants has been noted to stimulate luxurious plant growth with succulent leaves and shoots, delayed fruit maturity, reduced fruit yield and quality, and increased plants susceptible to pests and diseases (Gilliam, 2006). Increased canopy formation by dense foliage decrease light at the soil level (Hautier et al., 2009). Excessive nitrogen content in the soil may shift plant communities towards composition typical of higher N availability such as; less of nitrogen fixing leguminous plants, nitrophilous and acid tolerant plants (Brunet et al., 1998; Hautier et al., 2009) (since excess nitrogen result in nitrate leaching and base cation depletion, likely causing acidification of the soil (Lucas et al., 2011) and associated loss of biodiversity with significant ripple effect on the food chain.

A considerable change in the belowground populations of bacteria, mycorrhizal fungi, and non-mycorrhizal fungi have been reported with an increase in N deposition in soil (Clark et al., 2017; Lilleskov et al., 2008). Soil microorganisms use energy and nutrients from organic matter (OM) for growth and reproduction. However, if the soil's OM of more than the microorganisms' requirement, nutrients become available for plant uptake. In nutrient-deficient soil, microorganisms either do not grow or extract nutrients from the soil (nutrient immobilization) while digesting the OM, temporarily reducing soil fertility and plant growth (Hamilton, 2012). The ratio of carbon to nitrogen, or C:N ratio, in organic matter determines how quickly nutrients are released. Materials with $C:N < 20$ release nutrients immediately upon addition, while materials with $C:N > 20$ immobilize soil nutrients for some time after addition (Hamilton, 2012).

As nitrate, nitrogen is highly mobile in soils and may leach into groundwater if loading rates are too high, leading to contamination and making the water unsafe for drinking. This is widespread in Europe and has occurred in several places in Australia (Bowmer & Laut, 1992) and New Zealand (Keeley & Quin, 1979). Grazing of irrigated land has been found to increase the rate of nitrate leaching since animals recycle 90% of ingested nitrogen to soils (Cooper & Russell, 1988).

Continuous and indiscriminate discharge of wastewater on soils could lead to soil salinization, characterized by the gradual accumulation of water-soluble salts, including potassium (K^+), magnesium (Mg^{2+}), calcium (Ca^{2+}), chloride (Cl^-), sulfate (SO_4^{2-}), carbonate (CO_3^{2-}), bicarbonate (HCO_3^-), and sodium (Na^+), particularly in the topsoil as a result of high rates of evaporation. Sodification, and sodium accumulation in the soil, in particular, have detrimental effects on soil structure. The excess sodium ions can displace other essential cations like calcium and magnesium from soil particles, leading to the breakdown of soil aggregates and the development of a compacted, impermeable layer. This impairs both water infiltration and drainage, resulting in reduced aeration and root growth. An increase in soil pH and elevated levels of exchangeable sodium percentage (ESP) in highly saline soil has been reported with a subsequent decrease in soil organic matter (Yang et al., 2020; Zhang et al., 2015). The decrease in soil organic matter has been linked to poor or no plant growth on the soil surface (Yang et al., 2020).

Understanding factors that influence soil microbial composition is essential since they drive nutrient cycles by affecting soil organic matter, soil aggregation, porosity and nutrient availability. Aside from pH, salinity is suggested as an important driver of soil microbial community composition and their global distribution (Auguet et al., 2010; Lozupone & Knight, 2007; Yang et al., 2020). A study by Yang et al. (2020)

discovered that high soil salinity increased the relative abundance of Gemmatimonadetes and Bacteroidetes, two phyla known to tolerate a wide range of salinity gradients, indicating a shift in the microbial community towards more salt-tolerant microbial species, while decreasing the abundance of Proteobacteria and Firmicutes. In another study (Zhang et al., 2019), soil salinity decreased the relative abundance of Planctomyces and Archangium, while increasing the relative abundance of Piscinibacter and Aquamicrobium, suggesting that soil salinization decreased and altered the distribution of soil bacteria that facilitate soil organic matter input.

Although, high soil salinity increased the relative abundance of the fungal phylum, Ascomycota, a significantly low richness and diversity of soil fungi in extreme salinity are reported by many studies (Yang et al., 2020; Zhang et al., 2019). These authors showed that the Nectriaceae and Plectosphaerellaceae fungi families were the most abundant microflora in non-saline soil, while salinization dramatically increased the relative abundance of Cladosporiaceae, Bulleribasidiaceae, and Aspergillaceae from non-saline to extreme saline soil. In addition, plant saprotrophs and litter saprotrophs were significantly lower in extremely saline soil than in non-saline soil (Yang & Sun, 2020), an indication that the rate of plant litter decomposition will be reduced in saline soil with a consequent reduction in soil organic matter. High soil salinity increased the relative abundance of Hydropisphaera, Arbuscular Mycorrhiza (AM), and Rhizophagus while decreasing the relative abundance of Trichoderma and the AM, Sclerocystis, Septoglomus and Glomus (Zhang et al., 2019).

Zhang et al. (2019) reported that the presence of high salt concentrations (exceeding 4 g/kg) in soil resulted in a notable decline in microbial biomass carbon and nitrogen, measuring 177.3 ± 2.8.1 mg/kg and 5.1 ± 0.9 mg/kg, respectively, when compared to the control condition, where salt content was less than 1 g/kg, with considerably higher levels of 288.8 ± 56.6 mg/kg and 23.7 ± 3.3 mg/kg for microbial carbon and nitrogen, respectively. The low soil microbial carbon could be attributed to the low soil organic carbon input due to scarce vegetation under high salt conditions and a low abundance of saprophytic organism availability in the soil.

Saline soil often impedes the regular physiological activity and productive potential of plants, particularly those that are sensitive to it (Kim et al., 2016; Plaut et al., 2013). This impediment results from hindrances to growth, reduced leaf area expansion, and disruptions in primary carbon metabolism, due to imposed ion toxicity, osmotic effect, water deficit, nutritional imbalance, and oxidative stress (Kim et al., 2016; Zhai et al., 2015). In a three-year study, Zhai et al. (2015) found that 5.5 dS/m saline treatment reduced the maximum leaf area index (LAI_m) and chlorophyll content of tomato with concomitant reduction in yield by 22.4~31.1%, 12.6~28.0% and 11.7~27.3%, respectively in 2012, 2013 and 2014. Miransari and Smith (2007) reported a decline of 5.3 and 29.7% in soybean yield at 640 and 1280 kg/ha NaCl, respectively. According to Kim et al. (2016), the yield of lettuce and Chinese cabbage in a greenhouse experiment were more affected by salinity than their morphological characteristics, such as the number of leaves, leaf length, and leaf width, with yield declining at 0.9 and 1.5 dS/m, respectively.

Pulp and paper mill effluent are reported to cause phytotoxicity and chromosomal aberration in plants. The effluent at 100% significantly inhibited the germination of green gram (*Phaseolus mungo*) (37.5%) and wheat (*Triticum aestivum*) (34%) while onion root tips showed abnormal metaphase, disturbed anaphase, anaphase bridge, chromosome laggards at the anaphase which may be linked to alteration of the spindle by toxic pollutants, sticky anaphase, c-mitosis, ring chromosome, and vacuolated nucleus (Yadav & Chandra, 2018).

Aniline, which is a precursor in many dyes is produced in the process of decolourization of dye wastewater and could impose long-term risks to the ecosystem (Rawat et al., 2016). Aniline is adsorbed easily to soil particles. Tao et al. (2017) reported that aniline decreased the mitotic index of wheat root-tip cells at test concentrations higher than 10 mg/L with an increase in the frequency of micronucleus and chromosomal aberrations at aniline concentrations, ranging between 5 and 100 mg/L. This was also followed by a corresponding decrease in root and shoot length, which were more pronounced at the 200 mg/L concentration. At this concentration (200 mg/L) there was a concomitant black rot of the roots, yellowing of seedlings and final death and decay of the plants (Tao et al., 2022). Tao et al. (2022) demonstrated that aniline at a concentration $\geq$ 25 mg/L significantly inhibited the germination rate, amylase activity and lipase activity during the germination of rice. Moreover, the study showed that aniline significantly caused a reduction in the length of the root and shoot of rice seedlings with blackened and rotted roots observed with increasing aniline concentration (Tao et al., 2022).

3.3 Human

The colour imparted by the dye is the first observable problem, as it reduces the transparency of the water and aesthetic value (Brillas & Martínez-Huitle, 2015). The organic substances in rural industry wastewater are released rapidly, reducing the amount of dissolved oxygen in receiving water bodies and leading to anaerobic conditions that release a foul odour. Additionally, these water bodies may become breeding grounds for mosquitoes and flies, increasing the risk of diseases like malaria, chikungunya, and dengue (Kumar & Desai, 2011).

The major concern of nitrate contamination arises from the species of ammonia, nitrite, atmospheric nitrogen and bounded forms of nitrogen. The presence of nitrate in high concentrations in surface and groundwater triggers several health problems. For instance, methemoglobinemia (blue baby syndrome) is a potentially fatal condition that occurs when the concentration of nitrate increases in the bloodstream of infants or adults in humans, resulting in the conversion of nitrates to nitrites. These nitrites bind with the haemoglobin (Fe^{2+}) of the blood cell, oxidizing it to ferric ion (Fe^{3+}), resulting in the formation of methemoglobin (MHb), a substance that does not enable oxygen transport to tissues, thereby causing asphyxia (lack of oxygen) and cyanosis of body tissues, resulting in the colour change of the skin to blue (Bulu & Saliu, 2021; Ludlow et al., 2019). Besides this, the formation of nitrosoamines from nitrite can give rise to cancers of the digestive tract since nitrosamines are the most efficacious carcinogens in mammals.

Due to their wide-ranging of toxic impact and carcinogenicity, wastewater containing aniline compound is classified as hazardous waste (Ahmed et al., 2010). The main toxic effects of aniline start with inducing methemoglobinemia by oxidizing the hemoglobin in the blood into MetHb (Kharapkar, 2022; Lee et al., 2013; Zhang et al., 2021) which impairs the delivery of oxygen to tissues. Aniline converts the Fe^{2+} in hemoglobin to Fe^{3+} (Kharapkar, 2022) which impairs its oxygen transport capacity. Insufficient delivery of oxygen and destruction of red blood cells may cause cardiopulmonary complaints with the development of ischemia, arrhythmia and shock. Aniline may also cause the destruction of red blood cells, which manifests as acute or delayed hemolytic anaemia.

Aniline exhibits significant toxic, carcinogenic and mutagenic effects on both wildlife and humans with long-term exposure suggested to increase the incidence of bladder or prostate cancer (Zhang et al., 2021) Moreover, it can persist in nature and accumulate in living organisms (Cui et al., 2017). Moreover, aniline generates more toxic secondary intermediates such as phenylhydroxylamine which could also impact the oxygen-carrying capacity of haemoglobin and induce tissue hypoxia, which will harm the nervous and cardiovascular systems (Zhang et al., 2021).

A further concern for the management of rural industry wastewater in humans is exposure to potentially pathogen-contaminated water. The diary wastewater includes a variety of microorganisms due to its highly dense organic nature. The predominant species in heterotrophic bacterial strains are *Acinetobacter* spp., *Aeromonas* spp., *Alcaligenes* spp., *Comamonas* spp., *Enterobacter* spp., *Flavobacterium* spp., *Klebsiella* spp., *Moraxella* spp., *Pseudomonas* spp., *Sphingomonas* spp., *Stenotrophomonas* spp., *Mycobacterium* spp., *Bacillus* spp., and *Nocardia* spp. These bacteria are described as opportunistic pathogens that may be discovered in heterotrophic plate count microbiota, including strains of *Pseudomonas aeruginosa* spp., *Acinetobacter* spp., *Aeromonas* spp., and *Klebsiella pneumoniae* spp. (Afsharnia et al., 2018; Bartram et al., 2003). According to Dungan et al. (2012), putative zoonotic pathogens in dairy wastewaters include *Campylobacter jejuni*, stx1- and eaeA-positive *Escherichia coli, Listeria onocytogenes, Mycobacterium avium* subsp. *paratuberculosis*, and *Salmonella enterica.* Campylobacteriosis or Campylobacter enteritis, is an infection of the intestine by Campylobacter bacteria and is characterized by acute diarrhea and abdominal pain. Similarly, Guillain-Barré syndrome is associated with Campylobacter infection (Finsterer, 2022; Scallan Walter et al., 2020), a condition where a person's immune system attacks their nerves, leading to muscle weakness and sometimes paralysis. The pathogenic *E. coli* includes the Shiga toxin-producing *E. coli* (STEC), enteropathogenic *E. coli* (EPEC), enterotoxigenic *E. coli* (ETEC), enteroaggregative *E. coli* (EAEC), diffusely adherent *E. coli* and enteroinvasive *E. coli*, including Shigella strains, all causing a variety of human diseases such as diarrhoea, haemolytic uraemic syndrome and haemorrhagic colitis (Kaper et al., 2004) leading to acute renal failure and often death. In humans, the presence of salmonella species can lead to gastroenteritis associated with vomiting, bloody diarrhoea and abdominal cramps. Human enteric viruses are causative agents of many non-bacterial gastrointestinal tract infections, respiratory infections, conjunctivitis, hepatitis and other serious infections such as meningitis, encephalitis and paralysis (Okoh et al., 2010).

Heavy metals are associated with various toxicity and the incidence of numerous diseases in humans, including Alzheimer's disease, and different forms of cancer (Bakulski et al., 2020; Huat et al., 2019; Vellingiri et al., 2022). Besides, unique metals show specific indications of their toxicity. For example, arsenic exposure causes black foot disease, hyperglycemia, hyperkeratosis, and immune system dysfunction (Dani & Walter, 2018; Garcia-Vargas & Cebrian, 2023).

Furthermore, intermediate byproducts of some azo dye compounds like benzidine and 2-naphthylamine are suggested to be more harmful and toxic than the original dye compound and responsible for a significant increase in the occurrence of cancer (Christie, 2007).

4 Conclusion

Rural industries are integral to the economic well-being of rural areas, providing employment and income to local communities. However, the environmental cost associated with these industries, particularly in terms of water pollution and consumption, poses significant challenges. The lack of proper water management facilities and technical know-how has led to the indiscriminate discharge of wastewater into water bodies and soil, resulting in serious contamination of local ecosystems, and threatening the sustainability of the rural communities. The large volume of wastewater generated by these industries is rich in nutrients, organic carbon, and various pollutants, which may alter the ecological balance of the environment with further effects on humans.

Aside from reducing the aesthetic values of environmental media, due to colour and foul smell, the continuous discharge of wastewater from rural industries, without prior treatment predisposes soil and water bodies to nutrient enrichment. This causes a shift in the co-existence and stability of the ecological system and long-term hazards like the accumulation of heavy metals and carcinogenic compounds. Accumulation of contaminants, such as heavy metals, biomagnifies and bioaccumulates in biotic organisms, thereby moving along the food chain.

The effect of rural industries wastewater also transcends humans, as significant health risks are associated with the consumption of contaminated water or use of contaminated soil. Two of such contaminants are elevated nitrate and aniline levels in groundwater, which pose significant health risks, such as methemoglobinemia, a condition that interferes with oxygen transport in the body, leading to oxygen deprivation and a bluish discolouration of the skin. Diverse pathogenic organisms that survive outside their hosts have also been identified in wastewater and linked to different diseases in humans.

References

Ågren, G. I., Wetterstedt, J. Å. M., & Billberger, M. F. K. (2012). Nutrient limitation on terrestrial plant growth–modeling the interaction between nitrogen and phosphorus. *New Phytologist, 194*(4), 953–960.

Ahmaruzzaman, M., & Gupta, V. K. (2011). Rice husk and its ash as low-cost adsorbents in water and wastewater treatment. *Industrial & Engineering Chemistry Research, 50*(24), 13589–13613.

Ahmad, T., Aadil, R. M., Ahmed, H., ur Rahman, U., Soares, B. C., Souza, S. L., Pimentel, T. C., Scudino, H., Guimarães, J. T.,Esmerino, E. A., Freitas, M. Q., Almada, R. B., Vendramel, S. M. R., Silva, M. C., & Cruz, A. G. (2019). Treatment and utilization of dairy industrial waste: A review. *Trends in Food Science & Technology, 88*, 361–372.

Ahmed, S., Nisar, M. F., Hussain, K., Majeed, A., Ghumroo, P. B., Afghan, S., Shahzad, A., Nawaz, K., & Ali, K. (2010). Isolation and characterization of a bacterial strain for aniline degradation. *African Journal of Biotechnology, 9*(8), 1173–1179.

Anderson, D. M., & Garrison, D. J. (1997). The ecology and oceanography of harmful algal blooms. *Limnology and Oceanography, 42*, 1009–1305.

Aniebo, A. O., Wekhe, S. N., & Okoli, I. C. (2009). Abattoir blood waste generation in rivers state and its environmental implications in the Niger Delta. *Toxicological and Environmental Chemistry, 91*, 619–625.

Afsharnia, M., Kianmehr, M., Biglari, H., Dargahi, A., & Karimi, A. (2018). Disinfection of dairy wastewater effluent through solar photocatalysis processes. *Water Science and Engineering, 11*(3), 214–219.

Akinwole, A. O., Dauda, A. B., & Ololade, A. O. (2016). Haematological response of Clarias gariepinus juveniles reared in treated wastewater after waste solids removal using alum or *Moringa oleifera* seed powder. *International Journal of Aquaculture, 6*(11), 1–8.

Arán, D. S., Harguinteguy, C. A., Fernandez-Cirelli, A., & Pignata, M. L. (2017). Phytoextraction of Pb, Cr, Ni, and Zn using the aquatic plant Limnobium laevigatum and its potential use in the treatment of wastewater. *Environmental Science and Pollution Research International, 24*(22), 18295–18308.

Auguet, J. C., Barberan, A., & Casamayor, E. O. (2010). Global ecological patterns in uncultured archaea. *ISME Journal, 4*, 182–190.

Badar, S., & Farooqi, I. H. (2012). Pulp and paper industry—Manufacturing process, wastewater generation and treatment. In *Environmental protection strategies for sustainable development* (pp. 397–436).

Bajpai, P., & Bajpai, P. (2018). Brief description of the pulp and papermaking process. In *Biotechnology for pulp and paper processing* (pp. 9–26).

Bakulski, K. M., Seo, Y. A., Hickman, R. C., Brandt, D., Vadari, H. S., Hu, H., & Park, S. K. (2020). Heavy metals exposure and Alzheimer's disease and related dementias. *Journal of Alzheimer's Disease, 76*(4), 1215–1242.

Bartram, J., Cotruvo, J. A., Exner, M., Fricker, C., & Glasmacher, A. (Eds.). (2003). *Heterotrophic plate counts and drinking-water safety*. IWA Publishing.

Bowmer, K. H., & Laut, P. (1992). Wastewater management and resource recovery in intensive rural industries in Australia. *Water Research, 26*, 201–208.

Boyd, C. E., & Boyd C. E. (2020). Suspended solids, color, turbidity, and light. In *Water quality: An introduction* (pp. 119–133).

Brillas, E., & Martínez-Huitle, C. A. (2015). Decontamination of wastewaters containing synthetic organic dyes by electrochemical methods. An updated review. *Applied Catalysis b, Environmental, 166*, 603–643.

Brunet, J., Diekmann, M., & Falkengren-Grerup, U. (1998). Effects of nitrogen deposition on field layer vegetation in south Swedish forests. *Environmental Pollution, 102*, 35–40.

Budiyono, B., Seno, J., & Sunarso, S. (2011). Study on slaughterhouse waste potency and characteristic for biogas production. *International Journal of Waste Resources (IJWR), 1*(2), 4–7.

Bulu, Y. I., & Oladoja, N. A. (2023). Process variables that defined the phytofiltration efficiency of invasive macrophytes in aquatic system. *International Journal of Phytoremediation, 25*(1), 1–19.

Bulu, Y. I., & Saliu, T. D. (2021). Ecological impacts of oxyanion in aqua systems. In *Progress and Prospects in the Management of Oxyanion Polluted Aqua Systems* (pp. 33–67).

Bustillo-Lecompte, C. F., Mehrvar, M., & Quiñones-Bolaños, E. (2014). Cost effectiveness analysis of TOC removal from slaughterhouse wastewater using combined anaerobic—Aerobic and UV/ H_2O_2 Processes. *Journal of Environmental Management, 134*, 145–152.

Bustillo-Lecompte, C. F., & Mehrvar, M. (2015). Slaughterhouse wastewater characteristics, treatment, and management in the meat processing industry: A review on trends and advances. *Journal of Environmental Management, 161*, 287–302.

Cardoso, L. G., Lemos, P. V. F., de Souza, C. O., Oliveira, M. B. P. P., & Chinalia, F. A. (2022). Current advances in phytoremediation and biochemical composition of Arthrospira (Spirulina) grown in aquaculture wastewater. *Aquaculture Research, 53*(14), 4931–4943.

Chandra, R., & Abhishek, A. (2011). Bacterial decolorization of black liquor in axenic and mixed condition and characterization of metabolites. *Biodegradation, 22*, 603–611.

Chapman, P. M., Hayward, A., & Faithful, J. (2017). Total suspended solids effects on freshwater lake biota other than fish. *Bulletin of Environmental Contamination and Toxicology, 99*, 423–427.

Chen, M., Ding, S., Chen, X., Sun, Q., Fan, X., Lin, J., Ren, M., Yang, L., & Zhang, C. (2018). Mechanisms driving phosphorus release during algal blooms based on hourly changes in iron and phosphorus concentrations in sediments. *Water Research, 133*, 153–164.

Choudhury, A., Lepine, C., Witarsa, F., & Good, C. (2022). Anaerobic digestion challenges and resource recovery opportunities from land-based aquaculture waste and seafood processing byproducts: A review. *Bioresource Technology, 354*, Article 127144.

Christie, R. (2007). *Environmental aspects of textile dyeing*. Elsevier.

Chung, W. J., Shim, J., & Ravindran, B. (2022). Characterization of cheese processed wastewater and treatment using calcium nanoparticles synthesised by Senna auriculata L flower extract. *Journal of King Saud University-Science, 34*(2), Article 101793.

Clark, C. M., Bell, M. D., Boyd, J. W., Compton, J. E., Davidson, E. A., Davis, C., Fenn, M. E., Geiser, L., Jones, L., & Blett, T. F. (2017). Nitrogen-induced terrestrial eutrophication: Cascading effects and impacts on ecosystem services. *Ecosphere, 8*(7), Article e01877.

Cooper, R. N., & Russell, J. M. (1988). *Effluent treatment in the New Zealand export meat industry* (Technical Report 854). MIRINZ, Hamilton, New Zealand.

Coppola, D., Lauritano, C., Palma, E. F., Riccio, G., Rizzo, C., & de Pascale, D. (2021). Fish waste: From problem to valuable resource. *Marine Drugs, 19*(2), 116. https://doi.org/10.3390/md19020116

Cui, D., Shen, D., Wu, C., Li, C., Leng, D., & Zhao, M. (2017). Biodegradation of aniline by a novel bacterial mixed culture AC. *International Biodeterioration and Biodegradation, 25*, 86–96.

Cui, J., Jin, Z., Wang, Y., Gao, S., Fu, Z., Yang, Y., & Wang, Y. (2021). Mechanism of eutrophication process during algal decomposition at the water/sediment interface. *Journal of Cleaner Production, 309*, Article 127175.

Dani, S. U., & Walter, G. F. (2018). Chronic arsenic intoxication diagnostic score (CAsIDS). *Journal of Applied Toxicology, 38*(1), 122–144.

Das, S., Goswami, S., & Talukdar, A. D. (2014). A study on cadmium phytoremediation potential of water lettuce, *Pistia stratiotes* L. *Bulletin of Environment Contamination and Toxicology, 92*(2), 169–174.

Dauda, A. B., Ajadi, A., Tola-Fabunmi, A. S., & Akinwole, A. O. (2019). Waste production in aquaculture: Sources, components and managements in different culture systems. *Aquaculture and Fisheries, 4*, 81–88.

Devine, R. (2003). Meat consumption trends in the world and the European Union. *Productions Animales, 16*, 325–327.

Dungan, R. S., Klein, M., & Leytem, A. B. (2012). Quantification of bacterial indicators and zoonotic pathogens in dairy wastewater ponds. *Applied and Environmental Microbiology, 78*(22), 8089–8095.

Farzadkia, M, Vanani, F., Golbaz, S., Sajadi, S., & Bazrafshan, E. (2016). Characterization and evaluation of treatability of wastewater generated in Khuzestan livestock slaughterhouses and assessing of their wastewater treatment systems. *Global NEST Journal, 18*, 1–10.

Fearon, J., Mensah, S. B., & Vivian, B. (2014). Abattoir operations, waste generation and management in the Tamale metropolis: Case study of the Tamale slaughterhouse. *Journal of Public Health and Epidemiology, 6*(1), 14–19.

Finsterer, J. (2022). Triggers of Guillain-Barré syndrome: Campylobacter jejuni predominates. *International Journal of Molecular Sciences, 23*(22), 14222.

Foyle, L., Burnett, M., Creaser, A., Hens, R., Keough, J., Madin, L., Price, R., Smith, H., Stone, S., & Kinobe, R. T. (2023). Prevalence and distribution of antimicrobial resistance in effluent wastewater from animal slaughter facilities: A systematic review. *Environmental Pollution, 318*, Article 120848.

Ganapuram, B. R., Alle, M., Dadigala, R., Dasari, A., Maragoni, V., & Guttena, V. (2015). Catalytic reduction of methylene blue and Congo red dyes using green synthesized gold nanoparticles capped by *Salmalia malabarica* gum. *International Nano Letters, 5*, 215–222.

Garcia-Vargas, G. G., & Cebrian, M. E. (2023). Health effects of arsenic. In *Toxicology of Metals* (Vol. I, pp. 423–438). CRC Press.

Garcha, S., Verma, N., & Brar, S. K. (2016). Isolation, characterization and identification of microorganisms from unorganized dairy sector wastewater and sludge samples and evaluation of their biodegradability. *Water Resources and Industry, 16*, 19–28.

Gilliam, F. S. (2006). Response of the herbaceous layer of forest ecosystems to excess nitrogen deposition. *Journal of Ecology, 94*(6), 1176–1191.

Gita, S., Shukla, S. P., Prakash, C., Saharan, N., & Deshmukhe, G. (2018). Evaluation of toxicity of a textile dye (Optilan Red) towards a green microalga *Chlorella vulgaris*. *International Journal of Current Microbiology and Applied Sciences, 7*, 3346–3355.

Ghaly, A. E., Kamal, M., & Mahmoud, N. S. (2005). Phytoremediation of aquaculture wastewater for water recycling and production of fish feed. *Environment International, 31*(1), 1–13.

Hamilton, D. (2012). *Organic matter content of wastewater and manure*. Oklahoma Cooperative Extension Service.

Hernández-Zamora, M., Martínez-Jerónimo, F., Cristiani-Urbina, E., & Cañizares-Villanueva, R. O. (2016). Congo red dye affects survival and reproduction in the cladoceran Ceriodaphnia dubia. Effects of direct and dietary exposure. *Ecotoxicology, 25*, 1832–1840.

Hautier, Y., Niklaus, P. A., & Hector, A. (2009). Competition for light causes plant biodiversity loss after eutrophication. *Science, 324*, 636–638.

Higgins, M. J., Adams, G., Chen, Y.-C., Erdal, Z., Forbes, R. H., Glindemann, D., Hargreaves, J. R., McEwen, D., Murthy, S. N., Novak, J. T., & Witherspoon, J. (2008). Role of protein, amino acids, and enzyme activity on odor production from anaerobically digested and dewatered biosolids. *Water Environment Research, 80*(2), 127–135.

Higgins, M. J., Yarosz, D. P., Chen, Y. C., Murthy, S. N., Maas, N., Cooney, J., Glindemann, D., & Novak, J. T. (2006). Cycling of volatile organic sulfur compounds in anaerobically digested biosolids and its implications for odors. *Water Environment Research, 78*, 243–252.

Houser, J. N. (2006). Water color affects the stratification, surface temperature, heat content, and mean epilimnetic irradiance of small lakes. *Canadian Journal of Fisheries and Aquatic Sciences, 63*(11), 2447–2455.

Hossain, K., & Ismail, N. (2015). Bioremediation and detoxification of pulp and paper mill effluent: A review. *Research Journal of Environmental Toxicology, 9*(3), 113.

Hu, T. L., & Wu, S. C. (2001) Assessment of the effect of azo dye RP2B on the growth of a nitrogen fixing cyanobacterium—Anabaena sp. *Bioresource Technology, 77*(1), 93–95.

Huat, T. J., Camats-Perna, J., Newcombe, E. A., Valmas, N., Kitazawa, M., & Medeiros, R. (2019). Metal toxicity links to Alzheimer's disease and neuroinflammation. *Journal of Molecular Biology, 431*(9), 1843–1868.

Husam, A. N., & Nassar, A. (2019). Slaughterhouses wastewater characteristics in the Gaza strip. *Journal of Water Resource and Protection, 11*, 844–885.

Jang, J., Hur, H. G., Sadowsky, M. J., Byappanahalli, M. N., Yan, T., & Ishii, S. (2017). Environmental Escherichia coli: Ecology and public health implications—A review. *Journal of Applied Microbiology, 123*(3), 570–581.

Jiang, J., Wang, Y. P., Yang, Y., Yu, M., Wang, C., & Yan, J. (2019). Interactive effects of nitrogen and phosphorus additions on plant growth vary with ecosystem type. *Plant and Soil, 440*, 523–537.

Kaper, J. B., Nataro, J. P., & Mobley, H. L. (2004). Pathogenic *Escherichia coli. Nature Reviews Microbiology, 2*(2), 123–140.

Kanan, S., & Samara, F. (2018). Dioxins and furans: A review from chemical and environmental perspectives. *Trends in Environmental Analytical Chemistry, 17*, 1–13.

Keeley, G. M., & Quin, B. F. (1979). The effects of irrigation with meatworks-fellmongery effluent on water quality in unsaturated zone and shallow aquifer. *Progress in Water Technology, 11*, 369–386.

Kehinde, F., & Aziz, H. A. (2014). Textile wastewater and the advanced oxidative treatment process, an overview. *International Journal of Innovative Research in Science, Engineering and Technology, 3*(8), 15310–15317.

Khan, S., & Malik, A. (2014). Environmental and health effects of textile industry wastewater. In: *Environmental deterioration and human health: Natural and anthropogenic determinants* (pp 55–71).

Kharapkar, P. (2022). Acquired Methemoglobinemia—A case of aniline dye exposure. *Indian Practitioner, 75*(10).

Kim, H., Jeong, H., Jeon, J., & Bae, S. (2016). Effects of irrigation with saline water on crop growth and yield in greenhouse cultivation. *Water, 8*(4), 127.

Kolhe, A. S., & Powar, V. P. (2011). Physico-chemical analysis of effluents from dairy industry. *Recent Research in Science and Technology, 3*, 29–32.

Kolhe, A. S., Ingale, S. R., & Bhole, R. V. (2009). Effluent of dairy technology. *Shodh Samiksha Aur Mulyankan (International Research Journal), 2*, 459–461.

Kritzberg, E. S., Hasselquist, E. M., Škerlep, M., Löfgren, S., Olsson, O., Stadmark, J., Valinia, S., Hansson, L. A., & Laudon, H. (2020). Browning of freshwaters: Consequences to ecosystem services, underlying drivers, and potential mitigation measures. *Ambio, 49*, 375–390.

Kumar, D., & Desai, K. (2011). Pollution abatement in milk and dairy industry. *Journal of Current Pharma Research, 1*, 45–152.

Kupusovic, T., Midzic, S., Silajdzic, I., & Bjelavac, J. (2007). Cleaner production measures in small-scale slaughterhouse industry—Case study in Bosnia and Herzegovina. *Journal of Cleaner Production, 15*, 378–383.

Kushwaha, J. P., Srivastava, V. C., & Mall, I. D. (2011). An overview of various technologies for the treatment of dairy wastewaters. *Critical Reviews in Food Science and Nutrition, 51*(5), 442–452.

Laspidou, C. S., & Rittmann, B. E. (2002). A unified theory for extracellular polymeric substances, soluble microbial products, and active and inert biomass. *Water Research, 36*(11), 2711–2720.

Lee, C. H., Kim, S. H., Kwon, D. H., Jang, K. H., Chung, Y. H., & Moon, J. D. (2013). Two cases of methemoglobinemia induced by the exposure to nitrobenzene and aniline. *Annals of Occupational and Environmental Medicine, 25*, 1–7.

Lee, S., Maniquiz, M. C., & Kim, L. H. (2010). Characteristics of contaminants in water and sediment of a constructed wetland treating piggery wastewater effluent. *Journal of Environmental Sciences, 22*(6), 940–945.

Lilleskov, E. A., Wargo, P. M., Vogt, K. A., & Vogt, D. J. (2008). Mycorrhizal fungal community relationship to root nitrogen concentration over a regional atmospheric nitrogen deposition gradient in the northeastern USA. *Canadian Journal of Forest Research, 38*, 1260–1266.

Lin, G. (1997). Transformation of a rural economy in the Zhujiang delta. *The China Quarterly, 149*, 56–80.

Lozupone, C. A., & Knight, R. (2007). Global patterns in bacterial diversity. *Proceedings of the National Academy of Sciences, 104*, 11436–11440.

Lucas, R. W., Klaminder, J., Futter, M. N., Bishop, K. H., Egnell, G., Laudon, H., & Högberg, P. (2011). A meta-analysis of the effects of nitrogen additions on base cations: Implications for plants, soils, and streams. *Forest Ecology and Management, 262*(2), 95–104.

Ludlow, J. T., Wilkerson, R. G., & Nappe, T. M. (2019). *Methemoglobinemia.*

MacDonald, J. M., & Ollinger, M. E. (2000). Scale economies and consolidation in hog slaughter. *American Journal of Agricultural Economics, 82*(2), 334–346.

Maddhinni, V. L., Vurimindi, H. B., & Yerramilli, A. (2006). Degradation of azo dye with horse radish peroxidase (HRP). *Journal of the Indian Institute of Science, 86*(5), 507.

Malollari, I., Pinguli, L., Buzo, R., Lajqi, V., Makolli, S., & Cani, X. H. (2019). Actual situation of wastewater from food industry and a case study of their treatment. *Journal of Environmental Protection and Ecology, 20*(1), 432–438.

Mboowa, D. (2021). A review of the traditional pulping methods and the recent improvements in the pulping processes. *Biomass Conversion and Biorefinery, 14*(1), 1–12.

Mishra, P., Naik, S., Babu, P. V., Pradhan, U., Begum, M., Kaviarasan, T., Vashi, A., Bandyopadhyay, D., Ezhilarasan, P., Panda, U. S., & Murthy, M. R. (2022). Algal bloom, hypoxia, and mass fish kill events in the backwaters of puducherry, southeast coast of India. *Oceanologia, 64*(2), 396–403.

Mittal, G. S. (2004). Characterization of the effluent wastewater from abattoirs for land application. *Food Reviews International, 20*(3), 229–256.

Miransari, M., & Smith, D. L. (2007). Overcoming the stressful effects of salinity and acidity on soybean nodulation and yields using signal molecule genistein under field conditions. *Journal of Plant Nutrition, 30*(12), 1967–1992.

Mook, W. T., Chakrabarti, M. H., Aroua, M. K., Khan, G. M. A., Ali, B. S., Islam, M. S., & Abu Hassan, M. A. (2012). Removal of total ammonia nitrogen (TAN), nitrate and total organic carbon (TOC) from aquaculture wastewater using electrochemical technology: A review. *Desalination, 285*, 1–13.

Moorthy, A. K., Rathi, B. G., Shukla, S. P., Kumar, K., & Bharti, V. S. (2021). Acute toxicity of textile dye Methylene blue on growth and metabolism of selected freshwater microalgae. *Environmental Toxicology and Pharmacology, 82*, Article 103552.

Mostofa, K. M. G., Honda, Y., & Sakugawa, H. (2005). Dynamics and optical nature of fluorescent dissolved organic matter in river waters in Hiroshima Prefecture, Japan. *Geochemical Journal, 39*, 257–271.

Mudhoo, A., Thayalan, G., Muthoora, N. J., Muthoora, M. N., Oozeer, B. Z., Rago, Y. P., Ramphul, M. P., Valaydon, A. K., & Kumar, S. (2013). Dioxins and furans: sources, impacts and remediation. In *Pollutant diseases, remediation and recycling* (pp. 479–541).

Mulani, M. S., Kamble, E. E., Kumkar, S. N., Tawre, M. S., & Pardesi, K. R. (2019). Emerging strategies to combat ESKAPE pathogens in the era of antimicrobial resistance: A review. *Frontiers in Microbiology, 10*, 539.

Munavalli, G. R., & Saler, P. S. (2009). Treatment of dairy wastewater by water hyacinth. *Water Science and Technology, 59*(4), 713–722.

Nandomah, S., & Tetteh, I. K. (2023). Potential ecological risk assessment of heavy metals associated with abattoir liquid waste: A narrative and systematic review. *Heliyon, 9*(8), Article e17359.

Nyanti, L., Soo, C. L., Ahmad-Tarmizi, N. N., Ling, T. Y., Sim, S. F., Grinang, J., & Ganyai, T. (2018). Effects of water temperature, dissolved oxygen and total suspended solids on juvenile Barbonymus schwanenfeldii (Bleeker, 1854) and Oreochromis niloticus (Linnaeus, 1758). *Aquaculture, Aquarium, Conservation & Legislation, 11*(2), 394–406.

Okoh, A. I., Sibanda, T., & Gusha, S. S. (2010). Inadequately treated wastewater as a source of human enteric viruses in the environment. *International Journal of Environmental Research and Public Health, 7*(6), 2620–2637.

Olawuyi, K. A., Enitan, S. S., Akele, R. Y., Alim, S., Junaid, I. N. A., Avwioro, G. O., Itodo, G. E., Engom, O., Alaba, G., Ileoma, E., & Enitan, C. B. (2020). A review on the state of abattoirs in Nigeria during COVID-19 pandemic era: Potential threats and public health interventions. *Journal of Public Health and Diseases, 3*(6), 105–118.

Osuolale, O., & Okoh, A. (2017). Human enteric bacteria and viruses in five wastewater treatment plants in the Eastern Cape, South Africa. *Journal of Infection and Public Health, 10*(5), 541–547.

Pavlaki, M. D., Morgado, R. G., Ferreira, V., Rocha, R. J., Soares, A. M., Calado, R., & Loureiro, S. (2021). Cadmium accumulation and kinetics in *Solea senegalensis* tissues under dietary and water exposure and the link to human health. *Water, 13*(4), 522.

Piedrahita, R. H. (2003). Reducing the potential environmental impact of tank aquaculture effluents through intensification and recirculation. *Aquaculture, 226*(1–4), 35–44.

Plaut, Z., Edelstein, M., & Ben-Hur, M. (2013). Overcoming salinity barriers to crop production using traditional methods. *Critical Reviews in Plant Sciences, 32*(4), 250–291.

Prasongsuk, S., Lotrakul, P., Imai, T., & Punnapayak, H. (2009). Decolourization of pulp mill wastewater using thermotolerant white rot fungi. *Science Asia, 35*, 37–41.

Pulido, O. M. (2008). Domoic acid toxicologic pathology: A review. *Marine Drugs, 6*(2), 180–219.

Raghunath, B. V., Punnagaiarasi, A., Rajarajan, G., Irshad, A., Elango, A., & Mahesh Kumar, G. (2016). Impact of dairy effluent on environment—A review. In: *Integrated waste management in India: Status and future prospects for environmental sustainability* (pp. 239–249).

Ram, A., Jaiswar, J. R. M., Rokade, M. A., Bharti, S., Vishwasrao, C., & Majithiya, D. (2014). Nutrients, hypoxia and mass fishkill events in Tapi Estuary, India. *Estuarine, Coastal and Shelf Science, 148*, 48–58.

Ramos, S., Silva, V., Dapkevicius, M. D. L. E., Caniça, M., Tejedor-Junco, M. T., Igrejas, G., & Poeta, P. (2020). Escherichia coli as commensal and pathogenic bacteria among food-producing animals: Health implications of extended spectrum β-lactamase (ESBL) production. *Animals, 10*(12), 2239.

Ratia, H., Vuori, K., & Oikari, A. (2012). Caddis larvae (Trichoptera, Hydropsychidae) indicate delaying recovery of a watercourse polluted by pulp and paper industry. *Ecological Indicators, 15*(1), 217–226.

Rawat, D., Mishra, V., & Sharma, R. S. (2016). Detoxification of azo dyes in the context of environmental processes. *Chemosphere, 155*, 591–605.

Sampaio, E., Santos, C., Rosa, I. C., Ferreira, V., Pörtner, H. O., Duarte, C. M., Levin, L. A., & Rosa, R. (2021). Impacts of hypoxic events surpass those of future ocean warming and acidification. *Nature Ecology and Evolution, 5*(3), 311–321.

Sand-Jensen, K., Bruun, H. H., & Baastrup-Spohr, L. (2017). Decade-long time delays in nutrient and plant species dynamics during eutrophication and re-oligotrophication of Lake Fure 1900–2015. *Journal of Ecology, 105*(3), 690–700.

Sandhya, S. (2010). Biodegradation of azo dyes under anaerobic condition: Role of azoreductase. In *Biodegradation of azo dyes* (pp. 39–57).

Sarayu, K., & Sandhya, S. (2012). Current technologies for biological treatment of textile wastewater—A review. *Applied Biochemistry and Biotechnology, 167*, 645–661.

Sauvey, A., Claquin, P., Le Roy, B., Jolly, O., & Fauchot, J. (2023). Physiological conditions favorable to domoic acid production by three Pseudo-nitzschia species. *Journal of Experimental Marine Biology and Ecology, 559*, Article 151851.

Savin, M., Bierbaum, G., Hammerl, J. A., Heinemann, C., Parcina, M., Sib, E., Voigt, A., & Kreyenschmidt, J. (2020). Antibiotic-resistant bacteria and antimicrobial residues in wastewater and process water from German pig slaughterhouses and their receiving municipal wastewater treatment plants. *Science of the Total Environment, 727*, Article 138788.

Scallan Walter, E. J., Crim, S. M., Bruce, B. B., & Griffin, P. M. (2020). Incidence of Campylobacter-associated Guillain-Barre Syndrome estimated from health insurance data. *Foodborne Pathogens and Disease, 17*(1), 23–28.

Sew, G., Calbet, A., Drillet, G., & Todd, P. A. (2018). Effects of concentration and size of suspended particles on the ingestion, reproduction and mortality rates of the copepod, *Acartia tonsa. Marine Environmental Research, 140*, 251–264.

Schlink, A. C., Nguyen, M. L., & Viljoen, G. J. (2010). Water requirements for livestock production: A global perspective. *Revue Scientifique Et Technique, 29*(3), 603–619.

Sharma, J., Sharma, S., & Soni, V. (2021). Classification and impact of synthetic textile dyes on Aquatic Flora: A review. *Regional Studies in Marine Science, 45*, Article 101802.

Sharma, P., & Singh, S. P. (2021). Pollutants characterization and toxicity assessment of pulp and paper industry sludge for safe environmental disposal. In *Emerging treatment technologies for waste management* (pp. 207–223).

Sharma, P., Iqbal, H. M., & Chandra, R. (2022). Evaluation of pollution parameters and toxic elements in wastewater of pulp and paper industries in India: A case study. *Case Studies in Chemical and Environmental Engineering, 5*, Article 100163.

Shen, G. Q., Lu, Y. T., Wang, M. N., & Sun, Y. Q. (2005). Status and fuzzy comprehensive assessment of combined heavy metal and organochlorine pesticide pollution in the Taihu Lake region of China. *Journal of Environmental Management, 76*(4), 355–362.

Shete, B., & Shinkar, N. P. (2013). Dairy industry wastewater sources, characteristics and its effects on environment. *International Journal of Current Engineering and Technology, 3*, 1611–1615.

Shivsharan, V. S., Kulkarni, S. W., & Wani, M. (2013). Physicochemical characterization of dairy effluents. *International Journal of Life Science and Pharma Research, 2*, 182–191.

Singh, J. K., Kumar, A., Bilal, M., & Chandra, R. (2021). Organometallic pollutants of paper mill wastewater and their toxicity assessment on Stinging catfish and sludge worm. *Environmental Technology and Innovation, 24*, 101831.

Singh, A. K., Kumar, A., & Chandra, R. (2022). Environmental pollutants of paper industry wastewater and their toxic effects on human health and ecosystem. *Bioresource Technology Reports, 20*, Article 101250.

Skarp, C. P. A., Hänninen, M. L., & Rautelin, H. I. K. (2016). Campylobacteriosis: The role of poultry meat. *Clinical Microbiology and Infection, 22*(2), 103–109.

Soh, Y. N. A., Kunacheva, C., Webster, R. D., & Stuckey, D. C. (2020). Identification of the production and biotransformational changes of soluble microbial products (SMP) in wastewater treatment processes: A short review. *Chemosphere, 251*, Article 126391.

Srivastava, A. K., Rana, S. V., Mehrotra, T., & Singh, R. (2016). Characterization and immobilization of bacterial consortium for its application in degradation of dairy effluent. *J Pure Appl Microbiol., 10*(3), 2199–2209.

Swanson, K. E., Kuhn, R. G., & Xu, W. (2001). Environmental policy implementation in rural China: A case study of Yuhang, Zhejiang. *Environmental Management, 27*, 481–491.

Tang, Y., Jiang, Q., Tang, H., Wang, Z., Yin, Y., Ren, F., Kong, L., Jiao, X., & Huang, J. (2020). Characterization and prevalence of Campylobacter spp. from broiler chicken rearing period to the slaughtering process in Eastern China. *Frontiers in Veterinary Science, 7*, 227.

Tao, N., Zhang, W., Si, L., Zhang, R., Wang, D., & Guo, C. (2022). Effects of aniline on growth, oxidative and DNA damage of rice (Oryza sativa L.) seedlings. *Environmental Technology & Innovation, 28*, 102583.

Tao, N., Liu, G., Bai, L., Tang, L., & Guo, C. (2017). Genotoxicity and growth inhibition effects of aniline on wheat. *Chemosphere, 169*, 467–473.

Tara, N., Siddiqui, S. I., Rathi, G., Chaudhry, S. A., & Asiri, A. M. (2020). Nano-engineered adsorbent for the removal of dyes from water: A review. *Current Analytical Chemistry, 16*(1), 14–40.

Tchobanoglous, G., & Burton, F. (1991). *Wastewater engineering: Treatment, disposal, and reuse* (3rd ed.). McGraw-Hill Inc.

Valta, K., Kosanovic, T., Malamis, D., Moustakas, K., & Loizidou, M. (2014). Overview of water usage and wastewater management in the food and beverage industry. *Desalination and Water Treatment, 53*, 1–13.
Van Alfen, N. K. (Ed.). (2014). *Encyclopedia of Agriculture and food systems*. Elsevier.
Van Boeckel, T. P., Brower, C., Gilbert, M., Grenfell, B. T., Levin, S. A., Robinson, T. P., Teillant, A., & Laxminarayan, R. (2015). Global trends in antimicrobial use in food animals. *Proceedings of the National Academy of Sciences, 112*(18), 5649–5654.
Vellingiri, B., Suriyanarayanan, A., Selvaraj, P., Abraham, K. S., Pasha, M. Y., Winster, H., Gopalakrishnan, A. V., Singaravelu, G., Reddy, J. K., Ayyadurai, N., & Kumar, N. (2022). Role of heavy metals (copper (Cu), arsenic (As), cadmium (Cd), iron (Fe) and lithium (Li)) induced neurotoxicity. *Chemosphere, 301*, Article 134625.
Venugopal, V. (2006). *Seafood processing: Adding value through quick freezing*. CRC Press.
Venugopal, V., & Sasidharan, A. (2021). Seafood industry effluents: Environmental hazards, treatment and resource recovery. *Journal of Environmental Chemical Engineering, 9*(2), Article 104758.
Verma, A. K., Dash, R. R., & Bhunia, P. (2012). A review on chemical coagulation/flocculation technologies for removal of colour from textile wastewaters. *Journal of Environmental Management, 93*(1), 154–168.
Vitousek, P. M., & Howarth, R. W. (1991). Nitrogen limitation on land and in the sea: How can it occur? *Biogeochemistry, 13*, 87–115.
Wang, M., Webber, M., Finlayson, B., & Barnett, J. (2008). Rural industries and water pollution in China. *Journal of Environmental Management, 86*(4), 648–659.
Wheeler, D., Wang, H., & Dasgupta, S. (2000). *Can China grow and safeguard its environment? The case of industrial pollution* (Working Paper No. 68). Centre for Research on Economic Development and Policy Reform, Stanford University.
Yaakob, M. A., Mohamed, R. M. S. R., Al-Gheethi, A. A. S., & Kassim, A. H. M. (2018). Characteristics of chicken slaughterhouse wastewater. *Chemical Engineering Transactions, 63*, 637–642.
Yadav, S., & Chandra, R. (2018). Detection and assessment of the phytotoxicity of residual organic pollutants in sediment contaminated with pulp and paper mill effluent. *Environmental Monitoring and Assessment, 190*(10), 581.
Yang, C., Wang, X., Miao, F., Li, Z., Tang, W., & Sun, J. (2020). Assessing the effect of soil salinization on soil microbial respiration and diversities under incubation conditions. *Applied Soil Ecology, 155*, Article 103671.
Yang, C., & Sun, J. (2020). Soil salinity drives the distribution patterns and ecological functions of fungi in saline-alkali land in the Yellow River Delta, China. *Frontiers in Microbiology, 11*, Article 594284.
Yaseen, D. A., & Scholz, M. (2019). Textile dye wastewater characteristics and constituents of synthetic effluents: A critical review. *International Journal of Environmental Science and Technology, 16*, 1193–1226.
You, Y., & Silbergeld, E. K. (2014). Learning from agriculture: Understanding low-dose antimicrobials as drivers of resistome expansion. *Frontiers in Microbiology, 5*, 90890.
Zare, E. N., Motahari, A., & Sillanpää, M. (2018). Nanoadsorbents based on conducting polymer nanocomposites with main focus on polyaniline and its derivatives for removal of heavy metal ions/dyes: A review. *Environmental Research, 162*, 173–195.
Zhai, Y., Yang, Q., & Hou, M. (2015). The Effects of saline water drip irrigation on tomato yield, quality, and blossom-end rot incidence—A 3a case study in the South of China. *PLoS ONE, 10*(11), Article e0142204.
Zhang, W. W., Chong, W. A. N. G., Rui, X. U. E., & Wang, L. J. (2019). Effects of salinity on the soil microbial community and soil fertility. *Journal of Integrative Agriculture, 18*(6), 1360–1368.
Zhang, Y., Cao, C., Guo, L., Wu, Q., & Cui, Z. (2015). Soil properties, bacterial community composition, and metabolic diversity responses to soil salinization of a semiarid grassland in northeast China. *Journal of Soil and Water Conservation, 70*(2), 110–120.

Zhang, C., Chen, H., Xue, G., Liu, Y., Chen, S., & Jia, C. (2021). A critical review of the aniline transformation fate in azo dye wastewater treatment. *Journal of Cleaner Production, 321*, Article 128971.

Zhao, X., & Hardin, I. R. (2007). HPLC and spectrophotometric analysis of biodegradation of azo dyes by Pleurotus ostreatus. *Dyes and Pigments, 73*(3), 322–325.

Management of Wastewater from Aquaculture Operations

Brakemi Egbedi, Michael Kinsella, Shiau Pin Tan, and Helen Hughes

Abstract Aquaculture of farmed animals contributes to global food security, but the generation of large volumes of wastewater is a major threat to its operational sustainability. Low-cost wastewater treatment technologies, such as the use of adsorption-based techniques, constructed wetlands and phycoremediation have been investigated for the treatment of aquaculture wastewater. The use of industrial by-products, which are originally regarded as waste, makes adsorption technology a suitable alternative. However, the low removal efficiency of some adsorbents in their natural state is one of the limitations. Constructed wetlands, on the other hand, mimic natural processes and have low-energy requirement, low maintenance cost and are effective in removing both organic and inorganic contaminants. The land requirement, potential issue of clogging as well as the need for technical know-how to effectively design, manage and maintain the ecological system in constructed wetlands, are some of the limitations. Phycoremediation involves the use of algae in wastewater remediation. Since the algal biomass can be further harvested and used in various applications, it increases the process economic benefits. Overall, the use of these technologies for the treatment of aquaculture wastewater remediation should be encouraged, however, there is need to address some of the limitations to ensure their scalability as well as their integration with other wastewater treatment methods, plants and infrastructure.

B. Egbedi (✉) · M. Kinsella · S. P. Tan · H. Hughes
Department of Science, Pharmaceutical and Molecular Biotechnology Centre (PMBRC) and Eco-Innovation Research Centre (EIRC), South-East Technological University, Waterford Campus, Waterford, Ireland
e-mail: brakemi.egbedi@postgrad.setu.ie; begbedi@unilag.edu.ng

M. Kinsella
e-mail: Mike.Kinsella@setu.ie

S. P. Tan
e-mail: Graece.Tan@setu.ie

H. Hughes
e-mail: Helen.Hughes@setu.ie

B. Egbedi
Department of Marine Sciences, University of Lagos, Akoka, Lagos, Nigeria

N. A. Oladoja and E. I. Unuabonah (eds.), *Strategic Management of Wastewater from Intensive Rural Industries*, Springer Water,
https://doi.org/10.1007/978-3-031-90314-4_2

Keywords Aquaculture wastewater · Adsorption · Constructed-wetlands · Phycoremediation · Microalgae · Macroalgae

1 Introduction

Aquaculture is the farming of aquatic animals and aquatic plants in both inland (freshwater) and coastal (marine, brackish) water environment (Edwards & Demaine, 1997). Globally, aquaculture and industrial fisheries sector (capture fisheries) have contributed immensely to food security and nutrition (FAO, 2022). The global aquaculture production in 2022 was 130.9 million tonnes (~USD 313 billion), with 94.4 million tonnes attributed to aquatic animals and 36.5 million tonnes for algae production. About 62.6% of the aquaculture activities occurred in inland waters, while marine and coastal aquaculture accounted for 37.4% (FAO, 2024). With the increasing global population, expected to reach 9 billion in 2050, the demand for fish production to meet the dietary requirements of the people will be heightened (Mair et al., 2023).

The classification of aquaculture practice is based on several factors, such as the type of farming system adopted, culture intensity, culture structure employed, water exchange type (Soltan, 2016), integrated farming systems (Funge-Smith & Phillips, 2001), scale of production (Edwards & Demaine, 1997), climate (Bricknell et al., 2021) and type of aquatic environment/salinity. Owing to the variety of methods, intensity, species and the water environments, the classification of aquaculture is complex. An overview of the types of aquaculture, the culture systems and the production systems has been discussed by Ahmad et al. (2021), and a simplified classification is presented in Fig. 1.

Irrespective of the aquaculture system adopted, wastes are generated and these wastes are classified as solid, liquid and dissolved wastes (Ahmad et al., 2021; Dauda et al., 2019). Both aquaculture and industrial fishing practices use large volumes of water, thereby generating large amount of wastewater (Turcios & Papenbrock, 2014). The contribution of the aquaculture industry to global food security can only be environmentally and ecologically sustained via the responsible use of resources, the proper handling, treatment, disposal, and reuse of the wastes generated, including wastewater (Verdegem et al., 2023). Therefore, this chapter is aimed at discussing issues majorly related to inland aquaculture industry wastewater in terms of the sources, composition, environmental impacts, and management practices using low-cost technologies.

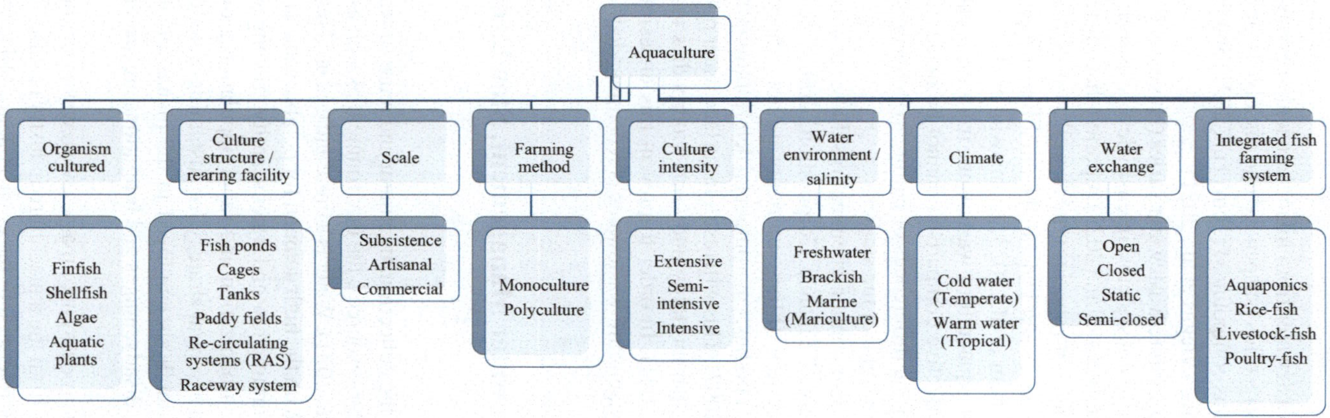

Fig. 1 Aquaculture classifications (adapted from Ahmad et al., 2021; Bricknell et al., 2021; Edwards & Demaine, 1997; Funge-Smith & Phillips, 2001; Soltan, 2016)

2 Aquaculture Wastewater Composition

The major components of wastewater from inland aquaculture consists of dissolved and particulate organic matter, solids (suspended and dissolved), and nutrients (such as phosphorus and nitrogenous compounds) (Tom et al., 2021) and this is heightened by the intensity of the practice (Viegas et al., 2021).

The characteristics of aquaculture wastewater varies (Hussan et al., 2019), and this can be associated with the different systems, intensity levels, as well as species cultivated as already highlighted. Bhatt et al. (2023) reported that wastewater in shrimp aquaculture wastewater, varied with species type as well as system and facility characteristics. The volumes of wastewater generated from different aquaculture systems have been documented (Kashem et al., 2023; Kurniawan et al., 2021). The physicochemical characteristics of aquaculture wastewater from different farm practices are shown in Table 1. An effective comparison cannot be made of the wastewater shown in Table 1 because of the differences of culture systems and species, as well as incomplete parameters for some of the studies.

Aquaculture wastewater also contain chemical residues, such as antibiotics and feed supplements (International Finance Corporation, 2007), all of which have adverse impacts on the ecosystem. If these nutrients are not removed, prior to their discharge into receiving water bodies, environmental deterioration will ensue through eutrophication (Crab et al., 2007). This will result in surface water pollution, decrease in dissolved oxygen levels, fish mortality and such waterbodies may become unfit for human use (Tan et al., 2021). In order to prevent this undesirable occurrence, effective management of aquaculture wastewater is recommended.

3 Aquaculture Wastewater Management, Standards, and Regulations

Wastewater management is the collection, treatment and reuse of wastewater and its overall aim is to protect water resources, the environment as well as human, animal and plant health (Roozbahani, 2021). Water is a finite and essential resource needed to sustain life in all its forms. With the high demand for potable water (Guimarães et al., 2018), freshwater reserves running low and water scarcity growing in certain regions because of climate change, urbanization and rapid population growth, wastewater management has become an essential strategy for sustaining water resources (Ding, 2023). Low-cost treatment methods employed in the management of aquaculture wastewater, the advantages, disadvantages and limitations are presented in Sect. 3.1 and Table 6 respectively.

Wastewater management process encompasses wastewater treatment (Lasisi et al., 2023), therefore wastewater from the aquaculture industry require treatment prior to their re-use, disposal, and/or release into the environment (Al-Dawery et al., 2023). The effectiveness of a wastewater treatment process is dependent on several factors

Table 1 Characteristics of wastewater from different aquaculture systems

Country	Nigeria	Nigeria	Nigeria	Nigeria	Thailand	USA
Species	Catfish	NS	Catfish	NS	Shrimp	Shrimp
System	Concrete fish pond	Fish pond	Fish pond	Pond	NS	Intensive re-circulating raceway system
pH	6.68	6.79–6.87	6.3	7.9	7.7	8.1
TSS (mg/L)	0.70	NS	787.5	45	188.9	NS
Temperature (°C)	27.80	27–29	NS	30.2	NS	NS
COD (mg/L)	108.70	NS	600	758	128	1593
BOD (mg/L)	46.80	4.0–6.0	120	317	NS	NS
NH_4^+–N (mg/L)	0.08	NS	NS	NS	NS	NS
NO_3^-–N (mg/L)	8.50	0.57–0.78 (NO^{3-}—2.53 to 3.45)	4.49 (NO^{3-}—19.88)	–	NS	7.07 (NO^{3-}—31.3)
PO_4^{3-}-P (mg/L)	2.50	0.17–0.42 (PO_4^{3-}—0.51 to 1.3)	1.13 (PO_4^{3-}—3.48)	N.S	11.85 (PO_4^{3-}—36.33)	NS
DO (mg/L)	4.17	5.8–7	NS	9.6	5.7	NS
TDS (mg/L)	NS	330–552	NS	650	400	NS
TS (mg/L)	NS	NS	NS	695	NS	33,100
Turbidity (NTU)	NS	130–300	NS	404	222.5	NS

(continued)

Table 1 (continued)

NO_2^-(mg/L)	NS	NS	NS	NS	NS	250
References	Omotade et al. (2019)	Musa et al. (2020)	Aderibigbe (2018)	Igwegbe and Onukwuli (2019)	Vijuksungsith et al. (2023)	Boopathy and Lyles (2008)

Country	Malaysia	Thailand	USA	Egypt	Canada
Species	Shrimp	Shrimp	Shrimp	Nile Tilapia	NS
System	Intensive shrimp culture	NS	NS	Re-circulation aquaculture system	Hydroponics
pH	7.95	8.41	7.85	NS	7.3
TSS (mg/L)	86	NS	NS	NS	100
Temperature (°C)	23.82	28	NS	NS	NS
COD (mg/L)	NS	NS	NS	NS	240
BOD (mg/L)	NS	NS	NS	NS	NS
NH_4^+–N (mg/L)	NS	3.12 (NH_4^+—4.02)	NS	20.96 (NH_4^+—27)	NS
NO_3^-–N (mg/L)	NS	1.11 (NO^{3-}—4.90)	0.24 (NO^{3-}—1.06)	2.71 (NO^{3-}—12)	110
PO_4^{3-}-P (mg/L)	3.71 (PO_4^{3-}—11.37)	5.716 (PO_4^{3-}—17.53)	3.766 (PO_4^{3-}—11.55)	22.175 (PO_4^{3-}—68)	2.152 (PO_4^{3-}—6.60)
DO (mg/L)	2.8	NS	NS	2.66	NS
TDS (mg/L)	NS	NS	NS	692	NS
TS (mg/L)	NS	NS	NS	NS	1000
Turbidity (NTU)	81	NS	NS	112.70	NS
NO_2^-(mg/L)	1.50	0.42	0.863	0.19	16.42 (NO_2^-–N-5)
References	Iber et al. (2023b)	Krasaesueb et al. (2023)	Bhatt et al. (2023)	El-Sherbiny et al. (2019)	Ghaly et al. (2005)

Note NS = Not stated, TSS = Total suspended solids, COD = Chemical oxygen demand, BOD = Biological oxygen demand, NH_4^+–N = Ammonium–nitrogen, NO_3^-–N = Nitrate–nitrogen, PO_4^{3-}–P = Phosphate as phosphorus, DO = Dissolved oxygen, TDS = Total dissolved solids, TS = Total solids, NO_2^- = Nitrite, NO_2^-–N = Nitrite–nitrogen. Conversions were rounded up which include the physicochemical constituents of the wastewater (Roozbahani, 2021). Since the treated wastewater must meet certain standards prior to its discharge (Rezai & Allahkarami, 2021), the physicochemical parameters after treatment is a quality indicator for assessing the efficacy of the treatment method.

Wastewater quality standards are made to prevent effluents from causing adverse effects on receiving waters (Boyd, 2003). The monitoring of effluent quality is usually done following environmental regulations, standards and/or directives set up by regulatory bodies (Yang et al., 2025). Different countries have their own regulations and guidelines to protect waterbodies. In USA for example, the guidelines are set by the Environmental Protection Agency (EPA) under the Clean Water Act. Furthermore, the Clean Water Act also mandates companies and aquaculture facilities to obtain the National Pollutant Discharge Elimination System (NPDES) permit which aids the monitoring, regulation and control of discharge into waterbodies (US EPA, 2024; Mugg et al., 2007).

For countries in the European Union, the EU Water Framework Directive is the primary legislation and included are directives to protect the quality of waterbodies (both groundwater and surface water) (Acierno et al., 2006; EU Commission, n.d.). In Canada, the Fisheries Act has guidelines to prevent pollution of waterbodies. The Wastewater systems Effluent Regulations, which is under the Fisheries Act, authorizes the release of wastewater effluents after meeting the effluent quality standard. It must be stated that this regulation does not apply in some territories in Canada. Furthermore, it applies to wastewater systems with an average daily volume of 100 m^3 or more, during a calendar year. It is also applicable if the final discharge point of the wastewater is in water frequented by fish or in any place that could reach these waters (Government of Canada, 2024). In Nigeria, the National Environmental standards and Regulations Enforcement Agency (NESREA) created in 2007, is the main environmental regulatory agency. It has several guidelines on wastewater management, one of which is the National Environmental (Surface and Ground Water Quality Control) regulations, 2010 (Kayode et al., 2018).

In addition to national/local effluent regulations and guidelines, international organizations such as the Food and Agriculture Organization (FAO) of the United Nations and the World Health Organization (WHO) also establish guidelines and standards, which encourage sustainable practices and minimize environmental risks. These guidelines also aid to shape or inform national polices and regulations in developing and developed countries globally (Yang et al., 2025). The FAO Code of Conduct for Responsible Fisheries is a framework that provides principles and standards for fisheries conservation, development and management (FAO, 1995). It encompasses all aspects of aquaculture, environmental sustainability as well as the socio-economic aspects of fisher folks and fisheries communities (Yang et al., 2025). FAO also has a guideline for interpreting water quality intended for irrigation titled 'Water quality for agriculture' (Ayers & Westcot, 1985). This guideline informs users who intend using treated aquaculture wastewater effluents for irrigation purposes. In 2023, a joint collaboration between FAO and the International Water Management Institute (IWMI) towards the 2030 Agenda for Sustainable Development, released an updated publication titled 'Water Quality in Agriculture: Risks and Risk Mitigation' which provides the latest information on water quality requirements for fish production, crop irrigation, livestock as well as environmental protection (Drechsel et al., 2023). This updated guideline draws on the previous FAO guideline documents: *Water Quality for Agriculture*, *Wastewater Treatment and Use in Agriculture* and *Water Desalination for Agricultural purposes* and is complemented by the *Brackish Water Use* and *Antimicrobial Resistance* guidelines of the FAO (IWMI, 2023). The World Health Organization (WHO) also develops international guidelines on water quality and human health. Some of these guidelines include Guidelines for drinking water quality (GDWQ) (WHO, n.d.a) and Guidelines on recreational water quality which is published in two volumes namely the 'Guidelines for safe recreational water quality: Volume 1 Coastal and freshwaters' and 'Guidelines for safe recreational water environments Volume 2: Swimming pools and similar environments' (WHO, n.d.b). These guidelines are indirectly applicable to aquaculture practices since their effluent discharge may influence coastal and freshwater environments as well as public health.

The United Nation Environment Programme (UNEP) is the world's leading authority on the environment. It aims to find solutions to the triple planetary crisis (crisis of climate change, nature and biodiversity loss and the crisis of pollution and waste) (UNEP, n.d.a). It also promotes the protection, conservation, restoration and management of waterbodies by collaborating with countries to achieve these aims (UNEP, n.d.b).The Global Programme of Action (GPA) focuses on ensuring that land-based activities do not degrade the marine environment. It has a founding document called 'The Global Programme of Action for the Protection of the Marine Environment from Land-based Activities (GPA/UNEP)' which is the only global environment initiative that directly addresses the connection between terrestrial and aquatic ecosystems (freshwater, coastal and marine) (UNEP, n.d.c). GPA is implemented through cooperation on a regional basis (Action guide, n.d.). It serves as a conceptual and practical guidance document for national and/or regional governments who wish to design and implement actions to prevent, reduce, or stop the degradation of marine environment from land-based activities. It directly helps to achieve the UN SDG Goal 14—'*Conserve and sustainably use the oceans, seas and marine resources for sustainable development*' (United Nations, n.d.).

The International Finance Corporation has developed effluent guidelines for the aquaculture sector. The International Finance Corporation is a member of the World Bank Group and it is its private-sector arm and the largest global development institution that focuses on providing funds and investment for private enterprises in developing countries. (International Finance Corporation, n.d.a). Environmental, Health, and Safety (EHS) Guidelines are technical reference documents accepted by the World Bank Group. The EHS guidelines set out the levels and measures, which must be adhered to by clients and/or borrowers of the World Bank Group. These guidelines, which contain information on environmental, health and safety issues applicable to industry sectors, are to be used in conjunction with the relevant sector guidelines (International Finance Corporation, n.d.b, n.d.c). These guidelines serve as a performance indicator and are useful for monitoring, pollution prevention and control, and they are also indicative of good industry practice (International Finance

Corporation, 2007). These effluent guidelines are applicable for the direct discharge of treated wastewater effluents to surface waters, which can then be used for general purposes (International Finance Corporation, 2007). Effluent discharge limits for some countries in Africa and Asia has been documented by Kayode et al. (2018) while water quality standards by the European Union, World bank (International Finance Corporation), U.S.A has been documented by Yang et al. (2025).

In order to ensure that these regulations are met, monitoring for compliance is a necessity. The frequency of monitoring has to be adequate to ensure that appropriate data is obtained for the parameters being monitored. When monitoring is done regularly and conscientiously, necessary corrective actions are taken to ensure that wastewater effluents do not constitute harm to the receiving water bodies and the ecosystem (International Finance Corporation, 2007). If effluents do not comply with standards, the appropriate treatment and/or management method has to be performed. In developed countries, the enforcement of these frameworks is possible; however, this is not the same with the developing countries. Furthermore, these countries do not have the financial capabilities to effectively regulate and enforce the effluent standards compliance (Boyd, 2003). Kayode et al. (2018) reported that developing countries have stipulated effluent standards, which were designed with a faulty premise to meet their counterpart developed nations with high level wastewater treatment technologies. The lack of facilities to meet the stipulated wastewater effluent guidelines, poor political willpower, poor collaboration among responsible agencies and few/weak institutions are some of the identified reasons for the non-compliance with these regulations. The use of low-cost wastewater technologies and strategies is thus essential, to alleviate many of these challenges which threaten the sustainability of aquaculture practices especially in developing countries.

3.1 Technological Solutions for Aquaculture Wastewater Management

Conventional methods such as physical, chemical and biological treatment have been employed in the treatment of wastewater from various aquaculture systems (Turcios & Papenbrock, 2014). Other methods, which include natural ecological treatment (Li et al., 2024) such as bioremediation, phyco-remediation (Jusoh et al., 2020), and the use of constructed and natural wetlands have also been studied. A summary of these processes, including their advantages and limitations have been discussed by Kashem et al. (2023). Owing to the presence of different types of pollutants in aquaculture wastewater, it is impossible to utilize the same treatment method for all types of wastewater generated from aquaculture practices (Devi & Gowri, 2007). A combination of processes is usually needed to meet the wastewater treatment needs. Some of the unit processes that have been investigated for the treatment of aquaculture wastewater are discussed below:

3.1.1 Adsorption

It is a phenomenon, whereby molecules or ions of target pollutants (adsorbate), adhere onto the surface of a solid material (adsorbent) through intermolecular forces (Satyam & Patra, 2024). Adsorption has been found to be effective in the treatment of aquaculture wastewater (Kashem et al., 2023). The major advantage is the use of low-cost materials, which reduces the overall cost of the technique. In addition, its simplicity of design, ease of operation, further makes it an attractive low-cost wastewater treatment technique (Bhatnagar & Sillanpää, 2011; Egbedi et al., 2023).

Various natural and synthetic materials have been used as adsorbents in wastewater treatment and they can be classified as conventional or non-conventional (Fig. 2). The classification of adsorbents into nanostructured and conventional materials is yet another way of categorizing adsorbents used in wastewater treatment (Tantalu et al., 2024). Some of the adsorbents used in wastewater treatment include mollusc shells (Egbedi et al., 2023), zeolite, sawdust, sawdust ash, granite powder, crushed granite (El-Sherbiny et al., 2019), activated carbon, chitosan, limestone, quicklime (Tantalu et al., 2024); composite materials such as chitosan-bentonite clay composites (Bensalem et al., 2024), clay, fly ash; agricultural wastes such as rice husk, fruit peels, biochar, spent coffee grounds, eggshells, nut shells as well as modified adsorbents such as magnetic adsorbents (Soni et al., 2020).

The key in selecting adsorbents for use in wastewater treatment is to ascertain if it has strong affinity for the target contaminant of interest, high porosity, large surface area, selective adsorption capacity (Tantalu et al., 2024) and of course its regeneration

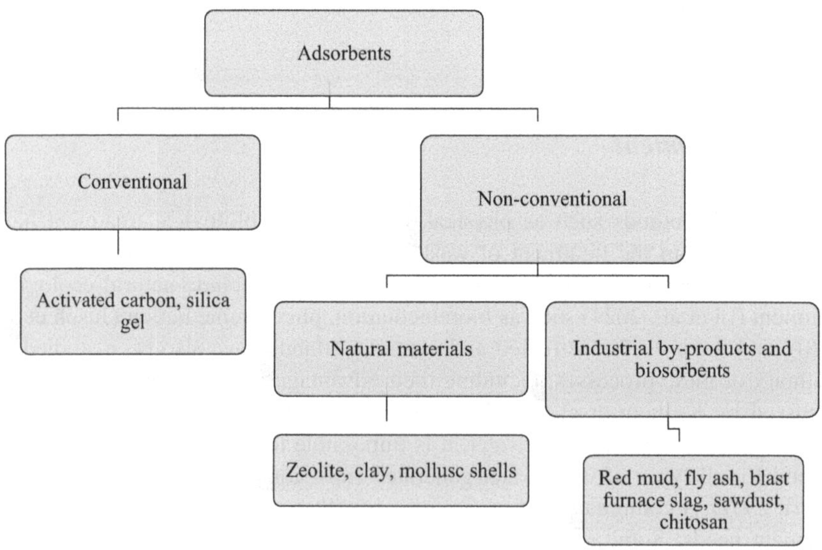

Fig. 2 Classification of adsorbent (adapted from Laiq et al., 2021 with slight modification)

or reusability capacity. Adsorbents exhibit varying physical and chemical properties (which contribute to their effectiveness in wastewater treatment for target pollutants) and they each have their respective advantages and disadvantages. Therefore, the selection of adsorbents is a crucial step that must be taken to ensure its effectiveness and efficiency in wastewater treatment (Li et al., 2024). For example, carbonaceous adsorbents, such as activated carbon is one of the commonly used adsorbents in wastewater treatment for the removal of various types of pollutants such as nutrients, pesticides, dyes, pharmaceuticals, heavy metals and so on (Laiq et al., 2021). This is because of its numerous advantages such as its high surface area, porosity and recyclability/re-use potentials (Kashem et al., 2023); surface reactivity, thermal stability and inertness (Soni et al., 2020), chemical stability, and its versatility. They can be produced from any waste material containing carbon such as bituminous coal, wood, shells of coconut, apricots, peach, and so on (Manoilova & Ruskova, 2019). They can also be developed into different forms (i.e., powdered activated carbon (PAC), granular activated carbon (GAC), and activated carbon fibres) (Taylor & Soltani, 2023). However, it also has some disadvantages such as difficulty to separate the powder from the effluent and high regeneration cost (Soni et al., 2020). According to Nguyen (2015), the low sorption capacity and inefficiency of some natural adsorbents used in their natural state without pre-treatment compared to commercial adsorbents birthed the idea of the modification of these adsorbents for an improved pollutant-removal capacity. Modification methods include but are not limited to thermal treatment and chemical activation.

In order to evaluate the effectiveness of adsorbents in removing target contaminants from wastewater, batch and column experiments are usually conducted (Nur, 2014). Batch studies are easy and quick to perform. They are often used to compare the adsorption capacities of various adsorbents and to investigate the adsorption mechanism (Loganathan et al., 2014). However, the results obtained under batch conditions are usually not directly applicable to most wastewater treatment systems because of the disparity in experimental conditions and real-life scenarios, where the contact time in batch studies do not replicate that of real-time wastewater treatment plants (Zulfadhly et al., 2001).

The sole use of adsorption as a treatment technology is often not sufficient for the treatment of aquaculture wastewater to meet regulatory standards. Therefore, adsorption is often used in combination with other treatment methods to enhance the overall treatment efficiency. In addition, there is paucity of information on the use of these arrays of investigated adsorbents in real wastewater treatment plants, because of scalability issues. Studies where adsorbents have been employed in the treatment of aquaculture wastewater are presented in Table 2. The advantage of adsorption is that in urban areas, where space requirement for the construction of wetlands may be insufficient, the use of removal technologies such as adsorption, which can be conducted in a more compact system with less space requirements can be employed (Nzubechi & Ikpemo, 2023). The fact that not all adsorbents can be effective in removing all types of pollutants, is one of the disadvantages of adsorption technology (Soni et al., 2020). Furthermore, majority of the studies involving adsorption are on experimental scale, while few studies have been conducted on pilot scale. Therefore,

Table 2 Adsorbents used in the treatment of aquaculture wastewater

Adsorbent	Wastewater source	Contaminants targeted	Removal efficiency	References
Activated carbon (Bamboo based)	Fish farm	Heavy metals (Cr^{2+}, Cu^{2+}, Fe^{2+}, Cd^{2+}, Pb^{2+}, Zn^{2+}, Ni^{2+})	65.94–100%	Akinbile et al. (2022)
Activated carbon (Bamboo based)	Fish farm	Bacteria, yeast and fungi	5.18–78.65%	Akinbile et al. (2022)
Activated carbon (Sawdust)	Catfish pond	COD, BOD, TSS, heavy metal (Pb), nitrate, phosphate	COD (83.3%), BOD (92.3%), TSS (93.2%), Pb (99.3%), nitrate (73.2%), phosphate (60%)	Nzubechi and Ikpemo (2023)
Zeolite	Tilapia pond	Heavy metals (As, Cd, Cr, Cu, Pb, Zn)	9.53–25.45%	Hasan et al. (2023)
Natural zeolite (Clinoptiolite)	Tilapia pond	Heavy metals (Cu, Pb, Fe, Mn, Cd, Zn), nitrate	Heavy metals (6.6–58.8%), nitrate (30.5–53.4%)	Hamed et al. (2022)
Chitosan	Shrimp farm	Turbidity and TSS	Turbidity (92.45%), TSS (69.49%)	Iber et al. (2023a)
Chitosan	Intensive shrimp culture	Turbidity, salinity	Turbidity (87.67%), salinity (21.43%)	Iber et al. (2023b)
Magnetite/Iron oxide (Fe_3O_4) nanoparticles	Fish farm	Nitrate, phosphate	Nitrate (74.25–80.76%), phosphate (78.17–80.55%)	Askari Hesni et al. (2019)

if adsorption technology is to be employed, there is need for the scale-up of this technology into real-time applications at the remediation sites.

3.1.2 Constructed Wetlands

Constructed wetlands (CWs) are man-made or engineered systems, which are designed to mimic the features and functions of natural processes (natural wetlands) involving soils, vegetation/plants, media or substrate and their associated microbial community, for the treatment of wastewater in controlled environmental conditions (Agaton & Guila, 2024; Sutanto & Bawole, 2021; Vymazal, 2007). Therefore, wastewater purification in CWs occurs through physical, biological and chemical processes (Omotade et al., 2019).

The use of CWs in aquaculture wastewater treatment has garnered attention because of its numerous benefits over conventional treatment methods. CWs offers a low-cost natural wastewater treatment alternative, which can be used in real-time and especially in developing countries where financial resources are a constraint (Agaton & Guila, 2024). Its low operational and maintenance cost, green technology, low energy requirements, easy operation and maintenance, effectiveness in removing contaminants from wastewater from different sectors also makes its use highly advantageous (Corbella & Puigagut, 2018; Ji et al., 2022).

Hydrology/water flow path and the type of vegetation, are the two most important parameters used in the classification of constructed wetlands (Vymazal, 2008). In terms of hydrology or water flow, CWs are classified into three, namely: Surface flow constructed wetland (SF CWs) or free water surface (FWS), subsurface flow constructed wetlands (SSF CWs) and Hybrid Constructed Wetlands. Subsurface flow CWs are further classified into subsurface vertical flow (VF) and horizontal flow (HF) based on effluent flow as shown in Fig. 3. Floating treatment wetlands (FTWs) are also a type of constructed wetland, which involves the use of naturally occurring floating plants on water surfaces. The roots of the plants are free and they act as filter (Colares et al., 2020).

Surface flow CWs is similar to natural wetlands (Wu et al., 2015) where the water flow is above the ground surface or substrate which is usually soil. They are flooded from the top, and water flows horizontally on top of the media (Abou-Elela, 2019) as shown in Fig. 3. Surface flow CWs usually maintains a shallow depth of water and the bottom of the soil is sealed to prevent wastewater seepage to the groundwater and to maintain water within its system (Abou-Elela, 2019; Hassan et al., 2021). Surface flow CWs are highly advantageous when compared with subsurface flow constructed wetlands (SSF CWs) based on their lower construction, operational and management cost. However, their large land area requirement and odour are some of the disadvantages (Hassan et al., 2021). The water level in subsurface flow constructed wetlands (SSF CWs) is always below the ground surface therefore SSF CWs offers good aesthetic properties and can be constructed below public parks unlike Surface flow CWs and they can be used even during cold climatic conditions (Hassan et al., 2021). Furthermore, there is less land area requirements, no odour and mosquitoes problems. In addition, subsurface flow constructed wetlands (SSF CWs) has a higher sorption/performance efficiency (owing to the variety of root-bed substrates used) despite their small land requirement (Abou-Elela, 2019).

In Horizontal subsurface flow constructed wetlands, the water flow and level are below the ground surface. To prevent clogging, solids are usually removed from the wastewater via the use of sedimentation tanks or filters (Hassan et al., 2021). The bed substrate is made up of selected types and sizes of gravel, sand or their combination, and a suitable plant such that the influent wastewater passes below the porous substrate media and plant roots (Abou-Elela, 2019; Hassan et al., 2021). The vertical water flow in vertical subsurface flow constructed wetlands which is achieved via gravity or by pumps, differentiates it from the horizontal subsurface flow constructed wetlands. Vertical subsurface flow CWs is useful in remediating wastewater that are high in organic matter and suspended solids. Furthermore, the

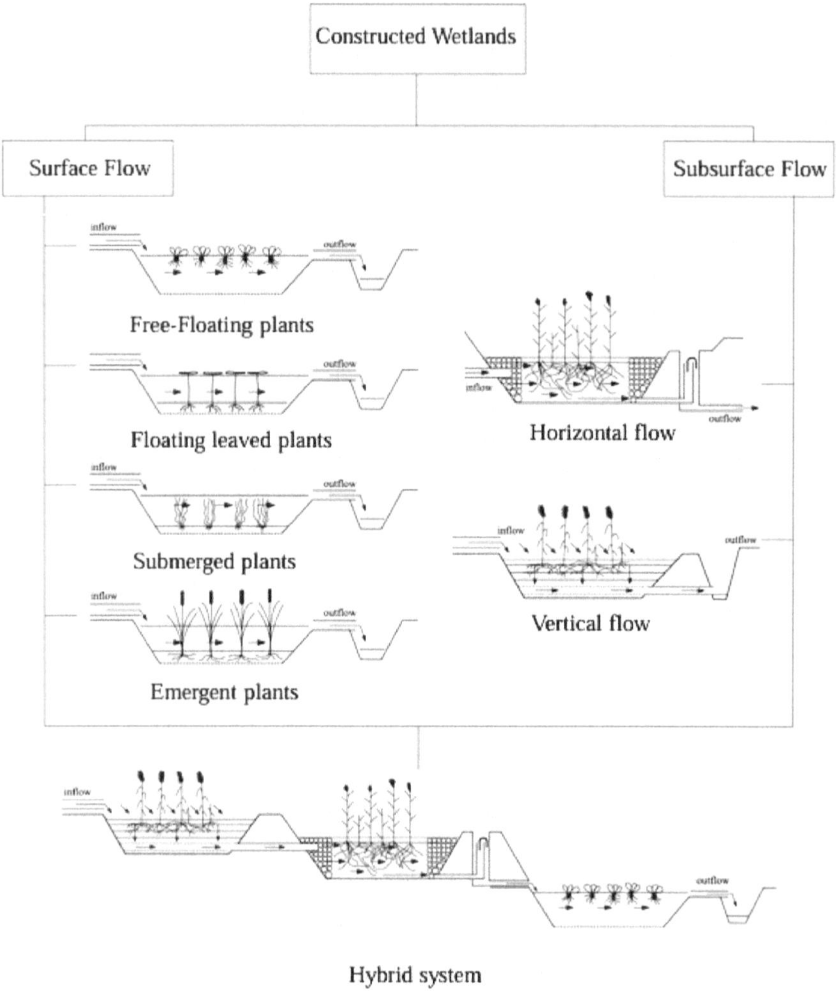

Fig. 3 Schematic diagram showing some of the types of CWs (Hiraishi et al., 2014)

higher oxygen level in vertical subsurface flow CWs when compared with horizontal subsurface flow CWs means that aerobic degradation is enhanced. Another advantage of vertical subsurface flow CWs is that it requires less land area requirement when compared with horizontal subsurface flow CWs and this makes it less expensive (Hassan et al., 2021).

Hybrid constructed wetlands is a combination of various types of constructed wetlands (Wu et al., 2015). The aim is to utilize the features of each of the combined system, in order to increase its efficiency in removing contaminants from wastewater (Swarnakar et al., 2022).

In FTWs, the plants are placed in a buoyant substrate such as a solid-core foam, which helps to position the plant crowns above the water surface. This facilitates the survival of both aquatic and semi-aquatic plant species and enables the use of a non-hydrophytic plant species. The plant's extensive root system in the water column provide a large surface area, which enhances its wastewater purification efficiency. The fact that FTWs can be made using already existing drainage canals or retention ponds is one of the advantages of its use in wastewater treatment (White, 2013).

The substrate media is also a key parameter in CWs. Substrate media can aid the removal of contaminants in wastewater through various processes, such as adsorption, precipitation and ion exchange. Therefore the substrate media must be properly selected to suit the features of the CW ecosystem, to enhance wastewater treatment, to prevent clogging issues (Mahmood et al., 2013) and to also serve as a suitable medium to support plant growth (Wu et al., 2015).

Wang et al. (2020) highlighted some of the factors that must be borne in mind when selecting substrate materials, which includes but is not limited to the availability; cost and source of the substrate; physicochemical properties; substrate life; re-use potentials; disposal issues; removal capacity for the targeted pollutants, etc. Hydraulic permeability is also an important factor when selecting substrate materials since poor hydraulic permeability may result in clogging issues and an overall reduction in the CW removal efficacy over time (Wang et al., 2010; Wu et al., 2015). Clogging occurs when the pores of the substrate used become filled with particles, such that the substrate's infiltration capacity and efficiency are negatively impacted, thus leading to a reduced treatment performance (Wang et al., 2010).

Various materials, which have been used as substrate can be classified as shown in Fig. 4. Mollusc shell can be classified under natural and/or agricultural or industrial by-product, but in this section, it has been classified under the latter. Other substrates which have been used are also highlighted by Wu et al. (2015).

In terms of vegetation type or macrophytes, CWs can be classified based on the type used such as emergent, submerged, free-floating and floating or leaved plants

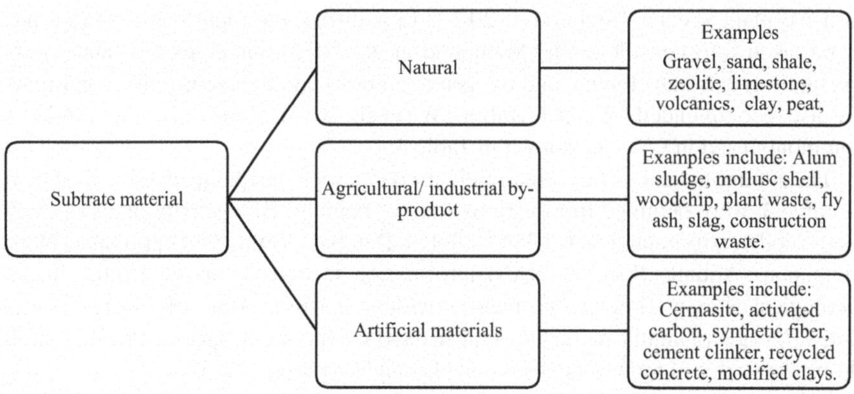

Fig. 4 Classification of substrate material used in CWs and some examples (Wang et al., 2020)

Table 3 Commonly used macrophytes/plants in CWs (Wu et al., 2015)

Type of plant used	Examples	Common name
Emergent plants	*Phragmites spp.*	Reed
	Typha spp.	Cattails
	Cyperus papyrus	Papyrus
	Iris spp.	Flags
	Juncus spp.	Rushes
	Eleocharis spp.	Spike-rush
	Scirpus spp.	Bulrush
Submerged plants	*Hydrilla verticillata*	Esthwaite waterweed
	Potamogeton crispus	Curled pondweed
	Myriophyllum verticillatum	Whorled water-milfoil
	Vallisneria natans	Eel grass
	Ceratophyllum demersum	Rigid hornwort
Floating leaved plants	*Nymphaea tetragona*	Pygmy waterlily
	Nymphoides peltata	Fringed waterlily
	Marsilea quadrifolia	Four-leaf clover
	Trapa bispinosa	Singhara nut/water chestnut
Free-floating plants	*Eirchhornia crassipes*	Common water hyacinth
	Salvinia natans	Salvinia/floating fern or moss
	Lemna minor	Common duckweed
	Hydrocharis dubia	Frogbit

(Gorgoglione & Torretta, 2018; Mahmood et al., 2013; Stefanakis, 2019). The choice of the plant species plays a pivotal role in CWs. The plant species selected should be suitable for the type of constructed wetland, wastewater and climatic conditions of the given locality. It should have a high tolerance to pollutant load and be non-invasive plant species (Stefanakis, 2019). In addition, the plant's growth rate has to be taken into consideration (Mahmood et al., 2013). Other factors such as its tolerance to extreme climatic and oxygen conditions and hypereutrophic conditions are also recommended for consideration (Wu et al., 2015). Some macrophytes/plants commonly used in CWs are shown in Table 3.

The common names of the commonly used macrophytes/plants in CWs as shown in Table 3, were obtained from various sources namely: Biodiversity maps (https://maps.biodiversityireland.ie/), EPPO Global Database (https://gd.eppo.int/), Non-indigenous Aquatic Species (NAS) information resources for the United States Geological Survey (https://nas.er.usgs.gov/default.aspx), Atlas of Florida Plants (https://florida.plantatlas.usf.edu/), Queensland Government Species Profile search (https://apps.des.qld.gov.au/species-search/) and Vymazal (2013).

CWs have been used in the treatment of various types of wastewaters such as agricultural and industrial wastewaters (Gorgoglione & Torretta, 2018). Zhimiao et al.

(2019) added ferrous ions (non-poisonous) into CWs to treat aquaculture wastewater, and 20 mg/L of ferrous ion was found as the optimal dose with the plant species *Canna indica*. After a week, the removal efficiencies of COD, TP and TN were 62%, 77% and 95% respectively.

Omotade et al. (2019) incorporated charcoal as a porous substrate in the CWs design. The efficiency of a charcoal-based constructed wetland planted with *Sacciolepsis africana* and *Commelina cyannae* was evaluated in the treatment of aquaculture wastewater from a catfish pond. After a 5-day retention time, the wetland removed 100% NO_3^-–N, 93% BOD_5, 88% COD and 50% TSS.

Bôto et al. (2016) evaluated the effectiveness of CWs planted with *Phragmites australis* for the removal of nutrients from a saline turbot fish aquaculture wastewater facility. The efficiency of the CWs to remove antibiotics was evaluated by doping the wastewater with two different antibiotics commonly used in aquaculture (i.e., enrofloxacin and oxytetracycline). A removal efficiency > 99.9% was observed for the antibiotics, and the initial BOD value which ranged between 7 and 43 mg/L was reduced to < 5 mg/L. A removal efficiency of > 99% was observed for NO_2^-, 86% removal for NO_3^-, > 94% for NH_4^+, and a PO_4^{3-} removal efficiency > 85%, were achieved after the 9-week experimental period.

Qu et al. (2020) constructed a compound wetland ecological system, which comprised of a fish pond, wastewater pipeline, sedimentation tank, ecological ditch, constructed wetland, multi-functional ecological pond and an intensive pond. The study showed a 73.03% removal for TN, 74.27% for NH_4^+–N, 64.29% for TP and a removal efficiency 91.89% for NO_2^-–N. Owing to the high performance of the compound wetland ecosystem, the authors recommended its use as a low-cost treatment for aquaculture wastewater. Furthermore, the treated water can be recycled and used in fishponds as a way to save water especially in areas where fresh water resources may be short in supply.

These research works have demonstrated the ability of CWs to contribute to environmental sustainability, due to its wastewater remediation potentials and the recirculation, reclamation and reutilization of the aquaculture wastewater (Bôto et al., 2016).

3.1.3 Phycoremediation

Phycoremediation involves the use of algae for the removal or biotransformation of pollutants in wastewater (Olguin, 2003). It does this by taking up nutrients from the wastewater for its growth (Nguyen et al., 2022). Microalgae use various mechanisms, such as biosorption, bioaccumulation and biodegradation for the remediation of wastewaters (Abdelfattah et al., 2023; Phang et al., 2015).

Algae are photosynthetic organisms and are important primary producers, which serve as food in the aquatic food web. Their high photosynthetic efficiency means they grow very fast, which is highly advantageous, since it will lead to an increase in biomass yield. This biomass can be used in a myriad of applications, such as in food, feed, cosmeceuticals, pharmaceuticals as well as feedstock for the production of biodiesel. Therefore, phycoremediation serves the dual purpose of wastewater treatment and the production of cheap algal biomass, which can be used for further applications (Phang et al., 2015).

Microalgae for Aquaculture Wastewater Treatment

Microalgae have been used in the treatment of wastewater from various sources, which include aquaculture wastewater (Ansari et al., 2017), fish processing wastewater (Khalatbari et al., 2024), sewage wastewater (Sharma et al., 2020; Vandana & Rakesh, 2023), wastewater from markets (Apandi et al., 2019), urban wastewater (Singh et al., 2017), textile wastewater (Soeprobowati & Hariyati, 2017).

There are several factors, which must be considered during phycoremediation and they can be broadly classified into abiotic, biotic, operational and economic factors as summarized in Fig. 5.

It is important to screen the microalgae for its ability to remove the target contaminant and to evaluate its intrinsic properties or chemical profile, so that they have high value for potential applications (Phang et al., 2015; Wilkie et al., 2011). In addition, the wastewater concentration profile as well as other biotic and abiotic factors can also influence the microalgae's mechanism and efficiency for contaminant removal in wastewater (Amaro et al., 2023).

Algae systems require a source of carbon, energy and nutrients for metabolism. Therefore, based on their energy and carbon source used in metabolism, they can be classified as photoautotrohic, heterotrophic or mixotrophic as shown in Fig. 6 (Nirmalakhandan et al., 2019).

Under varying external conditions microalgae can exhibit photoautotrophic, mixotrophic and heterotrophic growth and this in turn enables them to remove a variety of contaminants from wastewater (Amaro et al., 2023). Microalgae can use the nutrients from wastewater and CO_2 for their growth, therefore, its use in wastewater treatment aids the sequestration of CO_2, which in turn can contribute to the lowering of global warming (Li et al., 2023; Phang et al., 2015). The use of microalgae for wastewater treatment is a low-energy intensive wastewater treatment technology in terms of microalgae growth. This is because microalgae can derive energy directly from sunlight or from the wastewater (Nirmalakhandan et al., 2019). Therefore, those that are photoautotrophs, require less energy for growth. Microalgae can also be integrated with other waste water treatment methods for enhanced wastewater treatment (Abdelfattah et al., 2023).

Although the use of microalgae in phycoremediation offers numerous advantages or benefits, there are some limitations of it use. One of the limitations of using micro-algae in wastewater remediation is the need to ensure the continuous harvest of the algae biomass after wastewater treatment. This is to prevent high levels of biochemical oxygen demand (BOD) resulting from the biomass decay and the release of the adsorbed nutrients back into the waterbody (Phang et al., 2015). The associated high energy and cost requirement for the harvesting of the algae biomass especially

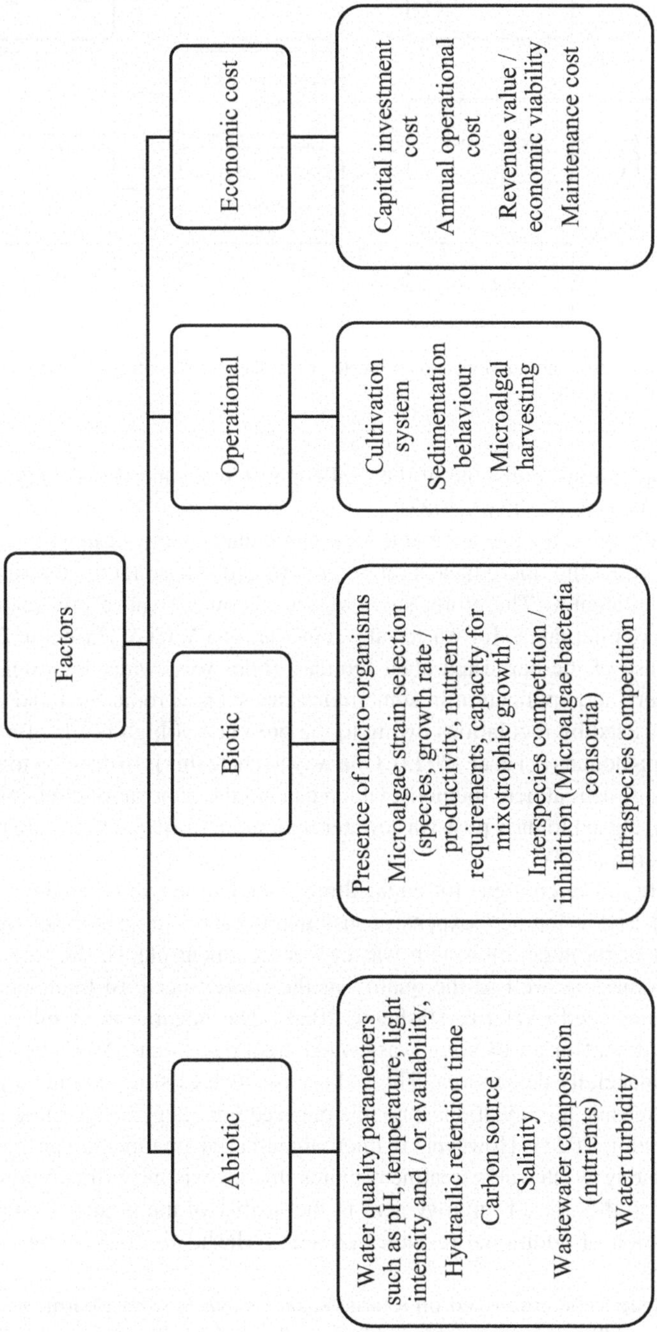

Fig. 5 Factors for Microalgae selection for use in wastewater remediation (adapted from Amaro et al., 2023; Nobre et al., 2024; Rani et al., 2021)

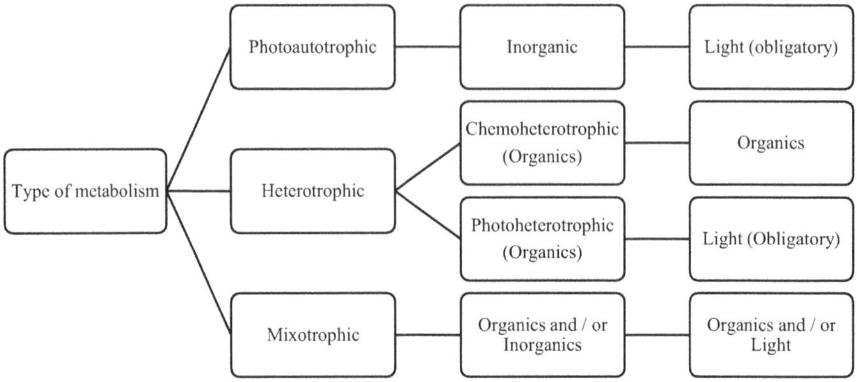

Fig. 6 The use of carbon and energy sources for algal metabolism classification (adapted from Nirmalakhandan et al., 2019)

at large scale operations, are some of the challenges which can pose a challenge to its scalability (Azizi & Hashemi, 2023).

Another challenge is the presence of toxic contaminants in the wastewater, which can negatively affect the microalgae strains' growth and subsequently, the wastewater treatment efficiency. Therefore, such wastewater may require pre-treatment, prior to the use of microalgae for wastewater remediation (Azizi & Hashemi, 2023). Secondly, the use of the harvested algal biomass (after wastewater treatment) for resource recovery and application in some industries such as food, feed and as bio fertilizer is restricted by regulations owing to the presence of high levels of metals, pollutants and pathogens (Li et al., 2023). One way to abate this is to develop low cost and low energy post-treatment methods, which can aid the removal of contaminants in the harvested algae biomass post-wastewater treatment, to ensure their safe re-use (Azizi & Hashemi, 2023).

The scalability of microalgae for aquaculture remediation can be attained if the challenges such as developing inexpensive sustainable harvesting methods to reduce its overall cost or its integration with wastewater treatment plants, the safe reuse of the algal biomass as well as the quality of the wastewater post-treatment, can be effectively managed (Azizi & Hashemi, 2023). The integration of microalgae wastewater treatment in an existing wastewater treatment plant (WWTP) can be complex and financially demanding. This is because such existing treatment plants may require extensive modifications, and this may require intensive investment cost (Azizi & Hashemi, 2023). However, if microalgae-based treatments can be integrated into existing wastewater treatment plants and/or existing infrastructures, it will contribute to the overall improvement in the quality of the treated wastewater and the generation of additional revenues (Azizi & Hashemi, 2023; Nobre et al., 2024).

For microalgae to be employed on a large scale for wastewater treatment, some of these challenges/limitations must be addressed (Li et al., 2023). One way to abate the problem is to use algae strains having special attributes, which include tolerance to extreme temperatures and harsh environments, quick sedimentation behaviour, capacity of mixotrophic growth (Olguin, 2003) and self-regulating capacity to suit fluctuations in environmental conditions (Amaro et al., 2023). The nutrient removal efficiency of microalgae strains may differ and some species may require additional nutrients/nutrient supplementation of the wastewater to improve their growth (Mahmoud, 2024).

Microalgal systems can be comprised of a single species of microalgae or multispecies or a consortium/symbiosis of microalgae and bacteria. Several microalgae strains, which include *Chlorella sorokiniana*, *Scenedesmus obliquus* and *Ankistrodesmus falcatus* (Ansari et al., 2017), *Platymonas subcordiformis*, have been used in aquaculture wastewater remediation (Guo et al., 2013), but *Chlorella* is among the species mostly used. It must be stated that microalgae can also be used in combination with other wastewater treatment methods. Case studies employing micro-algae in integrated wastewater treatment has been documented by Amaro et al. (2023). A summary of studies involving the use of microalgae for aquaculture wastewater remediation is presented in Table 4. The efficacy of this green technology for the remediation of various pollutants, such as heavy metals, nutrients from aquaculture wastewater are presented.

Macroalgae/seaweed for Aquaculture Wastewater Treatment

Seaweeds have been used in wastewater treatment, as they act as biofilters, and as an adsorbent to replace activated carbon (Arumugam et al., 2018). Seaweed harvested after the treatment of wastewater can be used in a plethora of applications, such as in bio-energy production, manure and as feedstock for the production of chemical and pharmaceuticals. Furthermore, the use of seaweed in aquaculture wastewater treatment is easy to adopt and shows high nutrient removal efficiency (Devi & Gowri, 2007), but the removal efficiency varies between seaweed species (Syahrir, 2024).

Syahrir (2024) observed that the seaweed species *G. verrucosa* showed the highest removal efficiency from a shrimp pond waste, when compared with the other species (*K. alvarezii* and *E. spinosum*). It was observed that different nutrient removal efficiencies were obtained for NH_3–N, NO_2^-–N, NO_3^-–N, PO_4^{3-}–P when different macroalgae species were used (Table 5). Cohen et al. (2022) observed that all the seaweed species evaluated removed ammonium and phosphate from the wastewater, but *Gracilaria tenuifrons* and *Gracilaria caudata* showed relatively higher nutrient removal, when compared with *Gracilaria domingensis*. Tanaka et al. (2020) observed that two seaweed species, among the eight species investigated, namely *Padina australis* and *Sargassum hemiphyllum* were most tolerant to low salinities. Therefore, the authors recommended their use in brackish and estuarine aquaculture systems.

Various macroalgal species have been used in wastewater remediation studies (Table 5). According to Copertino et al. (2009), *Ulva spp.* has one of the highest photosynthetic and nutrient uptakes, due to its large surface area to volume ratio,

Table 4 Aquaculture wastewater treatment using microalgae

Microalgae sp.	Wastewater source/type	Culture system	Pollutant reduction efficiency	Biomass productivity (mg/L/day)	Treatment time (days)	Amount of inoculation (g/L)	References
Chlorella sorokiniana	Fish farm in tarpaulin tanks	Experimental	Al = 55.54% Cd = 50.82% Mg = 21.71% Pb = 82.61% Zn = 84.06%	NS	28 days	NS	Carvalho et al. (2021)
Tetraselmis suecica	Turbot Fish farm	Tubular photobioreactor	N = 49.4% P = 99.0%	350	15 days	0.5	Michels et al. (2014)
Chlorella vulgaris	Aquaculture rearing tank of *Penaeus vannamei Boone*	Membrane photobioreactor (MBPR) continuous mode	TN = 86.1% TP = 82.7%	42.6	1 day	0.41	Gao et al. (2016)
Scenedesmus obliquus	Nile Tilapia Aquaculture wastewater	Batch	COD = 42% NO_3^-–N = 77.7% NO_2^-–N = 73.83% Total oxidizable nitrogen (TON) = 68.09% NH_4^+–N = 88.71% PO_4^{3-}–P = 100%	89.61	14 days	0.19	Ansari et al. (2017)

(continued)

Table 4 (continued)

Microalgae sp.	Wastewater source/type	Culture system	Pollutant reduction efficiency	Biomass productivity (mg/L/day)	Treatment time (days)	Amount of inoculation (g/L)	References
Chlorella sorokiniana	Nile Tilapia Aquaculture wastewater	Batch	COD = 69% NO_3^-–N = 75.76% NO_2^-–N = 81.79% TON = 67.89% NH_4^+–N = 98.21% PO_4^{3-}–P = 100%	107.85	14 days	0.22	Ansari et al. (2017)
Ankistrodesmus falcatus	Nile Tilapia Aquaculture wastewater	Batch	COD = 61% NO_3^-–N = 80.85% NO_2^-–N = 99.73% TON = 75.29% NH_4^+–N = 86.45% PO_4^{3-}–P = 98.52%	160.79	14 days	0.2	Ansari et al. (2017)

(continued)

Table 4 (continued)

Microalgae sp.	Wastewater source/type	Culture system	Pollutant reduction efficiency	Biomass productivity (mg/L/day)	Treatment time (days)	Amount of inoculation (g/L)	References
Chlorella vulgaris	Brown crab (*Cancer pagurus*) aquaculture wastewater	Batch	TN = 100% TP = 100% COD = 86.6% BOD_5 = 72.2%	146.4	22 days	4.9–5.15	Viegas et al. (2021)
Scenedesmus obliquus	Brown crab (*Cancer pagurus*) aquaculture wastewater	Batch	TN = 100% TP = 100% COD = 90.6% BOD_5 = 52.0%	NS	22 days	4.9–5.15	Viegas et al. (2021)
Isochrysis galbana	Brown crab (*Cancer pagurus*) aquaculture wastewater	Batch	TN = 100% TP = 100% COD = 83.9% BOD_5 = 68.2%	NS	33 days	4.9–5.15	Viegas et al. (2021)
Nannocloropsis salina	Brown crab (*Cancer pagurus*) aquaculture wastewater	Batch	TN = 100% TP = 100% COD = 71.8% BOD_5 = 77.7%	20.9	33 days	4.9–5.15	Viegas et al. (2021)
Spirulina major	Brown crab (*Cancer pagurus*) aquaculture wastewater	Batch	TN = 100% TP = 100% COD = 94.0% BOD_5 = 68.0%	NS	33 days	4.9–5.15	Viegas et al. (2021)

(continued)

Table 4 (continued)

Microalgae sp.	Wastewater source/type	Culture system	Pollutant reduction efficiency	Biomass productivity (mg/L/day)	Treatment time (days)	Amount of inoculation (g/L)	References
Chlorella vulgaris	Brown crab (*Cancer pagurus*) aquaculture wastewater	Semi-continuous	TN = 100% TP = 100% COD = 96.2% BOD_5 = 99.7%	879.8	34 days	4.9–5.15	Viegas et al. (2021)
Scenedesmus obliquus	Brown crab (*Cancer pagurus*) aquaculture wastewater	Semi-continuous	TN = 100% TP = 100% COD = 97.7% BOD_5 = 99.7%	811.7	34 days	4.9–5.15	Viegas et al. (2021)
Tetradesmus obliquus	Fish farm	Photobioreactor	NH_3 = 100% NO_3^- = 67.31% PO_4^{3-} = 54.97%	~ 11	15 days	NS	Enwereuzoh et al. (2021)
Chlamydomonas reinhardtii	Fish tank	Batch	NO_3^- = 82% TP = 100% COD = 62.3%	N.S	16 days	NS	Mahmoud (2024)
Chlorella vulgaris	Fish tank	Batch	NO_3^- = 80% TP = 100% COD = 48.1%	N.S	16 days	NS	Mahmoud (2024)

(continued)

Table 4 (continued)

Microalgae sp.	Wastewater source/type	Culture system	Pollutant reduction efficiency	Biomass productivity (mg/L/day)	Treatment time (days)	Amount of inoculation (g/L)	References
Haematococcus lacustris	Fish tank	Batch	NO_3^- = 78.4% TP = 100% COD = 58.97%	N.S	16 days	NS	Mahmoud (2024)
Chlorella sorokiniana	Aquaculture wastewater (*Orechromis nilotics* Recirculating aquaculture system)	Photobioreactor	N = 78% P = 77% Mg = 70% Zn = 90% Ni = 99%	140	NS	NS	Lugo et al. (2020)
Combined system (Mussel /microalgae (*Chlorella vulgaris*)/ bacteria)	Aquaculture wastewater	Batch	NH_3–N = 94.67% COD = 77.78% TP = 92.89%	NS	6 days	NS	Geng et al. (2022)

Note NS = Not stated

Table 5 Macroalgae used in aquaculture wastewater treatment

Seaweed sp.	Wastewater source/type	Scale/ application mode	Treatment time	Removal efficiency (%)	References
Gracilaria caudata + microcrustacean (*Artemia fransciscana*)	Aquaculture effluents	Lab study	72 h	NH_4^+ = 29.8% NO_2^- = 100% NO_3^- = 72.4%	Marinho-Soriano et al. (2011)
Gracilaria birdiae	Shrimp farm pond (*L. vannamei*) effluent	Lab study (Aquarium)	4 weeks	PO_4^{3-} = 93.5% NH_4^+ = 34% NO_3^- = 100%	Marinho-Soriano et al. (2009a)
Chlorella vulgaris	Fish farm	Semi-industrial reactor	3 weeks	PO_4^{3-} = 80.55% NO_3^- = 80.76%	Askari Hesni et al. (2019)
Gracilaria lemaneiformis	–	IMTA (cultivated with scallops (*Chlamys farreri*)	3 weeks	NH_4^+ = 83.7% P = 70.4%	Mao et al. (2009)
Gracillaria veruucosa	Shrimp pond	Lab scale	42 days	NH_3–N = 97.1% NO_2^-–N = 99.7% NO_3^-–N = 99.9% PO_4^{3-}–P = 99.7%	Syahrir (2024)
Kappaphycus alvarezii	Shrimp pond	Lab scale	42 days	NH_3–N = 81.6% NO_2^-–N = 94.6% NO_3^-–N = 94.5% PO_4^{3-}–P = 95.4%	Syahrir (2024)

(continued)

Table 5 (continued)

Seaweed sp.	Wastewater source/type	Scale/ application mode	Treatment time	Removal efficiency (%)	References
Eucheuma spinosum	Shrimp pond	Lab scale	42 days	NH_3–N = 90.5% NO_2^-–N = 93.9% NO_3^-–N = 96.4% PO_4^{3-}–P = 95.4%	Syahrir (2024)
Gracilaria caudata	Shrimp pond	Lab scale (conducted outside the lab)	4 h	NH_4^+–N = 59.5% NO_3^-–N = 49.6% PO_4^{3-}–P = 12.3%	Marinho-Soriano et al. (2009b)

which explains the choice as an ideal seaweed candidate for wastewater remediation/biofiltration. When used for wastewater treatment, certain factors have to be considered. The seaweed selected should have a wide range of salinity tolerance, high surface area to volume ratio in order to increase its nutrient uptake rate (Copertino et al., 2009). In addition, the seaweed should have some commercial value, to ensure a circular economy (Baruah et al., 2006).

Integrated multi-trophic aquaculture (IMTA) practice i.e. the integration of fish culture with that of extractive species such as seaweed, is one of the ways seaweeds have been used in aquaculture wastewater remediation (Copertino et al., 2009). This method is mutually beneficial/symbiotic, as the by-products from the culture of one aquatic species, such as fish, serves as a resource for the seaweed species, and this promotes environmental and economic benefits, as well as better management practices (Baruah et al., 2006). In this method, both inorganic and organic pollutants can be removed, and the system can be combined with aerobic—anaerobic treatment units.

Majority of studies have focused on the integration of seaweed and fish aquaculture (Copertino et al., 2009). The stocking density, nutrient in wastewater, the seaweed's ability to eco-physiologically adapt to the culture conditions and environmental factors are some factors, which can influence the seaweed's removal efficiency (Copertino et al., 2009). In Integrated Multiple-Trophic Aquaculture (IMTA), the suitability of the seaweed for integrated aquaculture as well as the economical/commercial value of the seaweed species must be considered (Al-Hafedh et al., 2015). For instance, *Ulva* sp. has lower economical value when compared with *Gracilaria* or other seaweed species, as the latter are commonly used as edible seaweeds/as human foods, as well as for phycocolloid production and other commercial valuable products (Al-Hafedh et al., 2015). Furthermore, its high yield and high bioremediation efficiency is also noteworthy (Chopin et al., 2001). Although a promising wastewater

Table 6 Characteristics, advantages and limitations of low-cost technologies for aquaculture wastewater treatment

Methods	Characteristics	Advantage	Limitations
Adsorption	It is a phenomenon, whereby molecules or ions of target pollutants (adsorbate), adhere onto the surface of a solid material (adsorbent) through intermolecular forces (Satyam & Patra, 2024).	1. Cheap, eco-friendly and can remove organic pollutants from wastewater (El-Sewify et al., 2022). 2. Waste materials can be utilized as adsorbents and the variety of adsorbents with high sorption capacities can render the treatment process effective (Younas et al., 2021).	1. Some of the adsorbents show limited adsorption capacity (El-Sewify et al., 2022) and varying efficiencies in aquaculture wastewater treatment (El-Sherbiny et al., 2019). 2. Adsorption can be physisorption or chemisorption (Artioli, 2008). In the case of the latter, desorption potential of spent adsorbents for re-use may be low thus limiting its reusability potential for further adsorption cycles (Egbedi et al., 2023).

(continued)

Table 6 (continued)

Methods	Characteristics	Advantage	Limitations
Constructed wetlands	They are engineered systems which use natural wetland processes under controlled conditions for the removal of pollutants from wastewater streams (Vymazal, 2022).	1. Low-cost and green wastewater treatment technology (Gorgoglione & Torretta, 2018). 2. It can be used in combination with other systems such as phytoremediation, to improve its efficiency in removing a wide range of contaminants (Li et al., 2022). 3. It can be used in both pilot/small scale and commercial scale (Li et al., 2022). 4. The operation and maintenance requirements are few (Gorgoglione & Torretta, 2018). 5. Can be readily applied in developing countries as a low-cost wastewater treatment option (Gorgoglione & Torretta, 2018).	1. The demand for land area is a limitation (Huang et al., 2013). 2. Vulnerable to temperature and climatic conditions (Huang et al., 2013). 3. The system is prone to clogging (Wang et al., 2021). 4. There is no standard design.

(continued)

Table 6 (continued)

Methods	Characteristics	Advantage	Limitations
Phyco-remediation (Microalgal phyco-remediation)	It involves the use of microalgae strains for the removal of pollutants from wastewater (Pradhan et al., 2022). Photosynthetic phytoplanktonic microalgae strains are usually employed (Sisman-Aydin, 2022).	1. It is an eco-friendly approach (Bhatt et al., 2023). 2. Value added products such as biofuel and micro-algal biomass can be derived and used in a plethora of applications which in turn can lead to a circular economy (Bhatt et al., 2023; Sisman-Aydin, 2022).	1. System requires large space for plant set-up and expensive bioreactors for installation (Bhatt et al., 2023). 2. High energy and cost requirement for the harvesting of the algae biomass pose a challenge to its scalability (Azizi & Hashemi, 2023). 3. All-year round light is needed and this would not be feasible for developing countries which may lack this facility (Bhatt et al., 2023). 4. Algal species selection is pivotal as its growth conditions and effectiveness is dependent on several factors such as nutrient level in the wastewater (Bhatt et al., 2023). 5. The use of the harvested algal biomass (post wastewater treatment) for resource recovery and application in some industries such as food, feed and as bio-fertilizer is restricted by regulations (Li et al., 2023).

(continued)

Table 6 (continued)

Methods	Characteristics	Advantage	Limitations
Phyco-remediation (Macroalgal phyco-remediation)	It involves the use of macroalgae/ seaweed for the removal of pollutants from wastewater (Arumugam et al., 2018).	1. The use of seaweed in aquaculture wastewater treatment is easy to adopt and shows high nutrient removal efficiency (Devi & Gowri, 2007). 2. It has a high biomass yield (Chopin et al., 2001). 3.Seaweed harvested after the treatment with wastewater can be used in various applications (Devi & Gowri, 2007). 4. Seaweed can be used in Integrated multi-trophic aquaculture (IMTA) practice for aquaculture wastewater remediation (Mao et al., 2009).	1. Removal efficiency of target pollutant varies between seaweed species (Syahrir, 2024). 2. Reuse of macroalgae biomass (post-wastewater treatment), for some applications may be restricted owing to regulations.

treatment technology, there is need for more research on pilot scale research as well as the regulatory acceptance of the biomass after wastewater treatment, for use in other applications.

Some of these low-cost technologies highlighted in Table 6 are nature-based solutions that offer a greener route, when compared with the conventional technologies, such as chemical treatments. The application of these technologies in developing countries is highly feasible because of the benefits they offer in terms of removal efficiency and low-cost.

In order to ensure that these technologies such as the use of constructed wetlands, adsorption and phycoremediation are embraced, there is need to raise awareness among farmers, local communities as well as consumers who are the end users. Engagement of local residents of communities in citizen science is important as it will allow for the full acceptance of these emerging technologies for wastewater treatment and management, and the enforcement of existing water regulations (Shahady & Boniface, 2018).

4 Conclusion

Aquaculture has a significant role to play in ensuring a food secure world for the growing human population. However, the issue of its wastewater poses a severe risk to its sustainability. The prevention of environmental degradation of receiving waterbodies, biodiversity loss and the negative impact of polluted waterbodies on human health can be attained through the effective management and treatment of wastewater generated from aquaculture practices. The use of adsorption technology, constructed wetlands (CWs) and phycoremediation has garnered interests owing to their numerous advantages in terms of cost, environmental friendliness, low-technical expertise, high efficiency in wastewater remediation over the conventional wastewater treatment methods. In developing countries, where finances may prevent the use of advanced wastewater treatment methods, these methods offer a viable and sustainable wastewater treatment option. In addition, the fact that these methods can be integrated in the form of a combined system, increases their potentials as alternative wastewater treatment technologies. There is a need for more pilot scale research where the integration of these systems in already existing wastewater treatment plants/facilities can be done, to evaluate their efficiency in real-time mode.

Acknowledgements This work was conducted at South East Technological University, Waterford, Ireland. The authors would like to thank the Irish Research Council and the HEA for the PhD funding.

References

Abdelfattah, A., Ali, S. S., Ramadan, H., El-Aswar, E. I., Eltawab, R., Ho, S.-H., Elsamahy, T., Li, S., El-Sheekh, M. M., Schagerl, M., Kornaros, M., & Sun, J. (2023). Microalgae-based wastewater treatment: Mechanisms, challenges, recent advances, and future prospects. *Environmental Science and Ecotechnology, 13*, Article 100205. https://doi.org/10.1016/j.ese.2022.100205

Abou-Elela, S. I. (2019). Constructed wetlands: The green technology for municipal wastewater treatment and reuse in agriculture. In A. Negem (Ed.), *Unconventional water resources and agriculture in Egypt. The handbook of environmental chemistry* (vol. 75, pp. 189–239). Springer. https://doi.org/10.1007/698_2017_69

Acierno, R., Blancheton, J. P., Bressani, G., Ceruti, L., Chadwick, D., Roque D'Orbcastel, E., Claricoates, J., & Donaldson, G. (2006). Manual on effluent treatment in aquaculture: Science and practice. Aquaetreat. 163 pp. https://archimer.ifremer.fr/doc/00000/6496/

Action Guide. (n.d.). *Global programme of action for the protection of marine environment (GPA/UNEP).* https://actionguide.info/m/orgs/209/

Aderibigbe, A. M. (2018). *Assessment of sawdust activated carbon (SAC) in the treatment of aquaculture effluent* (Master's Thesis) (104 pp). Ahmadu Bello University, Zaria, Nigeria.

Agaton, C. B., & Guila, P. M. (2024). Success factors and challenges: Implications of real options valuation of constructed wetlands as nature-based solutions for wastewater treatment. *Resources, 13*(1), 11. https://doi.org/10.3390/resources13010011

Ahmad, A., Sheikh Abdullah, S. R., Hasan, H. A., Othman, A. R., & Ismail, N. I. (2021). Aquaculture industry: Supply and demand, best practices, effluent and its current issues and treatment technology. *Journal of Environmental Management, 287*, Article 112271. https://doi.org/10.1016/j.jenvman.2021.112271

Akinbile, C. O., Epebinu, E. M., Olanrewaju, O. O., & Abolude, A. T. (2022). Performance analysis of bamboo-based (*Bambusa vulgaris*) activated carbon in aquaculture wastewater treatment. *Sustainable Water Resources Management, 8*, 178. https://doi.org/10.1007/s40899-022-00767-1

Al-Dawery, S. K., Al-Yaqoubi, G. E., Al-Musharrafi, A. A., Harharah, H. N., Amari, A., & Harharah, R. H. (2023). Treatment of fish-processing wastewater using polyelectrolyte and palm anguish. *Processes, 11*(7), 2124. https://doi.org/10.3390/pr11072124

Al-Hafedh, Y. S., Alam, A., & Buschmann, A. H. (2015). Bioremediation potential, growth and biomass yield of the green seaweed, *Ulva lactuca* in an integrated marine aquaculture system at the Red Sea coast of Saudi Arabia at different stocking densities and effluent flow rates. *Reviews in Aquaculture, 7*(3), 161–171. https://doi.org/10.1111/raq.12060

Amaro, H. M., Salgado, E. M., Nunes, O. C., Pires, J. C. M., & Esteves, A. F. (2023). Microalgae systems—Environmental agents for wastewater treatment and further potential biomass valorisation. *Journal of Environmental Management, 337*, Article 117678. https://doi.org/10.1016/j.jenvman.2023.117678

Ansari, F. A., Singh, P., Guldhe, A., & Bux, F. (2017). Microalgal cultivation using aquaculture wastewater: Integrated biomass generation and nutrient remediation. *Algal Research, 21*, 169–177. https://doi.org/10.1016/j.algal.2016.11.015

Apandi, N., Mohamed, R. M. S. R., Al-Gheethi, A., Gani, P., Ibrahim, A., & Kassim, A. H. M. (2019). *Scenedesmus* biomass productivity and nutrient removal from wet market wastewater, a bio-kinetic study. *Waste and Biomass Valorization., 10*, 2783–2800. https://doi.org/10.1007/s12649-018-0313-y

Artioli, Y. (2008). Adsorption. In: S. E. Jørgensen & B. D. Fath (Eds.), *Encyclopedia of ecology* (pp. 60–65). Oxford: Academic Press. https://doi.org/10.1016/B978-008045405-4.00252-4

Arumugam, N., Chelliapan, S., Kamyab, H., Thirugnana, S., Othman, N., & Nasri, N. S. (2018). Treatment of wastewater using seaweed: A review. *International Journal of Environmental Research and Public Health, 15*(12), 2851. https://doi.org/10.3390/ijerph15122851

Askari Hesni, M., Hedayati, S. A., Qadermarzi, A., Pouladi, M., Zangiabadi, S., & Naqshbandi, N. (2019). Comparison ability of algae and nanoparticles on nitrate and phosphate removal from aquaculture wastewater. *Environmental Health Engineering and Management, 6*(3), 171–177. https://doi.org/10.15171/EHEM.2019.19

Ayers, R. S., & Westcot, D. W. (1985). *Water quality for agriculture. Food and agriculture organization of the United Nations* (174 pp).

Azizi, M. S., & Hashemi, K. A. (2023). Grand challenges in application of microalgae for nutrient recovery. *Journal of Resource Recovery, 1*, 1019. https://doi.org/10.61186/jrr.2308.1019

Baruah, K., Norouzitallab, P., & Sorgeloos, P. (2006). Seaweeds: An ideal component for wastewater treatment for use in aquaculture. *Aquaculture Europe, 31*, 3–6.

Bensalem, S., Henni, A., Chouana, T., Hidouci, S., & Azib, S. (2024). Removal of pollutants from aquaculture wastewater using chitosan/bentonite composite beads. *Journal of Environmental Engineering and Science, 19*(3), 189–199. https://doi.org/10.1680/jenes.23.00012

Bhatnagar, A., & Sillanpää, M. (2011). A review of emerging adsorbents for nitrate removal from water. *Chemical Engineering Journal, 168*(2), 493–504. https://doi.org/10.1016/j.cej.2011.01.103

Bhatt, P., Huang, J. Y., Brown, P., Shivaram, K. B., Yakamercan, E., & Simsek, H. (2023). Electrochemical treatment of aquaculture wastewater effluent and optimization of the parameters using response surface methodology. *Environmental Pollution, 331*, Article 121864. https://doi.org/10.1016/j.envpol.2023.121864

Boopathy, R., & Lyles, C. N. (2008). Shrimp production and biological treatment of shrimp wastewater in the United States. In A. Pandy, C. Larroche, C. R. Soccol, & C. G. Dussap (Eds.), *New horizons in biotechnology* (pp. 235–252). Asia Tech Publishers.

Bôto, M., Almeida, C. M. R., & Mucha, A. P. (2016). Potential of constructed wetlands for removal of antibiotics from saline aquaculture effluents. *Water, 8*(10), 465. https://doi.org/10.3390/w8100465

Boyd, C. E. (2003). Guidelines for aquaculture effluent management at the farm-level. *Aquaculture, 226*(1–4), 101–112. https://doi.org/10.1016/S0044-8486(03)00471-X

Bricknell, I. R., Birkel, S. D., Brawley, S. H., Van Kirk, T., Hamlin, H. J., Capistrant-Fossa, K., Huguenard, K., Van Walsum, G. P., Liu, Z. L., Zhu, L. H., Grebe, G., Taccardi, E., Miller, M., Preziosi, B. M., Duffy, K., Bryon, C. J., Quigley, C. T. C., Bowden, T. J., Brady, D., … Moeykens, S. (2021). Resilience of cold water aquaculture: A review of likely scenarios as climate changes in the Gulf of Maine. *Reviews in Aquaculture, 13*(1), 460–503. https://doi.org/10.1111/raq.12483

Carvalho, E. M., Santos, C. R., Ansilago, M., Menegazzo, M. L., Nunes, N. S. P. (2021). Phycoremediation of fish farm wastewater by *Chlorella sorokiniana* and autochthonous microalgae. *Research, Society and Development, 10*(13), e259101320723. https://doi.org/10.33448/rsd-v10i13.20723

Chopin, T., Buschmann, A. H., Halling, C., Troell, M., Kautsky, N., Neori, A., Kraemer, G. P., Zertuche-González, J. A., Yarish, C., & Neefus, C. (2001). Integrating seaweeds into marine aquaculture systems: A key toward sustainability. *Journal of Phycology, 37*, 975–986. https://doi.org/10.1046/j.1529-8817.2001.01137.x

Cohen, F. P. A., Faria, A. V. F., Braga, E. S., Chiozzini, V. G., & Plastino, E. M. (2022). Gracilarioid algae (Rhodophyta) cultured in eutrophic synthetic seawater: Potential for growth and preliminary bioremediation assessment. *Journal of Applied Phycology, 34*, 2783–2791. https://doi.org/10.1007/s10811-022-02728-9

Colares, G. S., Dell'Osbel, N., Wiesel, P. G., Oliveira, G. A., Lemos, P. H. Z., da Silva, F. P., Lutterbeck, C. A., Kist, L. T., & Machado, Ê. L. (2020). Floating treatment wetlands: A review and bibliometric analysis. *Science of the Total Environment, 714*, Article 136776. https://doi.org/10.1016/j.scitotenv.2020.136776

Copertino, M. S., Tormena, T., & Seeliger, U. (2009). Biofiltering efficiency, uptake and assimilation rates of *Ulva clathrata* (Roth) J. Agardh (Clorophyceae) cultivated in shrimp aquaculture waste water. *Journal of Applied Phycology, 21*, 31–45. https://doi.org/10.1007/s10811-008-9357-x

Corbella, C., & Puigagut, J. (2018). Improving domestic wastewater treatment efficiency with constructed wetland microbial fuel cells: Influence of anode material and external resistance.

Science of the Total Environment, 631–632, 1406–1414. https://doi.org/10.1016/j.scitotenv.2018.03.084
Crab, R., Avnimelech, Y., Defoirdt, T., Bossier, P., & Verstraete, W. (2007). Nitrogen removal techniques in aquaculture for a sustainable production. *Aquaculture, 270*(1–4), 1–14. https://doi.org/10.1016/j.aquaculture.2007.05.006
Dauda, A. B., Ajadi, A., Tola-Fabunmi, A. S., & Akinwole, A. O. (2019). Waste production in aquaculture: Sources, components and managements in different culture systems. *Aquaculture and Fisheries, 4*(3), 81–88. https://doi.org/10.1016/j.aaf.2018.10.002
Devi, I. R. P., & Gowri, V. S. (2007). Biological treatment of aquaculture discharge waters by seaweeds. *Journal of Industrial Pollution Control, 23*(1), 135–140.
Ding, G. K. C. (2023). Managing water scarcity—The role of sustainable water management. In: K. Aviso, W. Deng, V. Strezov, Z. C. Dong, M. A. Panero, A. Guptill, J. Wu, G. P. Dillon, & G. P. Mendis (Eds.), *Encyclopedia of sustainable technogies* (2nd ed., pp. 766–789). Elsevier. https://doi.org/10.1016/B978-0-323-90386-8.00063-2
Drechsel, P., Marjani, Z. S., & Pedrero, F. (Eds.). (2023). *Water quality in agriculture: Risks and risk mitigation* (175 pp). FAO and IWMI. https://doi.org/10.4060/cc7340en
Edwards, P., & Demaine, H. (1997). *Rural aquaculture: Overview and framework for country reviews*. Regional Office for Asia and the Pacific, FAO, Thailand.
Egbedi, B., Kinsella, M., Tan, S. P., & Hughes, H. (2023). Phosphate removal from wastewater using scallop and whelk shells. *Journal of Water Process Engineering, 55*, Article 104159. https://doi.org/10.1016/j.jwpe.2023.104159
El-Sewify, I. M., Radwan, A., Shahat, A., El-Shahat, M. F., & Khalil, M. M. H. (2022). Superior adsorption and removal of aquaculture and bio-staining dye from industrial wastewater using microporous nanocubic Zn-MOFs. *Microporous and Mesoporous Materials, 329*, Article 111506. https://doi.org/10.1016/j.micromeso.2021.111506
El-Sherbiny, M., El-Chaghaby, G. A., & Abd El-Shafea, M. Y. (2019). Treatment of aquaculture waste effluent to be reused in fish culture in Egypt. *Egyptian Journal of Aquatic Biology and Fisheries, 23*(1), 233–243.
Enwereuzoh, U., Harding, K., & Low, M. (2021). Microalgae cultivation using nutrients in fish farm effluent for biodiesel production. *South African Journal of Chemical Engineering, 37*, 46–52. https://doi.org/10.1016/j.sajce.2021.03.007
EU Commission. (n.d.). *Water framework directive*. https://environment.ec.europa.eu/topics/water/water-framework-directive_en
FAO. (1995). *Code of conduct for responsible fisheries* (41 pp). FAO.
FAO. (2022). *The state of word fisheries and aquaculture 2022. Towards blue transformation* (236 pp). FAO. https://doi.org/10.4060/cc0461en
FAO. (2024). *The state of world fisheries and aquaculture 2024—Blue transformation in action* (232 pp). https://doi.org/10.4060/cd0683en
Funge-Smith, S., & Phillips, M. J. (2001). Aquaculture in the third millennium. In: R. P. Subasinghe, P. B. Bueno, M. J. Phillips, C. Hough, S. E. McGladdery, & J. R. Arthur (Eds.), *Technical proceedings of the conference on aquaculture in the third millennium* (pp. 129–135). NACA and FAO.
Gao, F., Li, C., Yang, Z. H., Zeng, G. M., Feng, L. J., Liu, J., Liu, M., & Cai, H. (2016). Continuous microalgae cultivation in aquaculture wastewater by a membrane photobioreactor for biomass production and nutrients removal. *Ecological Engineering, 92*, 55–61. https://doi.org/10.1016/j.ecoleng.2016.03.046
Geng, B., Li, Y., Liu, X., Ye, J., & Guo, W. (2022). Effective treatment of aquaculture wastewater with mussel/microalgae/bacteria complex ecosystem: A pilot study. *Science and Reports, 12*, 2263. https://doi.org/10.1038/s41598-021-04499-8
Ghaly, A. E., Kamal, M., & Mahmoud, N. S. (2005). Phytoremediation of aquaculture wastewater for water recycling and production of fish feed. *Environment International, 31*, 1–13. https://doi.org/10.1016/j.envint.2004.05.011

Gorgoglione, A., & Torretta, V. (2018). Sustainable management and successful application of constructed wetlands: A critical review. *Sustainability, 10*(11), 3910. https://doi.org/10.3390/su10113910

Government of Canada. (2024). *Overview of the wastewater systems effluent regulations*. Last updated June 14, 2024. https://www.canada.ca/en/environment-climate-change/services/wastewater/regulations.html

Guimarães, J. T., Souza, A. L. M., Brígida, A. I. S., Furtado, A. A. L., Chicrala, P. C. M., Santos, V. R. V., Alves, R. R., Luiz, D. B., & Mesquita, E. F. M. (2018). Quantification and characterization of effluents from the seafood processing industry aiming at water reuse: A pilot study. *Journal of Water Process Engineering, 26*, 138–145. https://doi.org/10.1016/j.jwpe.2018.10.006

Guo, Z., Liu, Y., Guo, H., Yan, S., & Mu, J. (2013). Microalgae cultivation using an aquaculture wastewater as growth medium for biomass and biofuel production. *Journal of Environmental Sciences, 25*, S85–S88. https://doi.org/10.1016/S1001-0742(14)60632-X

Hamed, M. M., Hussein, M. S., Abd-Eltwab, A. M., & Alsaiad, S. M. (2022). Using the natural zeolite (clinoptiolite) in removing ammonia, heavy metals and improving water quality in fish ponds. *Al-Azhar Journal of Agricultural Research, 47*, 79–88. https://doi.org/10.21608/ajar.2022.266486

Hasan, M. M., Haque, M. M., Hasan, N. A., Bashar, A., Ahammad, A. K. S., & Hossain, M. T. (2023). Assessing the impacts of zeolite on water quality, growth performance, heavy metal content and health condition of farmed tilapia (*Oreochromis niloticus*). *Aquaculture Reports, 31*, Article 101678. https://doi.org/10.1016/j.aqrep.2023.101678

Hassan, I., Chowdhury, S. R., Prihartato, P. K., & Razzak, S. A. (2021). Wastewater treatment using constructed wetland: current trends and future potential. *Processes., 9*(11), 1917. https://doi.org/10.3390/pr9111917

Hiraishi, T., Krug, T., Tanabe, K., Srivastava, N., Baasansuren, J., Fukuda, M., & Troxler, T. G. (2014). 2013 supplement to the 2006 IPCC guidelines for national greenhouse gas inventories: Wetlands (354 p.). IPCC, Switzerland.

Huang, J. L., Chen, Q., & Xu, L. H. (2013). Problems and countermeasures in the application of constructed wetlands. *Huan jing ke xue, 34*(1), 401–408.

Hussan, A., Adhikari, S., Das, A., Hoque, F., & Pillai, B. R. (2019). Fish culture without water!!! Does aquaculture contribute to water scarcity? Is aquaculture without water possible. *International Journal of Fisheries and Aquatic Studies, 7*(5), 39–48.

Iber, B. T., Torsabo, D., Engku, C. C. E. N. C., Wahab, F., Abdullah, S. R. S., Hassan, H. A., & Kasan, N. A. (2023a). A study on the recovery and characterization of suspended solid from aquaculture wastewater through coagulation/flocculation using chitosan and its viability as organic fertilizer. *Journal of Agriculture and Food Research, 11*, Article 100532. https://doi.org/10.1016/j.jafr.2023.100532

Iber, B. T., Torsabo, D., Chik, C. E. N. C. E., Wahab, F., Abdullah, S. R. S., Hassan, H. A., & Kasan, N. A. (2023b). Response surface methodology (RSM) approach to optimization of coagulation-flocculation of aquaculture wastewater treatment using chitosan from carapace of giant freshwater prawn *Macrobrachium rosenbergii*. *Polymers (Basel)., 15*(4), 1058. https://doi.org/10.3390/polym15041058

Igwegbe, C. A., & Onukwuli, O. D. (2019). Removal of total dissolved solids (TDS) from aquaculture wastewater by coagulation-flocculation process using Sesamum indicum extract: Effect of operating parameters and coagulation-flocculation kinetics. *Pharmaceutical Chemistry Journal, 6*(4), 32–45.

International Finance Corporation. (2007). *Environmental, health and safety guidelines for aquaculture* (19 pp).

International Finance Corporation. (n.d.a). *Who we are*. https://www.ifc.org/en/about

International Finance Corporation. (n.d.b). *Environmental, health, and safety (EHS) guidelines*. https://sustainable-infrastructure-tools.org/tools/environmental-health-and-safety-ehs-guidelines

International Finance Corporation. (n.d.c). *Manual or guideline-environmental, health, and safety guidelines*. https://www.ifc.org/en/insights-reports/2000/general-environmental-health-and-safety-guidelines

IWMI. (2023). *FAO and IWMI's milestone book gives actionable guidelines on water quality for agriculture*. https://www.iwmi.cgiar.org/news/fao-and-iwmis-milestone-book-gives-actionable-guidelines-on-water-quality-for-agriculture

Ji, Z., Tang, W., & Pei, Y. (2022). Constructed wetland substrates: A review on development, function mechanisms, and application in contaminants removal. *Chemosphere, 286*, Article 131564. https://doi.org/10.1016/j.chemosphere.2021.131564

Jusoh, A., Nasir, N. M., Yunos, F. H. M., Jusoh, H. H. W., & Lam, S. S. (2020). Green technology in treating aquaculture wastewater. *AIP Conference Proceedings, 2197*, Article 020001. https://doi.org/10.1063/1.5140892

Kashem, A. H. M., Das, P., Hawari, A. H., Mehariya, S., Thaher, M. I., Khan, S., Abduquadir, M., & Al-Jabri, H. (2023). Aquaculture from inland fish cultivation to wastewater treatment: A review. *Reviews in Environmental Science and Bio/Technology, 22*, 969–1008. https://doi.org/10.1007/s11157-023-09672-1

Kayode, O. F., Luethi, C., & Rene, E. R. (2018). Management recommendations for improving decentralized wastewater treatment by the food and beverage industries in Nigeria. *Environments, 5*(3), 41. https://doi.org/10.3390/environments5030041

Khalatbari, S., Sotaniemi, V. H., Suokas, M., Taipale, S., & Leiviskä, T. (2024). Microalgae technology for polishing chemically-treated fish processing wastewater. *Groundwater for Sustainable Development, 24*, Article 101074. https://doi.org/10.1016/j.gsd.2023.101074

Krasaesueb, N., Boonnorat, J., Maneeruttanarungroj, C., & Khetkorn, W. (2023). Highly effective reduction of phosphate and harmful bacterial community in shrimp wastewater using short-term biological treatment with immobilized engineering microalgae. *Journal of Environmental Management, 325*, Article 116452. https://doi.org/10.1016/j.jenvman.2022.116452

Kurniawan, S. B., Abdullah, S. R. S., Imron, M. F., Ahmad, A., Mohd Said, N. S., Mohd Rahim, N. F., Alnawajha, M. M., Abu Hasan, H., Othman, A. R., & Purwanti, I. F. (2021). Potential of valuable materials recovery from aquaculture wastewater: An introduction to resource reclamation. *Aquaculture Research, 52*(7), 2954–2962. https://doi.org/10.1111/are.15180

Laiq, E., Jabin, S., & Gupta, P. (2021). Role of adsorbents in treatment of pollutants from aqueous medium (A-review). *Oriental Journal of Chemistry, 37*(4), 868–879. https://doi.org/10.13005/ojc/370414

Lasisi, K. H., Ajibade, F. O., Idowu, T. E., Ajibade, T. F., Adelodun, B., Ojo, A. O., Fadugba, O. G., Olanrewaju, O. O., Adewumi, J. R. (2023). Tidal coastal wetlands for wastewater management In: L. F. R. Ferreira, A. Kumar, & M. Bilal (Eds.), *Recent advancements in wastewater management: Implications and biological solutions* (pp. 263–284). Elsevier. https://doi.org/10.1016/bs.apmp.2022.11.002

Li, B., Jia, R., Hou, Y., & Zhu, J. (2022). Treating performance of a commercial-scale constructed wetland system for aquaculture effluents from intensive inland *Micropterus salmoides* farm. *Frontiers in Marine Science, 9*, 1000703. https://doi.org/10.3389/fmars.2022.1000703

Li, J., Liu, Y., Wang, J., Liu, Y., Zhang, M., Zhao, L., Gu, S., Lin, R., & Chen, L. (2024). Research progress on the application of natural adsorbents in the treatment of livestock wastewater. *Desalination and Water Treatment, 317*, Article 100018. https://doi.org/10.1016/j.dwt.2024.100018

Li, L., Gao, K., Yang, M., Zheng, Q., Zhang, M., & Deng, X. (2023). Challenges and potential solutions of microalgae-based systems for wastewater treatment and resource recovery. *Frontiers in Bioengineering and Biotechnology, 11*, 1210228. https://doi.org/10.3389/fbioe.2023.1210228

Loganathan, P., Vigneswaran, S., Kandasamy, J., & Bolan, N. S. (2014). Removal and recovery of phosphate from water using sorption. *Critical Reviews in Environment Science and Technology, 44*(8), 847–907. https://doi.org/10.1080/10643389.2012.741311

Lugo, L. A., Thorarinsdottir, R. I., Bjornsson, S., Palsson, O. P., Skulason, H., Johannsson, S., & Brynjolfsson, S. (2020). Remediation of aquaculture wastewater using the microalga *Chlorella sorokiniana. Water, 12*(11), 3114. https://doi.org/10.3390/w12113144

Mahmood, Q., Pervez, A., Zeb, B. S., Zaffar, H., Yaqoob, H., Waseem, M., Zahidullah, & Afsheen, S. (2013). Natural treatment systems as sustainable ecotechnologies for the developing countries. *BioMed Research International*, 796373. https://doi.org/10.1155/2013/796373

Mahmoud, R. H. (2024). Assessment of phytoremediation efficiency and valuable biomass-yielding of different microalgal strains grown on aquaculture wastewater. *Egyptian Journal of Aquatic Biology and Fisheries, 28*(2), 883–902. https://doi.org/10.21608/ejabf.2024.351926

Mair, G. C., Halwart, M., Derun, Y., & Costa-Pierce, B. A. (2023). A decadal outlook for global aquaculture. *Journal of the World Aquaculture Society, 54*(2), 196–205. https://doi.org/10.1111/jwas.12977

Manoilova, L., & Ruskova, K. (2019). Activated carbon obtaining from various raw material/via chemical activation for the purpose of environmental purification–overview. *Mechanization in Agriculture & Conserving of the Resources, 65*(6), 204–206.

Mao, Y., Yang, H., Zhou, Y., Ye, N., & Fang, J. (2009). Potential of the seaweed *Gracilaria lemaneiformis* for integrated multi-trophic aquaculture with scallop *Chlamys farreri* in North China. *Journal of Applied Phycology, 21*, 649–656. https://doi.org/10.1007/s10811-008-9398-1

Marinho-Soriano, E., Azevedo, C. A. A., Trigueiro, T. G., Pereira, D. C., Carneiro, M. A. A., & Camara, M. R. (2011). Bioremediation of aquaculture wastewater using macroalgae and *Artemia. International Biodeterioration & Biodegradation, 65*, 253–257. https://doi.org/10.1016/j.ibiod.2010.10.001

Marinho-Soriano, E., Nunes, S. O., Carneiro, M. A. A., & Pereira, D. C. (2009a). Nutrients' removal from aquaculture wastewater using the macroalgae *Gracilaria birdiae. Biomass and Bioenergy, 33*, 327–331. https://doi.org/10.1016/j.biombioe.2008.07.002

Marinho-Soriano, E., Panucci, R. A., Carneiro, M. A. A., & Pereira, D. C. (2009b). Evaluation of *Gracilaria caudata* J. Agardh for bioremediation of nutrients from shrimp farming wastewater. *Bioresource Technology, 100*(24), 6192–6198. https://doi.org/10.1016/j.biortech.2009.06.102

Michels, M. H. A., Vaskoska, M., Vermuë, M. H., & Wijffels, R. H. (2014). Growth of *Tetraselmis suecica* in a tubular photobioreactor on wastewater from a fish farm. *Water Research, 65*, 290–296. https://doi.org/10.1016/j.watres.2014.07.017

Mugg, J., Serrano, A., Liberti, A., & Rice, M. A. (2007). *Aquaculture effluents: A guide for water quality regulators and aquaculturists* (No. 00-003, 12 pp). NRAC Publication.

Musa, J. J., Dada, P. O. O., Adewumi, J. K., Akpoebidimiyen, O. E., Musa, E. T., Otache, M. Y., & Yusuf, S. (2020). Fish pond effluent effect on physico-chemical properties of soils in Southern Guinea Savanna. *Open Access Library Journal, 7*, Article e5990. https://doi.org/10.4236/oalib.1105990

Nguyen, T. A. H. (2015). *Removal and recovery of phosphorus from municipal wastewater by adsorption coupled with crystallization* (235 pp, PhD thesis). University of Technology, Sydney, New South Wales, Australia.

Nguyen, L. N., Aditya, L., Vu, H. P., Johir, A. H., Bennar, L., Ralph, P., Hoang, N. B., Zdarta, J., & Nghiem, L. D. (2022). Nutrient removal by algae-based wastewater treatment. *Current Pollution Reports, 8*, 369–383. https://doi.org/10.1007/s40726-022-00230-x

Nirmalakhandan, N., Selvaratnam, T., Henkanatte-Gedera, S. M., Tchinda, D., Abeysiriwardana-Arachchige, I. S. A., Delanka-Pedige, H. M. K., Munasinghe-Arachchige, S. P., Zhang, Y., Holguin, F. O., & Lammers, P. J. (2019). Algal wastewater treatment: Photoautotrophic vs. mixotrophic processes. *Algal Research, 41*, 101569. https://doi.org/10.1016/j.algal.2019.101569

Nobre, M. L. F., Tavares, D., Fraga, C., Oliveira, B., Dias, M., Mesquita, S., Oliveira, C. M., & Pires, J. C. M. (2024). Techno-economic analysis of a circular microalgal approach for enhanced wastewater treatment and resource recovery in Northern Portugal. *Journal of Cleaner Production, 434*, Article 140389. https://doi.org/10.1016/j.jclepro.2023.140389

Nur, T. (2014). *Nitrate, phosphate and fluoride removal from water using adsorption process* (368 pp, PhD thesis). University of Technology, Sydney New South Wales, Australia

Nzubechi, A. O., & Ikpemo, O. C. (2023). Assessment of sawdust activated carbon in the treatment of aquaculture effluent. *Global Scientific Journals, 11*(11), 318–333.

Olguin, E. J. (2003). Phycoremediation: Key issues for cost-effective nutrient removal processes. *Biotechnology Advances, 22*(1–2), 81–91. https://doi.org/10.1016/S0734-9750(03)00130-7

Omotade, I. F., Alatise, M. O., & Olanrewaju, O. O. (2019). Recycling of aquaculture wastewater using charcoal based constructed wetlands. *International Journal of Phytoremediation, 21*(5), 399–404. https://doi.org/10.1080/15226514.2018.1537247

Phang, S. M., Chu, W. L., & Rabiei, R. (2015). Phycoremediation. In: D. Sahoo & J. Seckbach (Eds.), *The algae world. Cellular origin, life in extreme habitats and astrobiology* (pp. 357–389). Springer Netherlands. https://doi.org/10.1007/978-94-017-7321-8_13

Pradhan, B., Bhuyan, P. P., Nayak, R., Patra, S., Behera, C., Ki, J. S., Ragusa, A., Lukatkin, A. S., & Jena, M. (2022). Microalgal phycoremediation: a glimpse into a sustainable environment. *Toxics, 10*(9), 525. https://doi.org/10.3390/toxics10090525

Qu, J., Jia, C., Zhang, Q., Liu, P., Yang, M., & Zhao, M. (2020). Study on the purification of aquaculture wastewater by the compound wetland ecosystem. *IOP Conference Series: Earth and Environmental Science, 514*, 52023. https://doi.org/10.1088/1755-1315/514/5/052023

Rani, S., Gunjyal, N., Ojha, C. S. P., & Singh, R. P. (2021). Review of challenges for algae-based wastewater treatment: strain selection, wastewater characteristics, abiotic, and biotic factors. *Journal of Hazardous, Toxic, and Radioactive Waste, 25*(2), 3120004. https://doi.org/10.1061/(ASCE)HZ.2153-5515.0000578

Rezai, B., & Allahkarami, E. (2021). Wastewater treatment processes—techniques, technologies, challenges faced, and alternative solutions. In: R. R. Karri, G. Ravindran, & M. H. Dehghani (Eds.), *Soft computing techniques in solid waste and wastewater management* (pp. 35–53). Elsevier. https://doi.org/10.1016/B978-0-12-824463-0.00004-5

Roozbahani, A. (2021). Application of bayesian networks modelling in wastewater management. In: Karri, R. R., Ravindran, G., & Dehghani, M. H. (Eds.), *Soft computing techniques in solid waste and wastewater management* (pp. 111–130). Elsevier. https://doi.org/10.1016/B978-0-12-824463-0.00002-1

Satyam, S., & Patra, S. (2024). Innovations and challenges in adsorption-based wastewater remediation: A comprehensive review. *Heliyon, 10*(9), Article e29573. https://doi.org/10.1016/j.heliyon.2024.e29573

Shahady, T., & Boniface, H. (2018). Water quality management through community engagement in Costa Rica. *Journal of Environmental Studies and Sciences, 8*, 488–502. https://doi.org/10.1007/s13412-018-0504-7

Sharma, G. K., Khan, S. A., Shrivastava, M., Gupta, N., Kumar, S., Malav, L. C., Nogiya, M., & Dubey, S. K. (2020). Bioremediation of sewage wastewater through microalgae (*Chlorella minutissima*). *The Indian Journal of Agricultural Sciences, 90*(10), 2024–2028. https://doi.org/10.56093/ijas.v90i10.107985

Singh, R., Birru, R., & Sibi, G. (2017). Nutrient removal efficiencies of *Chlorella vulgaris* from urban wastewater for reduced eutrophication. *Journal of Environmental Protection, 8*, 1–11. https://doi.org/10.4236/jep.2017.81001

Sisman-Aydin, G. (2022). Comparative study on phycoremediation performance of three native microalgae for primary-treated municipal wastewater. *Environmental Technology and Innovation, 28*, Article 102932. https://doi.org/10.1016/j.eti.2022.102932

Soeprobowati, T. R., & Hariyati, R. (2017). The phycoremediation of textile wastewater discharge by *Chlorella pyrenoidosa* H. Chick, *Arthrospira platensis* Gomont, and *Chaetoceros calcitrans* (Paulson) H. Takano. *Aquaculture, Aquarium, Conservation & Legislation, 10*(3), 640–651.

Soltan, M. A. H. (2016). *Aquaculture systems* (27p). https://www.researchgate.net/publication/308928410

Soni, R., Bhardwaj, S., Shukla, D. P. (2020). Various water-treatment technologies for inorganic contaminants: current status and future aspects. In: P. Devi, P. Singh, & S. K. Kansal (Eds.), *Inorganic pollutants in water* (pp. 273–295). Elsevier. https://doi.org/10.1016/B978-0-12-818965-8.00014-7

Stefanakis, A. I. (2019). The role of constructed wetlands as green infrastructure for sustainable urban water management. *Sustainability, 11*(24), 6981. https://doi.org/10.3390/su11246981

Sutanto, H., & Bawole, P. (2021). Possibility study of implementing vertical constructed wetland for domestic waste water treatment in urban kampong. *Jurnal Ilmiah Sains dan Teknologi, 10*(2), 179–187. https://doi.org/10.22146/teknosains.63801

Swarnakar, A. K., Bajpai, S., & Ahmad, I. (2022). Various types of constructed wetland for wastewater treatment-a review. *IOP Conference Series: Earth and Environmental Science, 1032*, 012026. https://doi.org/10.1088/1755-1315/1032/1/012026

Syahrir, M. (2024). The most efficient seaweed species as a bioremediator of intensive pond waste. *Journal of Ecological Engineering, 25*(1), 108–118. https://doi.org/10.12911/22998993/174179

Tan, S. Y., Sethupathi, S., Leong, K. H., & Ahmad, T. (2021). Total phosphorus and reactive phosphate removal from aquaculture wastewater using calcined eggshell. *IOP Conference Series: Earth and Environmental Science, 945*, 12019. https://doi.org/10.1088/1755-1315/945/1/012019

Tanaka, Y., Ashaari, A., Mohamad, F. S., & Lamit, N. (2020). Bioremediation potential of tropical seaweeds in aquaculture: Low-salinity tolerance, phosphorus content, and production of UV-absorbing compounds. *Aquaculture, 518*, Article 734853. https://doi.org/10.1016/j.aquaculture.2019.734853

Tantalu, L., Hidayat, N., Suharto, B., & Nurika, I. (2024). Ammonia adsorbent development for white-shrimp ponds. *BIO Web of Conferences, 90*, 6002. https://doi.org/10.1051/bioconf/20249006002

Taylor, J. H., & Soltani, M. S. (2023). Carbonaceous adsorbents in the removal of aquaculture pollutants: A technical review of methods and mechanisms. *Ecotoxicology and Environmental Safety, 266*, Article 115552. https://doi.org/10.1016/j.ecoenv.2023.115552

Tom, A. P., Jayakumar, J. S., Biju, M., Somarajan, J., & Ibrahim, M. A. (2021). Aquaculture wastewater treatment technologies and their sustainability: A review. *Energy Nexus, 4*, Article 100022. https://doi.org/10.1016/j.nexus.2021.100022

Turcios, A. E., & Papenbrock, J. (2014). Sustainable treatment of aquaculture effluents—What can we learn from the past for the future? *Sustainability, 6*(2), 836–856. https://doi.org/10.3390/su6020836

UNEP. (n.d.a). *About the United Nations environment programme*. https://www.unep.org/who-we-are/about-us

UNEP. (n.d.b). *About ecosystem health*. https://www.unep.org/topics/fresh-water/ecosystem-health/about-ecosystem-health

UNEP. (n.d.c). *Global programme of action (GPA)*. https://www.unep.org/topics/ocean-seas-and-coasts/ecosystem-degradation-pollution/global-programme-action-gpa

United Nations. (n.d.). *Global programme of action for the protection of the marine environment from land-based activity (GPA) #SDGAction39890*. Department of Economic and Social Affairs-Sustainable Development. https://sdgs.un.org/partnerships/global-programme-action-protection-marine-environment-land-based-activity-gpa#description

US EPA. (2024). *Summary of the Clean Water Act*. Last updated June 12, 2024. https://www.epa.gov/laws-regulations/summary-clean-water-act

Vandana, R., & Rakesh, S. (2023). Phyco-remediation of sewage wastewater by microalgae. In: Taşeli, B. K. (Ed.), *Sewage management*. IntechOpen. https://doi.org/10.5772/intechopen.109257

Verdegem, M., Buschmann, A. H., Latt, U. W., Dalsgaard, A. J. T., & Lovatelli, A. (2023). The contribution of aquaculture systems to global aquaculture production. *Journal of the World Aquaculture Society, 54*(2), 206–250. https://doi.org/10.1111/jwas.12963

Viegas, C., Gouveia, L., & Gonçalves, M. (2021). Aquaculture wastewater treatment through microalgal. Biomass potential applications on animal feed, agriculture, and energy. *Journal of Environmental Management, 286*, 112187. https://doi.org/10.1016/j.jenvman.2021.112187

Vijuksungsith, P., Satapanajaru, T., Chokejaroenrat, C., Jarusutthirak, C., Sakulthaew, C., Kambhu, A., Yoo-iam, M., & Boonprasert, R. (2023). Removal and reuse of phosphorus from aquaculture water using activated carbon-based CaO_2 nanoparticles. *Environmental Technology and Innovation, 29*, Article 102990. https://doi.org/10.1016/j.eti.2022.102990

Vymazal, J. (2008). Constructed wetlands for wastewater treatment: A review. In: M. Sengupta & R. Dalwani (Eds.), *Proceedings of Taal2007: The 12th world lake conference* (pp. 965–980).

Vymazal, J. (2007). Removal of nutrients in various types of constructed wetlands. *Science of the Total Environment, 380*(1–3), 48–65. https://doi.org/10.1016/j.scitotenv.2006.09.014

Vymazal, J. (2013). Emergent plants used in free water surface constructed wetlands: A review. *Ecological Engineering, 61*, 582–592. https://doi.org/10.1016/j.ecoleng.2013.06.023

Vymazal, J. (2022). The historical development of constructed wetlands for wastewater treatment. *Land, 11*(2), 174. https://doi.org/10.3390/land11020174

Wang, H., Sheng, L., & Xu, J. (2021). Clogging mechanisms of constructed wetlands: A critical review. *Journal of Cleaner Production, 295*, Article 126455. https://doi.org/10.1016/j.jclepro.2021.126455

Wang, R., Korboulewsky, N., Prudent, P., Domeizel, M., Rolando, C., & Bonin, G. (2010). Feasibility of using an organic substrate in a wetland system treating sewage sludge: Impact of plant species. *Bioresource Technology, 101*, 51–57. https://doi.org/10.1016/j.biortech.2009.07.080

Wang, Y., Cai, Z., Sheng, S., Pan, F., Chen, F., & Fu, J. (2020). Comprehensive evaluation of substrate materials for contaminants removal in constructed wetlands. *Science of the Total Environment, 701*, Article 134736. https://doi.org/10.1016/j.scitotenv.2019.134736

White, S. A. (2013). Wetland technologies for nursery and greenhouse compliance with nutrient regulations. *HortScience Horts, 48*(9), 1103–1108. https://doi.org/10.21273/HORTSCI.48.9.1103

WHO. (n.d.a). *Water sanitation and health: Drinking-water quality guidelines*. https://www.who.int/teams/environment-climate-change-and-health/water-sanitation-and-health/water-safety-and-quality/drinking-water-quality-guidelines

WHO. (n.d.b). *Water sanitation and health: Recreational waters*. https://www.who.int/teams/environment-climate-change-and-health/water-sanitation-and-health/water-safety-and-quality/recreational-waters

Wilkie, A. C., Edmundson, S. J., & Duncan, J. G. (2011). Indigenous algae for local bioresource production: Phycoprospecting. *Energy for Sustainable Development, 15*(4), 365–371. https://doi.org/10.1016/j.esd.2011.07.010

Wu, H., Zhang, J., Ngo, H. H., Guo, W., Hu, Z., Liang, S., Fan, J., & Liu, H. (2015). A review on the sustainability of constructed wetlands for wastewater treatment: Design and operation. *Bioresource Technology, 175*, 594–601. https://doi.org/10.1016/j.biortech.2014.10.068

Yang, H., Tan, T., Du, X., Feng, Q., Liu, Y., Tang, Y., Bai, G., Liu, Z., Xia, S., Song, S., Quintana, M., Zhou, Q., Wu, Z., & Zhang, Y. (2025). Advancements in freshwater aquaculture wastewater management: A comprehensive review. *Aquaculture, 594*, Article 741346. https://doi.org/10.1016/j.aquaculture.2024.741346

Younas, F., Mustafa, A., Farooqi, Z. U. R., Wang, X., Younas, S., Mohy-Ud-Din, W., Hameed, M. A., Abrar, M. M., Maitlo, A. A., Noreen, S., & Hussain, M. M. (2021). Current and emerging adsorbent technologies for wastewater treatment: Trends, limitations, and environmental implications. *Water, 13*(2), 215. https://doi.org/10.3390/w13020215

Zhimiao, Z., Xiao, Z., Zhufang, W., Xinshan, S., Mengqi, C., Mengyu, C., & Yinjiang, Z. (2019). Enhancing the pollutant removal performance and biological mechanisms by adding ferrous ions into aquaculture wastewater in constructed wetland. *Bioresource Technology, 293*, Article 122003. https://doi.org/10.1016/j.biortech.2019.122003

Zulfadhly, Z., Mashitah, M. D., & Bhatia, S. (2001). Heavy metals removal in fixed-bed column by the macro fungus *Pycnoporus sanguineus*. *Environmental Pollution, 112*(3), 463–470. https://doi.org/10.1016/S0269-7491(00)00136-6

Strategic Management of Dairy Wastewater

Kehinde Iyanu Jimeto, Fidelis Odedishemi Ajibade, Temitope Fausat Ajibade, Chukwuemeka Kingsley John, Kayode Hassan Lasisi, Adedamola Oluwafemi Ojo, and James Rotimi Adewumi

Abstract The strategic management of dairy wastewater is critical for mitigating its environmental impact and process sustainability. Dairy wastewater is characterized by high organic content, significant levels of fats, proteins, and nutrients, which pose substantial environmental risks if improperly managed. Effective treatment is essential to address these challenges and to comply with stringent discharge standards. This chapter examines various treatment technologies for dairy wastewater, including biological methods (aerobic and anaerobic processes), chemical treatments, and electrochemical processes. Biological treatments involve aerobic and anaerobic systems that effectively reduce organic load and pollutants. Chemical treatments

K. I. Jimeto · F. O. Ajibade (✉) · T. F. Ajibade · K. H. Lasisi · J. R. Adewumi
Department of Civil and Environmental Engineering, Federal University of Technology, Akure, PMB 704, Nigeria
e-mail: foajibade@futa.edu.ng

K. I. Jimeto
e-mail: jimetokehindeiyanu@gmail.com

T. F. Ajibade
e-mail: tfotun@futa.edu.ng

K. H. Lasisi
e-mail: senserltd@gmail.com

J. R. Adewumi
e-mail: jradewumi@futa.edu.ng

C. K. John
Department of Construction and Civil Engineering, University Centre Jersey Highlands College, St Saviour Jersey, JE2 4FB, Jersey, UK
e-mail: chuks.john@highlands.ac.uk

K. H. Lasisi
College of Environmental Science and Engineering, Ocean University of China, Qingdao 266100, China

A. O. Ojo
Department of Civil Engineering, Yaba College of Technology, Lagos, Nigeria
e-mail: adedamola.ojo@yabatech.edu.ng

N. A. Oladoja and E. I. Unuabonah (eds.), *Strategic Management of Wastewater from Intensive Rural Industries*, Springer Water,
https://doi.org/10.1007/978-3-031-90314-4_3

focus on coagulation and oxidation to facilitate the removal of contaminants. Electrochemical methods, including electrocoagulation and electro-Fenton processes, offer advanced options for pollutant degradation. Additionally, phytoremediation and wetland systems provide natural treatment mechanisms, while constructed wetlands combine ecological benefits with treatment efficiency. Membrane technologies, such as microfiltration, ultrafiltration, nanofiltration, dialysis, electrodialysis, reverse osmosis, and membrane bioreactors, offer high-performance solutions for separating and concentrating pollutants. Integrated systems combining these technologies have shown promising results in optimizing treatment efficiency and meeting regulatory standards. By integrating multiple treatment approaches, these systems enhance overall effectiveness and sustainability. In conclusion, adopting a strategic, multi-faceted approach to dairy wastewater management can significantly reduce environmental impacts and improve compliance with discharge regulations. The development and implementation of integrated treatment systems are essential for advancing sustainable practices in the dairy industry.

Keywords Dairy industry · Environment · Integrated treatment systems · Wastewater

1 Introduction

Water is an indispensable natural resource for all human activities and is critical for the survival of all living organisms. Despite its significance, much of the earth's water remains inaccessible, locked in polar ice caps, glaciers, porous rocks, and subterranean reservoirs, leaving only a fraction readily available for human consumption (Joshiba et al., 2019). The unregulated utilization of water in industrial and agricultural sectors has precipitated a decline in water quality, exacerbating the scarcity of freshwater resources (Biswas et al., 2021; John & Pu, 2023). The escalating global population has further exacerbated this issue, with copious amounts of wastewater generated daily across residential, industrial, and agricultural domains, thus elevating concerns surrounding water pollution on both industrial and societal fronts (Al-Tayawi et al., 2023).

Industries are pivotal to economic growth and have paradoxically contributed to water shortages worldwide through unregulated discharge. This underscores the imperative for sustainable wastewater management practices (Joshiba et al., 2019). The discharge of untreated wastewater directly into natural water bodies emerges as a primary driver of water pollution, necessitating immediate attention from governmental bodies and industries alike (Ajibade et al., 2021a; El Khomri et al., 2022; El Messaoudi et al., 2022). Within urban areas, major contributors to wastewater generation include industries such as dairy, paper, metal, distillery, breweries, textile, tanning, leather manufacturing, chemical, beverage, and pharmaceutical industries. Wastewater from these industries is often characterized by high turbidity, chemical oxygen demand (COD), biochemical oxygen demand (BOD), toxic dyes, organic and persistent organic pollutants, inorganic nutrients, and heavy metal contamination. Treating such wastewater before discharge into natural reservoirs is essential to mitigate environmental impacts (Agrahari & Kumar, 2023; Azimi et al., 2017). However, the lack of proper technologies has resulted in a significant gap between the quantity of wastewater generated and the treatment capacity. This gap continues to grow over time, posing a serious challenge. If this trend persists, it could lead to severe water scarcity and toxicity, potentially affecting two-thirds of the world's population (Agrahari & Kumar, 2023; Yadav et al., 2020). Numerous factors including energy consumption, radioactive waste, urban expansion, sewage mismanagement, industrial effluents, mining activities, and agricultural runoff converge to perpetuate water pollution, underscoring the multifaceted nature of this global challenge (Al-Tayawi et al., 2023).

The dairy industry, known for its significant water requirements, generates substantial volumes of wastewater, primarily composed of chemically altered liquids and residues from cleaning processes (Arvanitoyannis & Giakoundis, 2006). The industry's water demand for its operations has led to the generation of abundant amounts of wastewater laden with proteins, salts, fats, lactose, and cleaning agents (Thassitou & Arvanitoyannis, 2001). According to Sarkar et al. (2006), substantial amounts of water are consumed throughout various processes. Therefore, the industry's requirements for water is regarded as one of the biggest sources of wastewater generation in many countries (Farizoglu & Uzuner, 2011; Shivayogimath & Naik, 2014).

The discharge of untreated dairy wastewater into natural water bodies poses grave threats to aquatic ecosystems, necessitating stringent treatment measures prior to disposal (Agrahari & Kumar, 2023; Kushwaha et al., 2011). Inadequate treatment systems in cheese dairies pose a significant environmental challenge due to the high organic matter content in the wastewater. However, the wastewater from cheese dairies can serve as a valuable alternative energy resource rather than as a pollutant source, if appropriately managed, given its high levels of COD and BOD (Alayu & Yirgu, 2018; Najafpour et al., 2008).

Various stakeholders have recognized the imperative of addressing industrial wastewater, particularly from dairy production, and they advocate for the development and implementation of efficient wastewater treatment technologies to alleviate environmental pollution burdens (Arvanitoyannis & Giakoundis, 2006; Tabelini et al., 2023). Effective treatment not only ensures environmental sustainability, but also promotes economic viability through the reduction of contaminant levels in the discharged effluents (Ajibade et al., 2021a).

Based on these premises, this chapter provides an overview of dairy wastewater, including its characteristics, environmental effects, and treatment methods. Treatments of dairy wastewater, using various methods and strategies (e.g., Biological, Mechanical, Chemical, Phytoremediation, constructed wetlands etc.), their advantages and disadvantages as well as the synergy achieved from the combination of strategies in treating dairy wastewater, are presented. It is expected that this will be a ready resource that will be useful to scientists and policy makers alike in their effort to develop effective strategies for the management of dairy wastewater.

2 The Dairy Industry

The dairy industry is recognized as a leading sector within the food industry that utilizes the most water and produces the most effluent per unit of production (Fig. 1). This industry's significant water consumption is not only due to producing various milk products, such as cheese, curd, butter, yogurt, and dry milk powder but also involves processes like sanitation, cooling of milk products, and cleaning and washing of processing equipment (Kaur, 2021). The dairy industry ranks third in water usage and effluent discharge volumes, after the chemical and refinery sectors (Briao & Tavares, 2007; Olmez & Kretzschmar, 2009). This industry is a global network responsible for the production and distribution of milk and its derivatives, ranging from cheese to yoghurt. Milk is regarded as a paramount source of essential nutrients among all food items. It serves as a nutritious fluid produced and discharged by the mammary glands of female mammals. With the rapid industrialization sweeping across the nation, there has been a marked increase in the number of dairies and conglomerates. Predominantly, milk is comprised of 87.5% water, with additional constituents, which include 3.9% fat, 4.8% lactose, 3.4% protein, 0.8% minerals, and 13% solid compounds. The dairy industry plays a pivotal role in generating copious amounts of wastewater from food processing activities in numerous nations. Due to the extensive water consumption in dairy product manufacturing, it remains the primary source of industrial wastewater from the food production sector (Joshiba et al., 2019). The inadequate treatment systems employed by cheese dairies has become a significant environmental concern because of the high organic matter they contain.

3 Characteristics of Dairy Wastewater

In a study by Agrahari and Kumar (2023), manufacturing of various products in the dairy industry through varying processes requires water for cleaning equipment, heating and cooling facilities. About 2% of the milk processed is wasted during the process, which leads to an increase in the wastewater quantity, and the contamination of the receiving water body. The high organic fraction of the wastewater discharged from the dairy industries poses a greater threat to the ecosystem, as they negatively impact water quality in various ways (Joshiba et al., 2019). Dairy wastewater is characterized by a white color (while whey tend to be yellowish-green), has an unpleasant odor and a turbid appearance (Slavov, 2017). The flow rate and composition of the effluents varies, based on the manufacturing and production processes. Dairy wastewater, like other effluents from the food industry, is characterized by its elevated levels of chemical oxygen demand (COD) and biological oxygen demand (BOD) (Agrahari & Kumar, 2023).

The characteristics of some dairy effluents, which represent the constituents of the wastewater are summarized in Table 1. The results presented in Table 1 shows that

Fig. 1 The dairy industry (Okojie & Bailey, 2019)

the wastewater typically has high COD value, ranging from 1–102.1 g/L, and BOD value, ranging from 0.22–60 g/L. Effluents from milk processing typically exhibit elevated temperatures, and significant variations in the values of pH, Total Suspended Solids (TSS), Total Dissolved Solids (TDS) BOD, COD, Total Nitrogen (TN), Total Phosphorus (TP), and Fat, Oil, and Grease (FOG) (Slavov, 2017).

When untreated dairy wastewater is released into water bodies, it introduces high organic content, which depletes the dissolved oxygen in the water, due to the ensuing putrefaction, and the water gets polluted. The average values of total BOD and COD indicate that dairy wastewaters contain significant organic matter concentration (Arvanitoyannis & Giakoundis, 2006). The specific characteristics of dairy effluents depend on various factors, which include industrial scale, processing methods and efficiency, process parameters, operational type, equipment selection for cleaning, type of waste discharged, and the cost of wastewater treatment. Constituents of dairy wastewater include milk fractions, accounting for approximately 2% of the total processed milk and its products, acidic compounds such as lactic acid, phosphoric acid, and citric acid, alkaline components like soda ash, trisodium phosphate, and sodium hydroxide, as well as cleaning detergents and sanitizing agents. (Agrahari & Kumar, 2023). The components of dairy wastewater vary, depending on factors such as the product being manufactured, seasonal variations like summer and winter, production cycles, and the use of different additives and cleaning agents (Bella & Rao, 2021; Ye & Li, 2023). Some of the stages involved in dairy production that generate wastewater are presented in Fig. 2.

Table 1 Characteristics of different wastewater originating from the dairy industry

Wastewater type	pH	COD (mg/L)	BOD_5 (mg/L)	FOG (mg/L)	TS (mg/L)	TSS (mg/L)	TP (mg/L)	TN (mg/L)	Reference
Cheese whey	3.92–6.5	50,000–102,100	27,000–60,000	900–14,000	55,000–70,900	1270–22,150	120–530	200–1760	Slavov (2017)
Mixed dairy processing	6–11	1150–9200	–	–	2705–3715	340–1730	8–68	–	Goli et al. (2019)
Butter	12.08	8930	220–2650	2880	–	700–5070	–	–	Slavov (2017)
Youghurt	4.53	6.5	–	–	–	–	–	–	
Raw wastewater	5.5–7.5	1500–3000	350–360	–	–	250–600	–	–	Joshiba et al. (2019)
Milk permeate	5.55–6.52	55,200–63,480	–	–	–	2670–3800	350–450	–	Goli et al. (2019)
Cheese/whey plant	5.2	189–6219	377–2214	–	–	188–2330	–	–	Britz et al. (2006)
Creamery	8–11	2000–6000	1200–4000	–	–	350–1000	–	–	Goli et al. (2019)
Cheese waste	4–12	650–3000	300–1400	–	–	250–2700	–	–	Joshiba et al. (2019)
Aavin dairy industry washwater	6.4–7.1	2500–3300	–	–	1300–1400	630–730	–	–	Raghunath et al. (2016)
Mixed dairy	4–11	500–10,400	240–5900	200–1920	710–7000	60–5800	100–600	10–660	Slavov (2017)
Dairy wastewater	7.2–7.5	2500–3000	1300–1600	–	8000–10,000	72,000–80,000	–	–	Raghunath et al. (2016)

(continued)

Table 1 (continued)

Wastewater type	pH	COD (mg/L)	BOD_5 (mg/L)	FOG (mg/L)	TS (mg/L)	TSS (mg/L)	TP (mg/L)	TN (mg/L)	Reference
Milk industrial wastewater	11.70	5000–14,250	3000–8910	–	–	1420–3540	–	–	Joshiba et al. (2019)
Milk and dairy products factory	8.34	10,251.2	4840.6			5802.6			Shete and Shinkar (2013)
Ice cream	5.1–6.96	5200	2450	–	3900	3100	14	–	Slavov (2017)
Butter/milk powder plant	10–11	–	1500	400	–	–	–	–	Britz et al. (2006)
Cheese	3.38–9.5	1000–63,300	590–5000	330–2600	1920–53,200	190–2500	5–280	18–830	Slavov (2017)

Note COD—chemical oxygen demand, BOD_5—5-day biochemical oxygen demand, FOG—fat, oil, and grease, TS—total solids, TSS—total suspended solids, TP—total phosphate, TN—total nitrogen

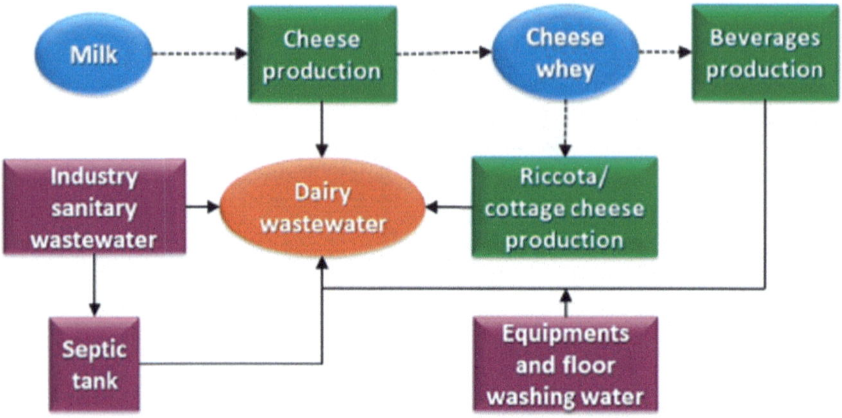

Fig. 2 Certain byproducts of milk and whey, along with the primary origins of dairy wastewater denoted by solid lines (Tabelini et al., 2023)

The impact of these chemicals and their additives on the COD of dairy wastewater is relatively small, when compared with the effect of other components like milk cream or whey (Arvanitoyannis & Giakoundis, 2006). The inadequate wastewater treatment from cheese dairies become a significant environmental concern because of the high organic matter contained in the wastewater (Stasinakis et al., 2022).

Cleaning wastewater typically originates from washing equipment that directly interacts with milk or dairy products, so it contains spillage of milk and its products, whey, and brines. More than 90% of the solid organic matters present in these effluents stem from milk and manufacturing residues, which include cheese remnants, whey, cream, waters from separation and clarification processes, starter cultures, yoghurt, fruit concentrates, and stabilizers.

Typically, dairy wastewater has an annual temperature of 17–25 °C, which results in faster biological degradation (Water Environment Federation, 2007). The average temperatures of industrial dairy effluents range from 17–18 °C in winter and 22–25 °C in summer. Using the Arrhenius equation, the biodegradation rates and oxygen consumption can be predicted to be 1.5 times higher in summer than in winter. The design winter temperature of 15 °C is adopted for this type of wastewater due to the utilization of hot water for washing and cleaning equipment.

4 Environmental Impact of Dairy Wastewater

The dairy industry produces between 0.2 to 10 L of wastewater for every litre of processed milk, with an average of 2.5 L produced per litre of processed milk. Waste from dairy processing is produced on an irregular basis, and their flow rates fluctuate a lot. The production schedule, operational procedures, plant design, level of water management implemented, and ultimately the quantity of water conserved affect the volume, concentration, and composition of the effluents generated in the dairy industry. Because of this, when a new technological cycle is applied in the processing line, the amount and quality of product content in the dairy effluent at a particular moment change (Shete & Shinkar, 2013).

As the nutrient components of the effluents decompose, the dissolved oxygen levels in water sources decrease, creating an anaerobic environment. The release of highly noxious odors from dairy effluents poses a significant challenge for residents living near water sources. Freshwater bodies contaminated by dairy waste become breeding grounds for various insects and pests. Due to their stagnant nature, these contaminated freshwater sources serve as breeding sites for mosquitoes, which transmit harmful diseases such as dengue, malaria, and yellow fever (Joshiba et al., 2019).

Wastewater discharged onto land can adversely impact soil quality and structure, while a portion of it may seep into underlying groundwater, affecting its quality as well. This issue becomes particularly severe when untreated wastewater is released from dairy or milk processing industries. A typical European dairy factory produces around 50 m^3 of wastewater daily, containing significant concentrations of organic matter (such as fats, proteins, and carbohydrates) and nutrients (primarily nitrogen and phosphorus) originating from milk and milk products (Demirel et al., 2005; Omil et al., 2002; Sirmacekic et al., 2022). The annual cost of treating and disposing of wastewater for such facilities is estimated to be in the millions of dollars (Tikariha & Sahu, 2014). The disposal of untreated wastewater is rapidly emerging as a significant economic and societal challenge for the dairy processing industry (Adewumi & Ajibade, 2015; Kesari et al., 2021; Vaishnav et al., 2023). Nearly all dairy factories are grappling with issues related to wastewater treatment, disposal, and utilization. Discharging untreated or partially treated wastewater into rivers, land, fields, or other water bodies, often in crude tanks, poses serious health and hygiene concerns (Tikariha & Sahu, 2014). The salinity levels in industrial dairy effluents contribute to an increase in non-volatile suspended solids in both primary and secondary sludge. Inorganic impurities found in dairy wastewater, such as Na^+, K^+, Ca^{2+}, and Cl^- ions, are primarily present in cheese and cottage cheese production, with concentrations ranging from 0.46 to 10%. These impurities mainly consist of NaCl, KCl (over 50%), and $Ca_3(PO_4)_2$, which is added as a preservative. Increased Na^+ levels indicate the use of alkaline cleaning agents in milk factories (Slavov, 2017), while the amount of Ca^{2+} in acidic whey is double that of sweet whey (Carvalho et al., 2013; Prazeres et al., 2012). Chlorides in dairy wastewater result from salt addition during brine and cooling liquor production, as well as from fresh water and milk. Chloride concentrations typically range from 0.15 to 0.2 g/L, with some reaching 0.8–1 g/L (Slavov, 2017).

5 Treatment of Dairy Wastewater

The manufacturing and processing sectors of the dairy industry consume large quantities of freshwater and produce considerable amounts of wastewater containing organic and nutrient materials, as byproducts, as well as extreme pH variations. Therefore, it is essential to treat this wastewater before disposal (Joshiba et al., 2019). Addressing this challenge necessitates the implementation of efficient and cost-effective wastewater treatment methods to safeguard freshwater resources. Various strategies and methods for treating dairy wastewater are discussed as follows.

5.1 Biological Treatment

Biological treatment, which is effective against organic matter, is one of the most efficient methods of treating dairy wastewater (Britz et al., 2006). It is an essential method for dairy wastewater treatment that purifies the wastewater from diverse sources and mitigates its adverse effects on the environment (Joshiba et al., 2019). It is an auspicious method, particularly for treating wastewater with high organic matter using microorganisms to break down the high organic load and treat harmful chemicals present. It is divided into two types, such as an aerobic and anaerobic method. The decomposition of organic compounds in the presence of oxygen is known as the aerobic treatment method (Joshiba et al., 2019) while anaerobic treatment involves the degradation of organic constituents with the help of microbes in the absence of oxygen (Kalat & Yüceer, 2017; Kaur, 2021). Both aerobic and anaerobic methods are utilized for efficient organic matter removal from dairy wastewater, but each has limitations that affect their effectiveness. Aerobic processes are preferred for less polluted effluents, while anaerobic processes are better suited for highly polluted ones. Aerobic systems are energy-intensive, posing a significant disadvantage compared to anaerobic methods (Goli et al., 2019). The comparison of the effectiveness of the two methods is presented in Table 2.

The effectiveness of a biological treatment hinges on two key factors: the capability of the microbes within the system to break down the organic matter in the effluent, and the efficiency of separating the biomass from the liquid phase during the last stage of the treatment process.

5.2 Aerobic Treatment Methods

Aerobic treatment depends on the microorganisms grown in an oxygen-rich environment to oxidize organic contaminants to carbon dioxide, water and cellular material (Britz et al., 2006). This treatment method eliminates BOD, and it is also utilized to reduce organic supplements like P and N in the wastewater (Joshiba et al., 2019). It's

Table 2 Comparison of the aerobic and anaerobic treatment methods (Goli et al., 2019)

Feature	Aerobic	Anaerobic
Organic removal efficiency	High	High
Wastewater quality	Great	Moderate to poor
Organic loading rate	Moderate	High
Sludge generation	High	Low
Nutrient demand	High	Low
Alkalinity demand	Low	High for certain industrial waste
Energy demand	High	Low to moderate
Temperature sensitivity	Low	High
Odor	Less opportunity for odors	Potential odor problems
Bioenergy and nutrient recovery	Less opportunity for odors	Yes
Mode of treatment	Total (rely on feedstock characteristics)	Essentially pretreatment

a highly challenging treatment method that occurs only in the presence of air, where microorganisms utilize oxygen to metabolize organic contents, converting them into biomass, CO_2, and H_2O (Das & Paul, 2022; Joshiba et al., 2019). Britz et al. (2006) concluded that extensive research on both laboratory and field-scale aerobic treatments has demonstrated the reliability and cost-effectiveness of aerobic treatment in generating high-quality effluent. The initial phase typically involves an acclimation period to facilitate the establishment of a competitive microbial community. Successful removal of ammonia-nitrogen is achievable to mitigate disposal challenges. Common issues related to aerobic processes include foaming and inadequate separation of solids from liquids. Due to the high organic content in dairy wastewater, biological treatment is considered the most suitable approach (Das & Paul, 2022; Gutiérrez et al., 1991). Aerobic treatment is typically categorized into two processes: suspended growth and attached growth processes. Traditionally, various aerobic treatment methods such as aerated lagoons, oxidation ponds, and activated sludge processes (ASP) have been successfully employed for treating organic industrial effluents (Joshiba et al., 2019). Currently, most dairy wastewater treatment facilities utilize aerobic processes, albeit with reduced efficiency, primarily due to filamentous growth and rapid acidification resulting from high lactose levels and limited water buffer capacity, respectively (Slavov, 2017).

5.2.1 Activated Sludge Process

The activated sludge process (ASP) is a highly preferred aerobic method for treating dairy wastewater because of its effectiveness in treating and removing the nutrients found in the water (Das & Paul, 2022). It uses a group of microorganisms that are suspended in wastewater in an aeration tank to absorb, adsorb, and biodegrade organic pollutants (Goli et al., 2019). The primary components of the ASP are protozoa and microorganisms that aid in breaking down organic substances and nutrients in wastewater (Das & Paul, 2022), converting it into activated sludge for recycling (Tocchi et al., 2012). Understanding microbial groups and their interaction with organic substances in biological wastewater treatment is crucial for optimizing treatment plant processes.

The microbes absorb and fully oxidize organic matter to harmless end products and convert inorganic substances into energy for microbial growth and biomass formation (flocs). To keep the flocs suspended, air is either diffused into the tank's bottom or mechanical aeration is used. Maintaining a dissolved oxygen level of 1–2 mg/L in the aeration tank is crucial, and the tank design should consider aeration duration and cell residence time. Afterwards, the mixture moves to the sedimentation tank, where activated sludge flocs agglomerate into larger particles and settles as sludge (Britz et al., 2006). These microorganisms also ingest suspended organic matter in dairy wastewater, transforming it into activated sludge, which is subsequently recycled and eliminated (Das & Paul, 2022). In this procedure, wastewater flows through an aerated tank, and the resulting activated sludge is separated from the dairy wastewater using a clarifier (Das & Paul, 2022; Patel et al., 2016), to produce clarified wastewater. This system is highly efficient in degrading complex substances like protein, lactose, fats, and oils with the help of microorganisms (Joshiba et al., 2019).

The biological aerobic metabolism process generates significant amounts of sludge, with approximately 0.6 kg of dry sludge produced per kg of BOD removed. While some sludge is recycled to the aeration tank, the remainder must be managed and disposed of in an environmentally responsible manner, which constitutes a major operational cost. In the U.S., ASP is widely used in municipal and industrial settings, including dairy effluent treatment, due to its effectiveness in reducing biochemical oxygen demand (BOD) and suspended solids (Grady et al., 2011).

5.2.2 Aerated Lagoons

Aerated lagoons offer efficient and straightforward means to reduce organic and inorganic loads from dairy effluent (Das & Paul, 2022; Renou et al., 2008). Depending on the wastewater type, aerated lagoons can function effectively in both aerobic and anaerobic processes. Except for aerated ponds, these methods typically operate without mechanical devices, relying on gravity for flow. Due to their simplicity and absence of a sludge recycling system, lagoons are favoured for efficient wastewater treatment. However, the uncontrolled environment leads to slower reaction times, resulting in retention times of up to 60 days. In warmer climates, operators may find lagoons to be a more suitable and economical option for wastewater treatment. In New Zealand, advanced pond systems are designed for the treatment of dairy farm effluent. These systems use natural processes such as sedimentation and biodegradation to treat wastewater in a controlled environment, ensuring effective treatment and nutrient removal (Craggs et al., 2004).

The cost-effectiveness and effectiveness of aerobic lagoons have positioned them as a viable method for treating dairy effluents in many developing countries. Researchers have noted that lagoons can effectively reduce nutrient and organic compound concentrations in industrial effluents. Ammonia (NH_3) is the predominant nitrogen compound in aerated lagoons, posing challenges to effluent treatment. Bacteria, a major group of prokaryotes in lagoons, significantly contribute to degrading organic loads in effluents. Various bacteria strains aid in converting complex organic and inorganic compounds into easily degradable forms. Understanding microbial parameters such as type, structure, morphology, and nutrient uptake rate enhances aerated lagoon design and operation. Temperature significantly impacts microbial metabolism in aerated lagoons, affecting overall efficacy (Joshiba et al., 2019). A significant drawback is the potential for hazardous waterborne diseases to occur in aerated lagoons. To mitigate this risk, precautionary measures should be taken by personnel working near the treatment plant, and maintenance of the treatment plant should be diligently performed (Alhamlan et al., 2013; Das & Paul, 2022).

The success of aerated lagoons hinges on the acclimatization of living organisms and their nutrient uptake rates. Maintaining removal efficiency in aerated lagoons is challenging due to microbial characteristics, prompting the implementation of new advancements for the efficient removal of toxic pollutants from dairy effluents (Joshiba et al., 2019). Numerous advancements have been introduced to enhance the treatment process, ensuring the efficient removal of toxic pollutants from the effluent (Das & Paul, 2022).

5.2.3 Rotating Biological Contactors

The rotating biological contactors (RBC) are a conventional aerobic biological wastewater treatment unit (Metcalf & Eddy, 2014) and utilize circular discs made of lightweight materials such as high-density plastic, with 40–60% of their surface area extending beyond the tank to facilitate oxygen transfer. These discs rotate at speeds of 1–3 rpm on a horizontal shaft, and the biofilm formed on the disc surface aids in the oxidation of organic compounds in the effluent. When the biofilm sludge becomes excessively thick, it is detached and removed from the sedimentation tank. Operational efficiency is measured in terms of g BOD/m^2/d (Goli et al., 2019). RBC is highly preferred for removing nitrogen compounds from industrial effluents, particularly those with high BOD and COD levels. It consists of rotating plastic discs, where microorganisms consume nutrients and organic matter, aided by aeration to promote their growth (Joshiba et al., 2019).

Das and Paul (2022) explained that the rotating biological contactor (RBC) functions primarily through adsorption and is commonly employed for treating both industrial and household wastewater. It is particularly preferred for its effectiveness in removing nitrogen compounds from wastewater (Mehrdadi et al., 2012). RBC operates as an attached growth process and is widely preferred due to its low energy consumption and high removal efficiency Aeration is provided to promote the growth of microorganisms, which consume nutrients and organic compounds in the biofilm, leading to an increase in biomass volume. As the biofilm layer thickens, the sludge, which has also thickened, falls from the carrier and is mixed with the sludge digester, eventually converting to water and gas (Kamath et al., 2018). RBC is preferred for treating various types of wastewaters due to its ease of operation, large surface area, simple construction, low energy consumption, minimal maintenance, cost-effectiveness, and resilience to shock. It has become the primary treatment system for aerobic wastewater treatment (Ebrahimi et al., 2010). The RBC process achieves an 85% removal efficiency for COD (Rusten et al., 1992), while a COD removal efficiency of 96% is achieved with a hydraulic retention time (HRT) of 36 h (Das & Paul, 2022; Ebrahimi et al., 2010).

In Canada, a laboratory-scale aerobic rotating biological contactor was tested using full-strength cheese whey, with HRT that varied between 11 and 5 days, and a temperature of 35 °C. The study observed methane (CH_4) production rates that ranged from 1.68 to 3.26 L of CH_4 per liter and achieved a COD reduction between 76 and 93%. However, stable operation was not possible at HRTs shorter than 5 days for reactors processing either full-strength or diluted whey. A two-stage fermentation system was also evaluated, demonstrating that stable operation and high treatment efficiency, with a COD removal rate of 89.5%, could be consistently achieved (Lo & Liao, 1986).

5.2.4 Sequencing Batch Reactor

A sequencing batch reactor (SBR) is a single-tank fill-and-draw system designed to aerate, settle, withdraw effluents, and recycle solids (Goli et al., 2019). According to Joshiba et al. (2019), the treatment of dairy wastewater is primarily accomplished through the utilization of SBRs as shown in Fig. 3.

This technology involves the addition of dairy wastewater into a single batch reactor, where it undergoes treatment to reduce the organic load before being discharged (Das & Paul, 2022; Kushwaha et al., 2011). Joshiba et al. (2019) reported that the sequencing batch reactor (SBR) operates through five distinct stages: fill, react, settle, decant, and idle. These stages are repeated cyclically as the SBR continues its treatment process. In the first stage, wastewater is mixed without aeration, as the tank fills to metabolize fermentable components. Aeration follows, aiding in oxidation and biomass formation. Sludge settles, and treated wastewater is separated to complete the cycle. SBR operation requires the site operator to adjust stage durations to match effluent composition fluctuations. SBRs are ideal for locations with limited land due to their small footprint and offer easy cycle modifications for nutrient separation when needed (Goli et al., 2019). The SBR seems to be a highly promising technology for dairy wastewater treatment, functioning as a 'fill and draw' activated sludge system. Its efficiency depends on various parameters including dissolved oxygen volume, retention time HRT, organic load composition, COD, nutrient composition, denitrification, and nitrification rates (Das & Paul, 2022). Its simplicity in construction and efficiency makes it a suitable treatment system

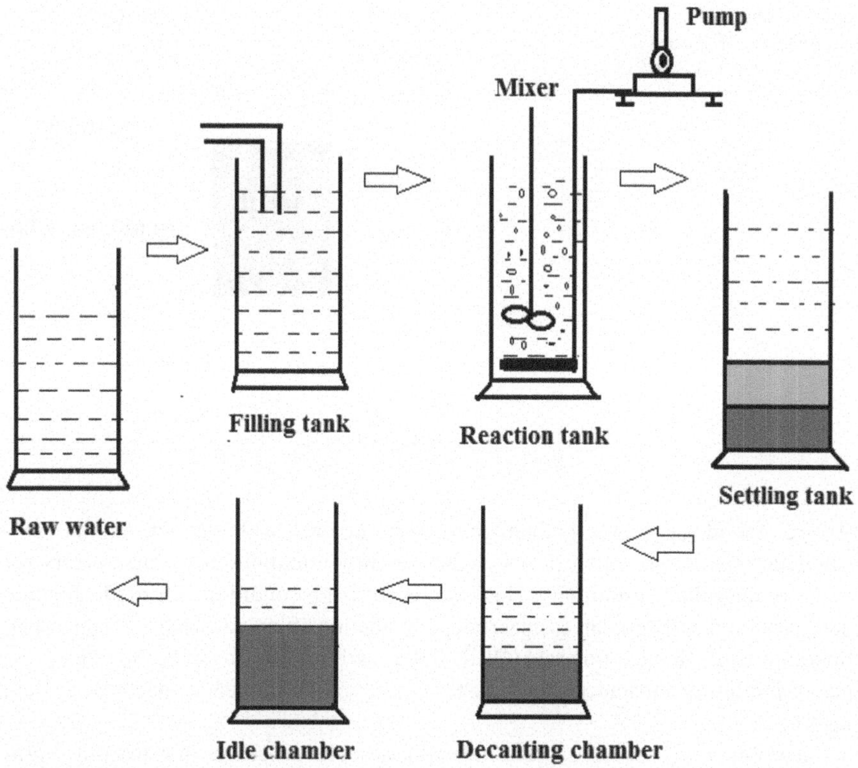

Fig. 3 Diagrammatic representation of sequence batch reactor (Joshiba et al., 2019)

for various industrial effluents (Joshiba et al., 2019). A bench-scale aerobic SBR effectively treated industrial milk factory wastewater in Iran, achieving over 90% COD removal for influent concentrations of 400–2500 mg/L. The SBR demonstrated cost-effective, easy operation with optimal conditions, proving its efficacy for dairy wastewater treatment (Mohseni-Bandpi & Bazari, 2004).

5.3 Anaerobic Treatment Methods

The anaerobic treatment (Fig. 4) of dairy wastewater, as described by, is a highly efficient process, where organic constituents are broken down through various intermediate processes with the help of microbes in the absence of oxygen (Das & Paul, 2022; Kalat & Yüceer, 2017). This method is recognized as crucial for mitigating pollution and managing waste (Joshiba et al., 2019). The method involves two steps stabilization and degradation with the help of microbes in the absence of oxygen (Chatzipaschali & Stamatis, 2012). Anaerobic degradation involves a

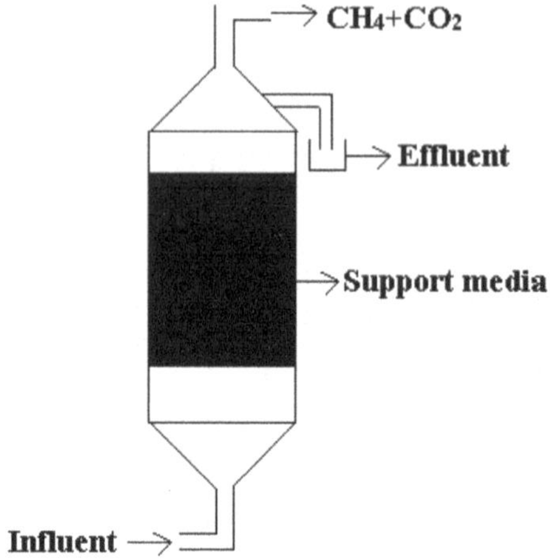

Fig. 4 Schematic representation of anaerobic filters (Joshiba et al., 2019)

mixture of acidogenic and methanogenic bacteria, each with distinct optimal environmental conditions which poses challenges in maintaining ideal parameters for the entire microbial community. However, optimizing conditions for methanogenic bacteria significantly enhances anaerobic degradation efficiency. Factors such as pH, temperature, organic loading rate (OLR), HRT, carbon-to-nitrogen (C/N) ratio, presence of inhibitory substances, and reactor configuration influence anaerobic system performance.

Anaerobic processes are preferred over aerobic processes for treating high-strength dairy wastewater due to their cost-effectiveness and non-production of unpleasant odors, when properly operated (Slavov, 2017). The major disadvantage of using aerobic treatment is the consumption of high energy through aeration, and reduced efficiency caused by microbial growth (Joshiba et al., 2019). Anaerobic systems are also preferred over aerobic methods for treating dairy waste due to their cost-effectiveness, methane production, lower sludge generation, reduced nutrient requirements, and capacity to handle high COD dairy waste without prior dilution. However, drawbacks include high initial costs, lengthy start-up periods, strict operational control, sensitivity to variable loads, and the discharge of ammonia nitrogen, necessitating additional treatment to meet discharge standards. The simplest anaerobic digester type, the anaerobic lagoon, is easier to construct but requires a large surface area (Britz et al., 2006). The biological treatment of dairy wastewater via anaerobic processes is crucial for producing renewable energy which is one reason it has gained considerable relevance due (Joshiba et al., 2019).

Anaerobic filters are suitable for low-strength wastewater (Karadag et al., 2015). These filters utilize packing materials to support microorganisms, enhancing surface area and porosity for better biomass attachment and reduced reactor volume (Das & Paul, 2022). Various types of anaerobic digesters, including anaerobic filters and up-flow anaerobic sludge blanket reactors, have been used in laboratories to treat dairy wastewater (Toumi et al., 2015). These systems rely on a consortium of microorganisms, and factors such as sludge retention and mass transfer are crucial for optimal efficiency. Researchers are increasingly interested in anaerobic treatment due to its ability to treat wastewater with minimal energy requirements and produce valuable by-products like biogas and nutrient-rich bio-fertilizers for agriculture (Joshiba et al., 2019).

The various materials employed in anaerobic filters include seashells, plastics, sintered glass, charcoal, ceramics, rocks, limestone, and clay (Das & Paul, 2022; Karadag et al., 2015). Various types of anaerobic digesters, including anaerobic filters and up-flow anaerobic sludge blanket reactors, utilize a range of materials for effective wastewater treatment. Anaerobic filters have shown an 80% COD removal efficiency (Loupasaki & Diamadopoulos, 2013). The up-flow anaerobic sludge blanket (UASB) reactor is popular due to its settling characteristics and cost-effectiveness. Continuous stirred tank reactors (CSTR) are commonly used in laboratory settings for dairy wastewater treatment than in large-scale treatments because of their HRT limitations (Carvalho et al., 2013), while researchers are exploring their potential for organic matter degradation and methane production (Morken et al., 2018). The efficiency of the anaerobic fluidized bed reactors (AFBRs) is hindered by biomass loss during operation and the presence of large amounts of solids the dairy wastewater (Kushwaha et al., 2011). AFBRs use various media for microbial attachment (Chowdhury et al., 2017) and offer advantages such as high mass transfer rates, low clogging, ability to handle high loads, high biomass, and low space requirements (Burghate & Ingole, 2013; Shete & Shinkar, 2013). The maximum COD removal efficiency of 84% has reported for fluidized bed reactors (Das & Paul, 2022; Joshiba et al., 2019).

Separated phase digesters are specialized systems that physically separate acid-forming bacteria from acid-consuming bacteria. They are effective for treating wastes with unbalanced carbon-to-nitrogen ratios, such as high-protein wastewater, or those that acidify rapidly like dairy wastewaters (IDF, 1990; Strydom et al., 2001). These digesters offer several advantages including handling wastewater with high organic loads, and it has short hydraulic retention times (Britz et al., 2006). A study used separated phase digesters, specifically anaerobic rotating biological contact reactors (AnRBC), to treat cheese whey. These digesters can be more accurately described as horizontally oriented tubular fixed-film digesters with internally rotating baffles. In the methane reactor, cedar wood baffles were used to promote rapid bacterial biofilm development. The acidogenic reactor, mixed by biogas recirculation, achieved only a 4% COD reduction but significantly increased volatile fatty acids from 168 to 1892 mg/L. This substrate led to an 87% COD reduction in the second phase. The original whey COD of 6720 mg/L was approximately diluted tenfold (Britz et al., 2006).

5.4 Chemical Treatment

Chemical treatment, which primarily involves the use of chemical reagents for oxidation or pH adjustment, removes colloids and soluble contaminants from milk processing effluents. For instance, when cheese wastewater reacts with $FeSO_4$ and H_2O_2, it can remove up to 80% of fat, even when starting with a concentration as high as 1.93 g/L (Vlyssides et al., 2012). Extreme pH of the dairy wastewater (below 6.5 or above 10) can accelerate pipe corrosion and harm the microbiological communities that are crucial for biological processes. Hence, pH correction is crucial to mitigate these adverse effects. If a Dissolved Air Flotation (DAF) unit is part of the treatment process, controlling pH also becomes essential for optimal coagulation conditions (Shammas et al., 2010). DAF systems in Canada use pressurized air to dissolve air into the water, which, when released at atmospheric pressure, forms bubbles that attach to suspended particles and float them to the surface for removal. This method is particularly effective for removing fats, oils, and greases from dairy effluent (Edzwald, 1995).

5.5 Electrochemical Methods

The electrocoagulation process is an exceptionally effective method for eliminating contaminants from wastewater. This technique operates on the principle of dissolving a metal electrode to release metal ions. Electrocoagulation (EC) is a promising, moderately eco-friendly method for wastewater treatment. Combining the benefits of coagulation, flotation, and electrochemistry (Babu et al., 2020; Moussa et al., 2017), EC has been extensively applied to various wastewaters, including those from tanneries (Deveci et al., 2019), textile mills (Khemila et al., 2018), laundries (Dimoglo et al., 2019), dairy plants (Reilly et al., 2019), and the steel industry (Changmai et al., 2020).

Key electrochemical techniques in dairy treatment include EC, electro-flotation, and anodic oxidation. In the EC process, metals like iron or aluminum serve as anodes, dissolving to release metal ions that act as coagulating agents. The system consists of metal plate anodes and cathodes immersed in wastewater. EC is valued for generating minimal sludge, not requiring additional chemicals, and having a compact footprint without sacrificing efficiency (Aswathy et al., 2016). Factors affecting EC performance include initial pH, conductivity, temperature, current density, effluent constituent concentrations, electrolyte concentration, electrode material, stirring speed, supporting electrolyte, and reaction time (Chezeau et al., 2020; Tahreen et al., 2020; Xu et al., 2021). The EC process's efficiency hinges on the maximization of floc yield, which enhances the removal efficiency of anions and organics (Xu et al., 2021). Floc yield is highest without aeration and decreases over time post-electrolysis. EC's future in wastewater treatment looks promising, especially with the potential of hybrid integration with other separation processes and the shift towards using solar power instead of conventional electricity for the EC process (Krishna et al., 2022).

Melchiors et al. (2016) evaluates the effectiveness of electroflocculation in treating dairy industry wastewater and recovering solid whey. Using an electrochemical apparatus with aluminum or iron electrodes, the study monitored the effects of electroflocculation time, initial pH, and applied potential intensity on chemical oxygen demand (COD), turbidity, and final pH. Optimal conditions (5 V potential, 2 cm electrode distance, 60 min, and pH 5.0) resulted in COD and turbidity removal rates of 97.0 and 99.6% with aluminum electrodes, and 97.4 and 99.1% with iron electrodes. The final pH was 6.72 for aluminum and 7.38 for iron electrodes. The results demonstrate that electroflocculation is a highly effective and sustainable method for dairy wastewater treatment, enabling water recovery for reuse and the recovery of solid whey for food supplement production.

In Iran, a study investigates the treatment of dairy wastewater using the electro-Fenton process with an iron electrode. The research focused on optimizing five key parameters: reaction time, current density, pH, H_2O_2/Fe^{2+} molar ratio, and H_2O_2/dairy wastewater (DW) volumetric ratio. The response surface methodology was employed to minimize experimental runs and determine optimal conditions. The study identified optimal conditions for maximum chemical oxygen demand (COD) removal (93.93%) as a 90-min reaction time, 56 mA/cm^2 current density, pH 7.52, H_2O_2/DW ratio of 0.898 ml/l, and H_2O_2/Fe^{2+} ratio of 3.965. For color removal (97.32%), optimal conditions were slightly adjusted to 86 min, 55.1 mA/cm^2, pH 7.48, H_2O_2/DW ratio of 0.907 ml/l, and H_2O_2/Fe^{2+} ratio of 3.987 (Davarnejad & Nikseresht, 2016).

5.6 Phytoremediation

Phytoremediation is the use of plants and biological processes to remove or eliminate contaminants and pollutants from polluted systems (Nwogwu et al., 2022). During phytoremediation, specific plant roots help in either breaking down soil contaminants or absorbing and storing them within the plant's tissues (Ajibade et al., 2013a, 2013b; NaIr & Kani, 2016). Phytoremediation is an economical and viable sustainable method for removing pollutants from wastewater. Additionally, it's environmentally friendly and poses minimal impact on nearby residents and workers since it relies on plants to clean the environment. Plants, along with their microbial communities, are utilized to absorb contaminants from soil or water. This remediation process involves retaining, eliminating, or degrading pollutants through natural mechanisms, similar to those found in ecosystems, where organic and inorganic cycles are involved. This leads to the development of cost-effective remediation technology. Numerous plant species inherently possess the ability to treat polluted soil, water, and air (Ajibade et al., 2021b). Heavy metals like cadmium and lead are not readily absorbed by microorganisms, therefore, utilizing plants to bioaccumulate these toxins through natural metabolic processes presents an optimal treatment option. Aquatic plants, in particular, demonstrate remarkable capacity in reducing the levels of toxic metals, BOD, and total solids from wastewater.

The use of azolla and water hyacinth to treat dairy wastewater showed that the treatment efficiency depends on climatic conditions, the concentration of contaminants, and temperature, among others (Nair & Kani, 2016). The efficiency in removing contaminants such as TSS, TDS, BOD, COD, Electrical Conductivity (EC), hardness, heavy metals, etc., varied from one plant species to another. The study showed that Azolla was more effective in reducing nearly all the parameters examined, and a decrease in the pH value of the dairy effluent.

According to Agrahari and Kumar (2023), increasing plant cover to remove harmful pollutants from the environment not only helps in carbon sequestration by reducing atmospheric carbon dioxide levels but also brings about various benefits for the economy, environment, and society through sustainable site management. This approach enhances ecosystem functions like nutrient cycling, fertility, carbon sequestration, and water purification. Moreover, this method proves to be more cost-effective than conventional treatment approaches, with lower operational costs and the added benefit of creating aesthetically pleasing surroundings. However, it is a time-consuming process, taking months to years to show significant results, and requires proper management and disposal of plants that may have absorbed hazardous contaminants.

5.7 Wetland

Wetland systems use natural processes, including self-sustaining microbial communities, to enhance wastewater treatment. It involves the arrangement of aquatic and terrestrial plants within an artificial system. Their simple construction and absence of sludge recycling make them particularly suitable for treating dairy effluent in developing countries (Slavov, 2017). This aquatic plant (macrophyte) is utilized for phytoremediation purposes, and these plants have demonstrated significant capability in accumulating harmful elements (Ajibade & Adewumi, 2017; Schwantes et al., 2019). Typically, dairy wastewater is treated in wetlands under aerobic conditions. In aerobic ponds containing milk wastes, it takes approximately five days to achieve an 85% reduction in BOD_5 at a temperature of 20 °C. However, for high-strength dairy wastewater, treatment is predominantly conducted in facultative wetlands (Slavov, 2017).

As highlighted by Das and Paul (2022), this method enhances wastewater treatment potential and improves its quality for safe discharge. Considered a sustainable wastewater treatment solution, wetland treatment offers efficiency comparable to conventional methods but with added eco-friendliness, cost-effectiveness, and energy savings (Bao et al., 2023; Lasisi et al., 2023). Plants in these systems, with their roots immersed in water for long periods, effectively remove contaminants from the wastewater. Wetland treatment exhibits high removal rates for various pollutants, including BOD, COD, suspended solids, nitrate, and coliforms, with moderate removal rates observed for total nitrogen and phosphorus. In constructed wetlands (CWs), the removal of nitrogen (N) is primarily achieved through several mechanisms, including physical settlement, denitrification, and uptake by plants and microbes (Ajibade et al., 2021c, 2023). Constructed wetlands in Canada use wetland vegetation, soil, and microbial activity to treat dairy effluent naturally. This method provides a low-cost, sustainable option for wastewater treatment, utilizing natural filtration and biodegradation processes (Kadlec & Wallace, 2008).

In wetland systems, dairy wastewater (DWW) is treated under aerobic conditions (Das & Paul, 2022). In the aerobic pond, Bitton (1994) reported that the removal efficiency of BOD_5 was 85% of milk wastes at 20 °C while facultative wetlands are used for high-strength dairy wastes (Sterritt & Lester, 1988). While wetland processes are effective at removing BOD and COD, they are less efficient at removing phosphorus and nitrogen, with phosphorus removal generally low and nitrogen removal above 50% (Verhoeven & Meuleman, 1998). Typically, phosphorus removal is low, while nitrogen removal exceeds 50%. Studies have shown that increasing the retention time of wastewater from 2 to 7 days enhances total nitrogen removal from 12 to 14% in unplanted wetlands and from 48 to 75% in planted wetlands. Similarly, total phosphorus removal efficiency increases from 12 to 36% in unplanted wetlands and from 37 to 74% in planted wetlands. In the wetland process, nitrogen removal primarily depends on microbial activities in the root zones, which are temperature-sensitive, whereas phosphorus removal is not directly temperature-dependent but is influenced by redox levels affected by oxygen availability. In aerobic wetland processes, effluents exhibit high quality, with removal efficiencies exceeding 90% for COD, TSS, BOD_5, organic nitrogen, and total coliforms (Das & Paul, 2022).

The wetland treatment process offers several advantages, including reduced energy consumption, minimal maintenance, and low operational demands (Brix, 1994; Das & Paul, 2022). However, drawbacks include the requirement for a sizable surface area, potential risks of surface and groundwater pollution, the presence of hazardous volatile substances, and susceptibility to insect infestations. Despite their easy operation, the complexity of the biological processes involved in wetland systems can surpass that of other wastewater treatment methods (Slavov, 2017).

5.8 *Membrane Technologies*

Membrane filtration is an advanced wastewater treatment process that enhances effluent quality to meet stringent discharge standards or enables the reuse of treated effluents, thereby conserving freshwater resources. This process involves removing or separating particulate and colloidal particles from a liquid by passing it through a selective barrier with a pore size ranging from 0.0001 to 1.0 μm. Membrane filtration technologies include microfiltration, ultrafiltration, nanofiltration, dialysis, electrodialysis, and reverse osmosis (Krishna et al., 2022). Membrane technologies can be categorized based on membrane type, pressure, configuration, and removal method. Ultrafiltration, a key technique within membrane technologies, is widely used in wastewater treatment (oily wastewater/oil and dye wastewater), bacterial and virus removal, and the textile industry (Ajibade et al., 2021d, 2022; Gentile et al., 2018; Lin et al., 2016; Santra et al., 2019). In the context of dairy effluent treatment, integrating membrane-based technologies with biological and chemical processes shows significant promise. However, challenges such as fouling, high investment, and operating costs have to be addressed (Kenari & Barbeau, 2016). To mitigate fouling, chemical coagulants or adsorbents like powdered activated carbon or zeolite are often used to enhance filtration efficiency by reducing suspended and colloidal particles in the feed solution.

MBR technology in Germany combines biological treatment with membrane filtration, allowing for high-quality effluent with a compact system footprint. The integration of microfiltration or ultrafiltration membranes ensures the removal of suspended solids and pathogens, making it highly effective for treating dairy effluent (Drews, 2020). In Japan, MBR systems are implemented for their efficiency in solid-liquid separation and their ability to produce high-quality effluent suitable for reuse. This technology addresses the challenges of high organic load and variable effluent quality in dairy processing facilities. Direct membrane separation using hollow fibers was applied in an activated sludge process, achieving over 95% COD removal and up to 60% nitrogen denitrification. Stable operation was maintained for 120 days with intermittent suction at low pressure, preventing membrane clogging. (Yamamoto et al., 1989). Anaerobic Membrane Bioreactor (AnMBR) systems in Spain integrate membrane filtration with anaerobic digestion, providing high-rate treatment of dairy wastewater. This combination allows for effective biomass retention and reduces membrane fouling, ensuring consistent treatment performance and high-quality effluent (Giménez et al., 2011). In China, AnMBR technology addresses the challenge of membrane fouling while offering energy-efficient treatment of dairy effluents. The system's ability to retain biomass and produce biogas as a renewable energy source makes it a sustainable solution for high-strength wastewater (Meng et al., 2017). This study by Fraga et al. (2017) in Uruguay assessed the performance of a pilot membrane bioreactor (MBR) for treating dairy wastewater to meet new discharge standards. Operated over four months at two locations with varying influent COD concentrations, the MBR achieved removal efficiencies of 94.1% for COD, 98.1% for BOD, and 99.6% for ammonium. It effectively removed total nitrogen and phosphorus when the COD/N ratio exceeded 10, meeting Uruguayan discharge standards and showing potential for water reuse.

6 Integrated System for Dairy Wastewater Treatment

Effective treatment of dairy wastewater is critical for environmental protection and resource recovery. Recent studies highlight various integrated advanced systems designed to enhance treatment efficiency and meet stringent discharge standards as follows.

Cervantes-Aviles et al. (2024) conducted a study in Mexico examining the treatment of dairy wastewater through a continuous flow system comprising two biological reactors: an Upflow Anaerobic Sludge Blanket (UASB) and a Hybrid Membrane

Bioreactor (H-MBR). Over 85 days, key parameters such as chemical oxygen demand (COD), volatile suspended solids (VSS), total nitrogen (TN), and phosphate were monitored. The UASB achieved a 97% COD removal, with an 18% increase under higher organic loading, which also boosted biogas production. The H-MBR further enhanced COD, TN, and phosphorus removal by 72%, 52%, and 19%, respectively, and ensured complete VSS removal. Effective sludge management was critical to preventing UASB destabilization and delaying membrane clogging. The study underscores the potential of the UASB and H-MBR system for efficient dairy wastewater treatment.

Another study by Ji et al. (2020) evaluated the engineering performance of a dairy wastewater treatment system integrating an anaerobic baffled reactor (ABR) with an upflow anaerobic sludge blanket (UASB). The ABR process facilitated the adsorption and decomposition of fats and the coagulation of proteins, creating favorable conditions for sludge retention in the UASB. The integrated system achieved significant outcomes, including a reduction in excess sludge from 3–5 t/d to 3 t/m, 98% COD removal, and compliance with effluent standards for other pollutants. Additionally, the system demonstrated excellent energy efficiency, with operating costs below 0.5 kWh/t, and maintained stable operations for over three years at ambient temperatures. The results affirm the technical and economic viability of this system for dairy wastewater treatment.

Moreover, Akansha et al. (2020) explores the treatment of dairy industry wastewater through a combined approach of aerated electrocoagulation and phytoremediation. Electrocoagulation was carried out using aluminum and iron electrodes, examining the effects of pH, electrode combinations, and voltage. The process was most effective at neutral pH, achieving a COD removal efficiency of 86.4% with the Al-Fe electrode combination under optimal conditions (pH 7, 5 V, 120 min). When paired with phytoremediation using *Canna indica*, COD removal increased to 97%, and the treated wastewater supported significant plant growth compared to the untreated water. Bacterial toxicity tests showed that both treated and untreated wastewater promoted bacterial growth. This combined method demonstrates high efficiency in treating dairy industry wastewater.

These studies underscore the effectiveness of integrating various treatment technologies for dairy wastewater management. Collectively, these approaches highlight the potential for advanced and sustainable solutions in dairy wastewater treatment, emphasizing the importance of optimizing treatment conditions and integrating technologies to meet environmental standards and enhance resource recovery.

7 Conclusion

Dairy wastewater presents a complex environmental challenge, demanding comprehensive understanding and effective solutions. Through the exploration of its concise overview and the identification of key characteristics, it becomes evident that dairy wastewater is a significant contributor to environmental pollution. The analysis of its environmental effects underscores the urgent need for action, as its discharge poses risks to aquatic ecosystems, soil quality, and public health. However, amidst these challenges, there is promise in the evaluation of various methods and technologies for dairy wastewater treatment. By assessing treatment options ranging from biological processes to advanced systems, opportunities arise for mitigating the adverse impacts of dairy wastewater discharge. This conclusion emphasizes the importance of holistic approaches in addressing dairy wastewater management, integrating knowledge of its characteristics and environmental effects with the implementation of efficient treatment methodologies. Ultimately, proactive measures are essential to safeguarding environmental sustainability and human well-being in the face of dairy wastewater pollution.

References

Adewumi, J. R., & Ajibade, F. O. (2015). The pollution effects of indiscriminate disposal of wastewater on soil in semi-urban area. *Journal of Applied Science and Environmental Management (JASEM), 19*(3), 412–419.

Agrahari, S., & Kumar, S. (2023). Phytoremediation: A shift towards sustainability for dairy wastewater treatment. *ChemBioEng Reviews, 11*(1):115–135. https://doi.org/10.1002/cben.202300038

Ajibade, F. O., Adeniran, K. A., & Egbuna, C. K. (2013). Phytoremediation efficiencies of water hyacinth in removing heavy metals in domestic sewage (A Case Study of University of Ilorin, Nigeria). *IJES,* 2(12), 16–27. ISSN(e): 2319-1813 ISSN(p): 2319-1805.

Ajibade, F. O., Nwogwu, N. A., Lasisi, K. H., Ajibade, T. F., Adelodun, B., Guadie, A., Ugya, A.Y., Adewumi, J. R., Wang, H. C., & Wang, A. (2021b). Removal of nitrogen oxyanion (nitrate) in constructed wetlands. In N. A. Oladoja & I. E. Unuabonah (Eds.), *Progress and prospects in the management of oxoanion polluted aqua systems*. Springer Nature. https://doi.org/10.1007/978-3-030-70757-6_12

Ajibade, F. O., Adelodun, B., Lasisi, K. H., Fadare, O. O., Ajibade, T. F., Nwogwu, N. A., Sulaymon I. D., Ugya, A. Y., Wang, H. C., Wang, A. (2021a). Environmental pollution and their socioeconomic impacts. In A. Kumar, V. K. Singh, P. K. Singh, & V. Mishra (Eds.), *Microbe mediated remediation of environmental contaminants*. Woodhead Publishing. https://doi.org/10.1016/B978-0-12-821199-1.00025-0

Ajibade, F. O., Adeniran, K. A., & Egbuna, C. K. (2013b). Phytoremediation efficiencies of water hyacinth in removing heavy metals in domestic sewage (A Case Study of University of Ilorin, Nigeria). *The International Journal of Engineering and Science, 2*(12), 16–27.

Ajibade, F. O., & Adewumi, J. R. (2017). Performance evaluation of aquatic macrophytes in a constructed wetland for municipal wastewater treatment. *FUTA Journal of Engineering and Engineering Technology, 11*(1), 1–11.

Ajibade, F. O., Wang, H., Guadie, A. A., Ajibade, T. F., Fang, Y., Sharif, H. M. A., Liu, W., & Wang, A. (2021c). Total nitrogen removal in biochar amended non-aerated vertical flow constructed wetlands for secondary wastewater effluent with low C/N ratio: Microbial community structure and dissolved organic carbon release conditions. *Bioresource Technology, 1*, 124430. https://doi.org/10.1016/j.biortech.2020.124430

Ajibade, T. F., Tian, H., Hassan Lasisi, K., Xue, Q., Yao, W., & Zhang, K. (2021d). Multifunctional PAN UF membrane modified with 3D-MXene/O-MWCNT nanostructures for the removal of complex oil and dyes from industrial wastewater. *Separation and Purification Technology, 275*, Article 119135. https://doi.org/10.1016/j.seppur.2021.119135

Ajibade, T. F., Tian, H., Lasisi, K. H., & Zhang, K. (2022). Bio-inspired PDA@WS2 polyacrylonitrile ultrafiltration membrane for the effective separation of saline oily wastewater and the removal of soluble dye. *Separation and Purification Technology, 299*, Article 121711. https://doi.org/10.1016/j.seppur.2022.121711

Ajibade, F.O., Yin, W., Guadie, A., Ajibade, T.F., Liu, Y., Kumwimba, M.N., Liu, W., Han, J., Wang, H., & Wang, A. (2023). Impact of biochar amendment on antibiotic removal and ARGs accumulation in constructed wetlands for low C/N wastewater treatment. *Chemical Engineering Journal, 459*, 141541. https://doi.org/10.1016/j.cej.2023.141541

Akansha, J., Nidheesh, P., Gopinath, A., Anupama, K., & Suresh Kumar, M. (2020). Treatment of dairy industry wastewater by combined aerated electrocoagulation and phytoremediation process. *Chemosphere, 253*, Article 126652. https://doi.org/10.1016/j.chemosphere.2020.126652

Alayu, E., & Yirgu, Z. (2018). Advanced technologies for the treatment of wastewaters from agro-processing industries and cogeneration of by-products: A case of slaughterhouse, dairy and beverage industries. *International Journal of Environmental Science and Technology, 15*, 1581–1596. https://doi.org/10.1007/s13762-017-1522-9

Alhamlan, F. S., Ederer, M. M., Brown, C. J., Coats, E. R., & Crawford, R. L. (2013). Metagenomics-based analysis of viral communities in dairy lagoon wastewater. *Journal of Microbiological Methods, 92*(2), 183–188. https://doi.org/10.1016/j.mimet.2012.11.016

Al-Tayawi, A. N., Sisay, E. J., Beszédes, S., & Kertész, S. (2023). Wastewater treatment in the dairy industry from classical treatment to promising technologies: An overview. *Processes, 11*, 2133. https://doi.org/10.3390/pr11072133

Arvanitoyannis, I. S., & Giakoundis, A. (2006). Current strategies for dairy waste management: A review. *Critical Reviews in Food Science and Nutrition, 46*(5), 379–390. https://doi.org/10.1080/10408390591000695

Aswathy, P., Gandhimathi, R., Ramesh, S. T., & Nidheesh, P. V. (2016). Removal of organics from bilge water by batch electrocoagulation process. *Separation and Purification Technology, 159*, 108–115. https://doi.org/10.1016/j.seppur.2016.01.001

Azimi, A., Azari, A., Rezakazemi, M., & Ansarpour, M. (2017). Removal of heavy metals from industrial wastewaters: a review. *Chem Bio Eng, 4*(1), 37–59. https://doi.org/10.1002/cben.201600010

Babu, D. S., Singh, T. S. A., Nidheesh, P. V., & Kumar, M. S. (2020). Industrial wastewater treatment by electrocoagulation process. *Separation Science and Technology, 55*, 3195–3227. https://doi.org/10.1080/01496395.2019.1671866

Bao, H., Yin, W., Wang, H., Lu, Y., Jiang, S., Ajibade, F. O., Ouyang, Q., Wang, Y., Nie, S., Bai, Y., Gao, H., & Wang, A. (2023). Automated machine learning-based models for predicting and evaluating antibiotic removal in constructed wetlands. *Bioresource Technology, 385*, Article 129436. https://doi.org/10.1016/j.biortech.2023.129436

Bella, K., Rao, P.V. (2021). Anaerobic digestion of dairy wastewater: effect of different parameters and co-digestion options—a review. *Biomass Conversion and Biorefinery 13*(93). http://dx.doi.org/%2010.1007/s13399-020-01247-2

Biswas, T., Bhushan, S., Prajapati, S. K., & Chaudhuri, S. R. (2021). An eco-friendly strategy for dairy wastewater remediation with high lipid microalgae-bacterial biomass production. *Journal of Environmental Management, 286*, Article 112196. https://doi.org/10.1016/j.jenvman.2021.112196

Bitton, G. (1994). *Wastewater microbiology*. Wiley Press.

Briao, V. B., & Tavares, C. R. G. (2007). Effluent generation by the dairy industry: Preventive attitudes and opportunities. *Brazilian Journal of Chemical Engineering, 24*, 487–497. https://doi.org/10.1590/S0104-66322007000400003

Britz, T. J., Van Schalkwyk, C., & Hung Y. (2006). Treatment of dairy processing wastewaters. In *Waste treatment on the food processing industry* (pp. 1–28).

Brix, H. (1994) Use of Constructed wetlands in water pollution control: historical development, present status, and future perspectives. *Water Science and Technology, 30*, 209–223.

Burghate, S. P., & Ingole, N. W. (2013). Fluidized bed biofilm reactor—A novel wastewater treatment reactor. *International Journal of Research in Environmental Science and Technology, 3*(4), 145–155.

Carvalho, F., Prazeres, A. R., & Rivas, J. (2013). Cheese whey wastewater: Characterization and treatment. *Science of the Total Environment, 445–446*, 385–396. https://doi.org/10.1016/j.scitotenv.2012.12.038

Cervantes-Avilés, P., Prado-Viramontes, E., Saber, A. N., Cuevas-Rodríguez, G. (2024). Treatment of dairy wastewater by a combined technologies approach: Anaerobic digestion and membrane technology. In M. Ksibi, A. Sousa, O. Hentati, H. Chenchouni, J. L. Velho, A. Negm, J. Rodrigo-Comino, R. Hadji, S. Chakraborty, & A. Ghorbal (Eds.), Recent advances in environmental science from the Euro-mediterranean and surrounding regions (4th Edition). EMCEI 2022. Advances in science, technology & innovation. Springer. https://doi.org/10.1007/978-3-031-51904-8_22

Changmai, M., Das, P. P., Mondal, P., Pasawan, M., Sinha, A., Biswas, P., Sarkar, S., & Purkait, M. K. (2022). Hybrid electrocoagulation–microfiltration technique for treatment of nanofiltration rejected steel industry effluent. *International Journal of Environmental Analytical Chemistry, 102*(1), 62–83.

Chatzipaschali, A. A., & Stamatis, A. G. (2012). Biotechnological utilization with a focus on anaerobic treatment of cheese whey: Current status and prospects. *Energies, 5*(9), 3492–3525. https://doi.org/10.3390/en5093492

Chezeau, B., Boudriche, L., Vial, C., & Boudjemaa, A. (2020). Treatment of dairy wastewater by electrocoagulation process: Advantages of combined iron/aluminum electrodes. *Separation Science and Technology, 55*, 2510–2527.

Chowdhury, M. M. I., Nakhla, G., & Zhu, J. (2017). Ultrasonically enhanced anaerobic digestion of thickened waste activated sludge using fluidized bed reactors. *Applied Energy, 204*, 807–818. https://doi.org/10.1016/j.apenergy.2017.07.057

Craggs, R. J., Sukias, J. P., Tanner, C. T., & Davies-Colley, R. J. (2004). Advanced pond system for dairy-farm effluent treatment. *Water Science and Technology, 48*(2), 291–297. https://doi.org/10.1080/00288233.2004.9513613

Das, P., & Paul, K. K. (2022). *Present scenario of dairy wastewater treatment: A state of art review*. https://doi.org/10.21203/rs.3.rs-1774888/v1

Davarnejad, R., & Nikseresht, M. (2016). Dairy wastewater treatment using an electrochemical method: Experimental and statistical study. *Journal of Electroanalytical Chemistry, 775*, 364–373. https://doi.org/10.1016/j.jelechem.2016.06.016

Demirel, B., Yenigun, O., & Onay, T. T. (2005). Anaerobic treatment of dairy wastewaters: A review. *Process Biochemical, 40*, 2583–2595.

Deveci, E. Ü., Akarsu, C., Gonen, Ç., & Ozay, Y. (2019). Enhancing treatability of tannery wastewater by integrated process of electrocoagulation and fungal via using RSM in an economic perspective. *Process Biochemistry, 84*, 124–133.

Dimoglo, A., Sevim-Elibol, P., Dinç, O., Gokmen, K., & Erdogan, H. (2019). Electrocoagulation/electroflotation as a combined process for the laundry wastewater purification and reuse. *Journal of Water Process Engineering, 31*, Article 100877.

Drews, A. (2020). Membrane fouling in membrane bioreactors—Characterisation, contradictions, cause and cures. *Journal of Membrane Science, 363*(1–2), 1–28.

Ebrahimi, A., Najafpour, G., Mohammadi, M., & Hashemiyeh, B. (2010). Biological treatment of whey in an UASFF bioreactor followed a three-stage RBC. *Chemical Industry and Chemical Engineering Quarterly*(2), 175–182. https://doi.org/10.2298/CICEQ100315025E

Edzwald, J.K. (1995) Principles and applications of dissolved air flotation. *Water Science & Technology, 31*, 1–23. https://doi.org/10.2166/wst.1995.0512

El Khomri, M., El Messaoudi, N., Dbik, A., Bentahar, S., Lacherai, A., Faska, N., & Jada, A. (2022). Regeneration of argan nutshell and almond shell using HNO_3 for their reusability to remove cationic dye from aqueous solution. *Chemical Engineering Communications, 209*, 1304–1315.

El Messaoudi, N., El Mouden, A., Fernine, Y., El Khomri, M., Bouich, A., Faska, N., Ciğeroğlu, Z., Américo-Pinheiro, J. H. P., Jada, A., & Lacherai, A. (2022). Green synthesis of Ag_2O nanoparticles using punica granatum leaf extract for sulfamethoxazole antibiotic adsorption: Characterization, experimental study, modeling, and DFT calculation. *Environmental Science and Pollution Research, 2022*, 1–18.
Farizoglu, B., & Uzuner, S. (2011). The investigation of dairy industry wastewater treatment in a biological high performance membrane system. *Biochemical Engineering Journal, 57*(1), 46–54. https://doi.org/10.1016/j.bej.2011.08.007
Fraga, F. A., García, H. A., Hooijmans, C. M., Míguez, D., & Brdjanovic, D. (2017). Evaluation of a membrane bioreactor on dairy wastewater treatment and reuse in Uruguay. *International Biodeterioration & Biodegradation, 119*, 552–564. https://doi.org/10.1016/j.ibiod.2016.11.025
Gentile, G. J., Cruz, M. C., Rajal, V. B., & de Cortalezzi, M. M. F. (2018). Electrostatic interactions in virus removal by ultrafiltration membranes. *Journal of Environmental Chemical Engineering, 6*, 1314–1321.
Giménez, J. B., Robles, Á., Carretero, L., Durán, F., Ruano, M. V., Gatti, M. N., Ferrer, J., & Seco, A. (2011). Experimental study of the anaerobic urban wastewater treatment in a submerged hollow-fibre membrane bioreactor at pilot scale. *Bioresource Technology, 102*(16), 8799–8806.
Goli, A., Shamiri, A., Khosroyar, S., Talaiekhozani, A., Sanaye, R., & Azizi, K. (2019). A review on different aerobic and anaerobic treatment methods in dairy industry wastewater. *Journal of Environmental Treatment Techniques, 6*(1), 113–141.
Grady, C. P. L., Daigger, G. T., Love, N. G., & Filipe, C. D. M. (2011). *Biological wastewater treatment* (3rd ed.). CRC Press.
Gutiérrez, J. L. R., Encina, P. A. G., & Fdz-Polanco, F. (1991). Anaerobic treatment of cheese-production wastewater using a UASB reactor. *Bioresource Technology, 37*(3), 271–276. https://doi.org/10.1016/0960-8524(91)90194-O
IDF. (1990). *Anaerobic treatment of dairy effluents—The present stage of development* (Doc., 252). Bulletin of the International Dairy Federation.
Ji, S., Ma, W., Wei, Q., Zhang, W., Jiang, F., & Chen, J. (2020). Integrated ABR and UASB system for dairy wastewater treatment: Engineering design and practice. *Science of the Total Environment, 749*, Article 142267. https://doi.org/10.1016/j.scitotenv.2020.142267
John, C. K., & Pu, J. H. (2023). *Household reusable rainwater technology for developing and under-developed countries*. Taylor and Francis Ltd.
Joshiba, J. G., Kumar, P. S., Femina, C. C., Jayashree, E., Racchana, R., & Sivanesan, S. (2019). Critical review on biological treatment strategies of dairy wastewater. *Desalination and Water Treatment, 160*, 94–109. https://doi.org/10.5004/dwt.2019.24194
Kadlec, R. H., & Wallace, S. D. (2008). *Treatment wetlands*. CRC Press.
Kalat, D. G., & Yüceer, A. (2017). Anaerobic mesophilic and thermophilic treatability of vegetable oil refining wastewater. *Process Safety and Environmental Protection, 109*, 151–157. https://doi.org/10.1016/j.psep.2017.04.001
Kamath, A., Kharat, O., Mehta, R., Kalsekar, S., & Patil, D. (2018). Treatment of dairy effluent using rotating biological contactors (RBC). *International Research Journal of Engineering and Technology, 5*(3), 2862–2866.
Karadag, D., Köroğlu, O. E., Ozkaya, B., & Cakmakci, M. (2015). A review on anaerobic biofilm reactors for the treatment of dairy indüstry wastewater. *Process Biochemistry, 50*(2), 262–271. https://doi.org/10.1016/j.procbio.2014.11.005
Kaur, N. (2021). Different treatment techniques of dairy wastewater. *Groundwater for Sustainable Development, 14*, Article 100640. https://doi.org/10.1016/j.gsd.2021.100640
Kenari, S. L. D., & Barbeau, B. (2016). Understanding ultrafiltration fouling of ceramic and polymeric membranes caused by oxidized iron and manganese in water treatment. *Journal of Membrane Science, 516*, 1–12.
Kesari, K. K., Soni, R., Jamal, Q. M. S., Tripathi, P., Lal, J. A., Jha, N. K., Siddiqui, M. H., Kumar, P., Tripathi, V., & Ruokolainen, J. (2021). Wastewater treatment and reuse: A review of its applications and health implications. *Water, Air, & Soil Pollution, 232*, 1–28.

Khemila, B., Merzouk, B., Chouder, A., Zidelkhir, R., Leclerc, J. P., & Lapicque, F. (2018). Removal of a textile dye using photovoltaic electrocoagulation. *Sustainable Chemistry and Pharmacy, 7*, 27–35.

Krishna, B. R., Bhuvaneshwari, S., Majeed, F., Jose, E., & Mohan, A. (2022). Different treatment methodologies and reactors employed for dairy effluent treatment—A review. *Journal of Water Process Engineering, 46*, 102622. https://doi.org/10.1016/j.jwpe.2022.102622

Kushwaha, J. P., Srivastava, V. C., & Mall, I. D. (2011). An overview of various technologies for the treatment of dairy wastewaters. *Critical Reviews in Food Science and Nutrition, 51*(5), 442–452. https://doi.org/10.1080/10408391003663879

Lasisi, K. H., Ajibade, F. O., Idowu, T. E., Ajibade, T. F., Adelodun, B., Ojo, A. O., Fadugba, O. G., Olanrewaju, O. O., & Adewumi, J. R. (2023). Tidal coastal wetlands for wastewater management. In A. Kumar (Ed.) *Advances in chemical pollution, environmental management and protection*. Elsevier. https://doi.org/10.1016/bs.apmp.2022.11.002

Lin, J., Ye, W., Baltaru, M. C., Tang, Y. P., Bernstein, N. J., Gao, P., Balta, S., Vlad, M., Volodin, A., & Sotto, A. (2016). Tight ultrafiltration membranes for enhanced separation of dyes and Na_2SO_4 during textile wastewater treatment. *Journal of Membrane Science, 514*, 217–228.

Lo, K. V., & Liao, P. H. (1986). Digestion of cheese whey with anaerobic rotating biological contact reactor. *Biomass, 10*, 243–252.

Loupasaki, E., & Diamadopoulos, E. (2013). Attached growth systems for wastewater treatment in small and rural communities: A review. *Journal of Chemical Technology and Biotechnology, 88*(2), 190–204. https://doi.org/10.1002/jctb.3967

Mehrdadi, N., Bidhendi, G. R. N., & Shokouhi, M. (2012). Determination of dairy wastewater treatability by bio-trickling filter packed with lava rocks—Case study PEGAH dairy factory. *Water Science and Technology, 65*(8), 1441–1447. https://doi.org/10.2166/wst.2012.032

Melchiors, M. S., Piovesan, M., Becegato, V. R., Becegato, V. A., Tambourgi, E. B., & Paulino, A. T. (2016). Treatment of wastewater from the dairy industry using electroflocculation and solid whey recovery. *Journal of Environmental Management, 182*, 574–580. https://doi.org/10.1016/j.jenvman.2016.08.022

Meng, F., Zhang, S., Oh, Y., Zhou, Z., Shin, H. S., & Yang, F. (2017). Fouling in membrane bioreactors: An updated review. *Water Research, 114*, 151–180. https://doi.org/10.1016/j.watres.2017.02.006

Metcalf & Eddy (2014). Wastewater engineering: treatment and resource recovery. (5th Ed.), McGraw-Hill, New York.

Mohseni-Bandpi, A., & Bazari, H. (2004). Biological treatment of dairy wastewater by sequencing batch reactor. *Iranian Journal of Environment Health Science Engineering, 1*(2), 65–69.

Morken, J., Gjetmundsen, M., & Fjørtoft, K. (2018). Determination of kinetic constants from the co-digestion of dairy cow slurry and municipal food waste at increasing organic loading rates. *Renewable Energy, 117*, 46–51. https://doi.org/10.1016/j.renene.2017.09.081

Moussa, D. T., El-Naas, M. H., Nasser, M., & Al-Marri, M. J. (2017). A comprehensive review of electrocoagulation for water treatment: Potentials and challenges. *Journal of Environmental Management, 186*, 24–41. https://doi.org/10.1016/j.jenvman.2016.10.032

Najafpour, G.D., Hashemiyeh, B.A., Asadi, M., & Ghasemi, M.B. (2008). Biological treatment of dairy wastewater in an up-flow anaerobic sludge-fixed film bioreactor. *American-Eurasian Journal of Agricultural & Environmental Sciences, 4*(2), 251–257.

Nair, C. S., & Kani, & K. M. (2016). Phytoremediation of dairy effluent using aquatic macrophytes. *International Journal of Scientific & Engineering Research, 7*, 253–259

Nwogwu, N. A., Ajala, O. A., Ajibade, F. O., Ajibade, T. F., Adelodun, B., Lasisi, K. H., Ugya, A. Y., Omotade, I. F., Adewumi, J. R., Ojo, O. M., Babalola, T. E., Akinbile, C. O. (2022). Phytoremediation mechanisms of heavy metals removal: A step towards a green and sustainable environment. In P. Singh (Ed.) *Innovative bio-based technologies for environmental remediation*. Taylor and Francis. https://doi.org/10.1201/9781003004684-11

Okojie, J., & Bailey, B. (2019). *Nigeria's dairy farmers suffer setback as lack of competitiveness stalls industry growth.* https://businessday.ng/exclusives/article/nigerias-dairy-farmers-suffer-setback-as-lack-of-competitiveness-stalls-industry-growth/. Accessed August 22, 2024.

Olmez, H., & Kretzschmar, U. (2009). Potential alternative disinfection methods for organic fresh-cut industry for minimizing water consumption and environmental impact. *LWT-Food Science and Technology, 42*, 686–693. https://doi.org/10.1016/j.lwt.2008.08.001

Omil, B., Mosquera, R., Reigueiro, A., & Merino A. (2002). Chemical and biological properties of an agroforestry soil treater with dairy-plant waste. In *Managing forest soils for sustainable productivity* (pp. 231–232).

Patel, A., Sharma, S., Mitra, S., & Shah, M. (2016). Performance and evaluation study of dairy waste water. *International Journal of Advanced Technology in Engineering and Science, 4*(4), 172–176.

Prazeres, A. R., Carvalho, F., & Rivas, J. (2012). Cheese whey management: A review. *Journal of Environmental Management, 110*, 48–68. https://doi.org/10.1016/j.jenvman.2012.05.018

Raghunath, B.V., Punnagaiarasi, A., Rajarajan G., Irshad, A., Elango, A. & Mahesh kumar G. (2016). Impact of dairy effluent on environment—a review. https://doi.org/10.1007/978-3-319-27228-3_22

Reilly, M., Cooley, P., Tito, D., Tassou, S. A., & Theodorou, M. K. (2019). Electrocoagulation treatment of dairy processing and slaughterhouse wastewaters. *Energy Procedia, 161*, 343–351. https://doi.org/10.1016/j.egypro.2019.02.106

Renou, S., Givaudan, J. G., Poulain, S., Dirassouyan, F., & Moulin, P. (2008). Landfill leachate treatment: Review and opportunity. *Journal of Hazardous Materials, 150*(3), 468–493. https://doi.org/10.1016/j.jhazmat.2007.09.077

Rusten, B., Odegaard, H., & Lundar, A. (1992). Treatment of dairy wastewater in a novel moving bed biofilm reactor. *Water Science and Technology, 26*(3–4), 703–711. https://doi.org/10.2166/wst.1992.0451

Santra, B., Kar, S., Ghosh, S., & Majumdar, S. (2019). An integrated process development for treatment of textile effluent involving ceramic membrane-driven ultrafiltration and biosorption. In *Waste water recycling and management* (pp. 75–84). Springer.

Sarkar, B., Chakrabarti, P. P., Vijaykumar, A., & Kale, V. (2006). Wastewater treatment in dairy industries—Possibility of reuse. *Desalination, 195*, 141–152.

Schwantes, D., Gonçalves, A. C., Schiller, A.D.P., Manfrin, J., Campagnolo, M.A., & Veiga, T.G. (2019). Salvinia auriculata in post-treatment of dairy industry wastewater. *International journal of Phytoremediation, 21*(13), 1368–1374. https://doi.org/10.1080/15226514.2019.1633260

Shammas, N., Wang, L., & Hahn, H. (2010). Costs of flotation. Flotation technology. Chapter 4. Fundamentals of wastewater flotation. In L. Wang, N. Shammas, W. Selke, D. Aulenbach (Eds.), *Handbook of environmental engineering* (vol. 12, p. 150). Humana Press. https://doi.org/10.1007/978-1-60327-133-2

Shete, B. S., & Shinkar, N. P. (2013). Dairy industry wastewater sources, characteristics and its effects on environment. *International Journal of Current Engineering and Technology, 3*(5), 1611–1615.

Shivayogimath, C. B., & Naik, V. R. (2014). Treatment of dairy industry wastewater using an electrocoagulation technique. *International Journal of Engineering Research & Technology, 3*(7), 971–974.

Sirmacekic, E., Atilgan, A., Rolbiecki, R., Jagosz, B., Rolbiecki, S., Gokdogan, O., Niemiec, M. & Kocięcka, J. (2022). Possibilities of using whey wastes in agriculture: Case of Turkey. *Energies, 15*(24), 9636. https://doi.org/10.3390/en15249636

Slavov, A.K. (2017). Dairy wastewaters—General characteristics and treatment possibilities—A review. *Food Technology and Biotechnology, 55*(1). https://doi.org/10.17113/ftb.55.01.17.4520

Stasinakis, A. S., Charalambous, P., & Vyrides, I. (2022). Dairy wastewater management in EU: Produced amounts, existing legislation, applied treatment processes and future challenges. *Journal of Environmental Management, 303*, Article 114152.

Sterritt, R. M., & Lester, J. N. (1988). *Microbiology for environmental and public health engineers*. E & FN Spon

Strydom, J. P., Mostert, J. F., & Britz, T. J. (2001). *Anaerobic digestion of dairy factory effluents* (WRC Report No 455/1/01). Water Research Commission. ISBN 1868457249.

Tabelini, D. G., Lima, J. P. P., Borges, A. C., & Aguiar, A. (2023). A review on the characteristics and methods of dairy industry wastewater treatment in the state of Minas Gerais, Brazil. *Journal of Water Process Engineering, 53*(2023), Article 103779. https://doi.org/10.1016/j.jwpe.2023.103779

Tahreen, A., Jami, M. S., & Ali, F. (2020). Role of electrocoagulation in wastewater treatment: A developmental review. *Journal of Water Process Engineering, 37*, Article 101440.

Thassitou, P.K., & Arvanitoyannis, I. (2001). Bioremediation: a novel approach to food waste management. *Journal of Food Science & Technology 12*(5–6):185–196. http://dx.doi.org/10.1016/S0924-2244(01)00081-4

Tikariha, A., & Sahu, O. (2014). Study of characteristics and treatments of dairy industry waste water. *Journal of Applied & Environmental Microbiology, 2*(1), 16–22.

Tocchi, C., Federici, E., Fidati, L., Manzi, R., Vincigurerra, V., & Petruccioli, M. (2012). Aerobic treatment of dairy wastewater in an industrial three-reactor plant: Effect of aeration regime on performances and on protozoan and bacterial communities. *Water Research, 46*(10), 3334–3344. https://doi.org/10.1016/j.watres.2012.03.032

Toumi, J., Miladi, B., Farhat, A., Nouira, S., Hamdi, M., Gtari, M., & Bouallagui, H. (2015). Microbial ecology overview during anaerobic codigestion of dairy wastewater and cattle manure and use in agriculture of obtained bio-fertilisers. *Bioresource Technology, 198*, 141–149. https://doi.org/10.1016/j.biortech.2015.09.004

Vaishnav, S., Saini, T., Chauhan, A., Gaur, G. K., Tiwari, R., Dutt, T., & Tarafdar, A. (2023). Livestock and poultry farm wastewater treatment and its valorization for generating value-added products: Recent updates and way forward. *Bioresource Technology, 382*, Article 129170.

Verhoeven, J. T., & Meuleman, A. F. (1998). Wetlands for wastewater treatment: Opportunities and limitations. *Ecological Engineering, 12*(1–2), 5–12. https://doi.org/10.1016/S0925-8574(98)00050-0

Vlyssides, A. G., Tsimas, E. S., Barampouti, E. M. P., & Mai, S. T. (2012). Anaerobic digestion of cheese dairy wastewater following chemical oxidation. *Biosystems Engineering, 113*, 253–258. https://doi.org/10.1016/j.biosystemseng.2012.09.001

Water Environment Federation (2007) Characterization and sampling of wastewater. In: *Operation of municipal wastewater treatment plants* (p. 5). WEF Press. http://www.wefnet.org/mopnew/Operation_of_Municipal_Wastewater_Treatment_Plants/Chapter%2017%20Revised_6th%20Edition.pdf

Xu, Q., Lu, J., Tegladza, I. D., Xua, K., Leng, X., & Dai, H. (2021). Could the optimization of electrocoagulation simply be based on the maximization of flocs yield? *Desalination and Water Treatment, 224*, 206–215.

Yadav, R.K., Chiranjeevi, P., & Patil, S.S.A. (2020). Integrated drip hydroponics-microbial fuel cell system for wastewater treatment and resource recovery. *Bioresource Technology Reports, 9*, 100392. https://doi.org/10.1016/j.biteb.2020.100392

Yamamoto, K., Hiasa, M., Mahmood, T., & Matsuo, T. (1989). Direct solid-liquid separation using hollow fiber membrane in an activated sludge aeration tank. *Water Science and Technology, 21*(4–5), 43–54.

Ye, M., & Li, Y. (2023). Methanogenic treatment of dairy wastewater: A review of current obstacles and new technological perspectives. *Science of the Total Environment, 866*, Article 161447. https://doi.org/10.1016/j.scitotenv.2023.161447.a2

Strategic Management of Wastewater from Dyeing and Textile Industries

Damilare Olorunnisola, Morenike O. Adesina, Chidinma G. Olorunnisola, Moses O. Alfred, and Emmanuel I. Unuabonah

Abstract The textile industry is one of the earliest and continuously growing industry in human civilization. This is owned to the essentiality of the need of man for clothing and cultural recognition. In spite of the advancement in technology and the textile production processes, the indigenous dyeing and textile industry still finds relevance and application till date. However, the large volume of water required for the process leads to the generation of high volume of wastewater. In addition, the poor knowledge of the environmental implications of the indigenous dyeing and textile production process as well as the need for the treatment of the wastewater by the local practitioners are some of the challenges of the industry. This is the reason, in most cases, why the waste streams are indiscriminately discharged into the environment without treatment. This chapter provides a general overview of textile industry, emphasizing on the indigenous dyeing processes. It summarizes the characteristics of wastewater generated from these processes, human and environmental impact, and the plausible management techniques. The chapter concludes with a roadmap connecting all stakeholders towards implementing a sustainable wastewater management plan for indigenous dyeing and textile industries.

Keywords Artisanal dyeing industries · Indigenous dyeing practices · Textile industry wastewater · Environmental impact · Sustainable wastewater management · Coagulation/flocculation

---

Damilare Olorunnisola and Morenike O. Adesina contributed equally to the chapter.

D. Olorunnisola · M. O. Adesina · C. G. Olorunnisola · M. O. Alfred (✉) · E. I. Unuabonah
African Centre of Excellence for Water and Environment Research (ACEWATER), Redeemer's University, Ede, Osun, Nigeria
e-mail: alfredm@run.edu.ng

D. Olorunnisola · M. O. Adesina · M. O. Alfred · E. I. Unuabonah
Department of Chemical Sciences, Redeemer's University, Ede, Osun, Nigeria

D. Olorunnisola · M. O. Adesina · C. G. Olorunnisola
Institute of Chemistry, University of Potsdam, Potsdam, Germany

D. Olorunnisola
Institute of Nutritional Science, University of Potsdam, Potsdam, Germany

N. A. Oladoja and E. I. Unuabonah (eds.), *Strategic Management of Wastewater from Intensive Rural Industries*, Springer Water,
https://doi.org/10.1007/978-3-031-90314-4_4

1 Introduction

Dyeing and textile production is one of the earliest industries in human civilization (Barber, 1991). Textile has not only adorned individuals and communities with exquisite regalia but have also served for the identification, spiritual, and communal purposes. Indigenous to several culture and communities are different colours and patterns of fabrics, and fabric weaving techniques. Such are the batik, indigenous Hausa indigo dyeing, Hij Mayek dyeing of Manipur India, the Adire of the indigenous Yoruba people (Fig. 1) etc. However, the evolution of these industries and cultures comes with its challenging impact on the environment. This is because several of these dyeing processes are conducted via crude structures, many of which are not environmentally friendly, and attention is not often paid to the treatment of the wastewater generated. The entrants of dye stuffs and textile wastewater into fresh water bodies constitute one of the critical challenges in water and wastewater treatment.

Dyeing and textile production processes utilize a huge volume of water, thereby generating a large volume of wastewater. As a result of the dyeing processes and the rising grown proportionally, creating significant environmental issues globally (Verma et al., 2012). The majority of the environmental issues connected with the textile industry are related to water contamination emanating from the release of untreated effluents from the use of hazardous chemicals during processing. The release of the untreated wastewater depletes the oxygen content of water bodies and blocks light passage (Rajagopalan, 2016). In addition, the presence of some recalcitrant functional groups on the dye molecules (e.g., Nitrogen and organic chlorine), makes a large number of dyes (about 40%) to be non-degradable and in some cases, are degraded to form toxic products (Khan & Malik, 2014; Rajagopalan, 2016). It is therefore important to treat wastewater from this industry before discharge into the environment.

The challenge with the indigenous dyeing and textile industry is the poor knowledge of the environmental implications of the processes and the need for the treatment of the wastewater generated by the local operators. This is the reason, in most cases, why the waste streams are indiscriminately discharged into the environment without treatment. This chapter endeavors to shed light on the imperative of addressing environmental pollution from indigenous dyeing and textile industries, offering a roadmap for their mitigation. The chapter provides a general overview of the dyeing and textile industry, emphasizing the characteristics and impact of the wastewater generated in their processes. It x-rays plausible management techniques that are currently being explored and capable of removing these contaminants from wastewater and concludes with a roadmap connecting all stakeholders towards implementing a sustainable wastewater management plan for indigenous dyeing and textile industries.

2 Overview of Indigenous Dyeing and Textile Industries

The textile industry can be considered as one of the fastest-growing industries in most countries. For example, in Nigeria, the textile sector has been reported to generate over N1 billion in revenue to the government (Owen et al., 2016). Also, in Bangladesh, the garment sector is linked to 77% of the country's foreign exchange (Hossain et al., 2018; Islam et al., 2011), while India has been reported to have more than 4,000 textile manufacturing units that exports their products to over 70 countries (Kumar et al., 2016).

The textile industry converts fibers into yarn and then transforms the yarn into fabrics or allied commodities. Thereafter, the materials are dyed and finished through various operations of production (Madhav et al., 2018). Textile production occurs in four stages, including yarn formation, fabric formation, fabric processing (i.e. wet processing), and textile fabrication (Madhav et al., 2018). Textile wet processing is a significant stage in textile fabrication as it gives the highest worth to the textiles by bestowing upon them their elegant and practical properties. Textile industries utilize different types of dyes, sizing materials and auxiliary chemicals during the wet processing (Saxena et al., 2017).

Indigenous dyeing and textile industries rely on natural materials (e.g., plants, insects, minerals) as raw materials for their operations (Fig. 2). Indigenous dyeing

Fig. 1 Adire textile dyeing of the Yoruba people of Nigeria

is often associated with the native culture, traditions and identity of the local people; hence, it is often used to characterize people along cultural lines. For example, the Lutuv Chin indigenous textile industries in Myanmar depend mainly on cloth woven from organic cotton and silk as primary sources of fabrics. Dyeing of these woven fabrics is done using different colours extracted from several plant parts. Different parts of the plants are sourced from the wild for dye stuffs for which trees (45.4%) were the most frequently reported species, followed by herbs (24.6%), shrubs (22.8%), and climbers (7.2%) (Ling et al., 2022). Leaves (28.5%) were the most used plant parts, followed by bark (25.7%), fruit (18.1% each), the whole plant (13.6% each), rhizome (7.0%), seed (4.2%), and flower (2.9%) (Ling et al., 2022). The sources of dyes used for indigenous Hausa dyeing in Northern Nigeria include indigo and other local indigenous dyes. The indigo dye and the dyeing process were introduced to Hausa land by the Berbers and Arabs from North Africa through the Trans-Sahara trade routes (Dutsenwai, 2010).

Dye-pits, indigo, wood ash, sediment from the dye-pit (dead remains of used indigo known as 'katsi' in Hausa), long stick, pail, wood, yellow powder from locust bean pod, dry cow dung and guinea corn husk are the common materials used by the Hausa dyers. Although, the process of preparing the dye solution slightly varies from one Hausa community to another, the equipment and ingredients are basically the same. Before the advent of foreign fabrics and yarns, indigenous hand spun yarns,

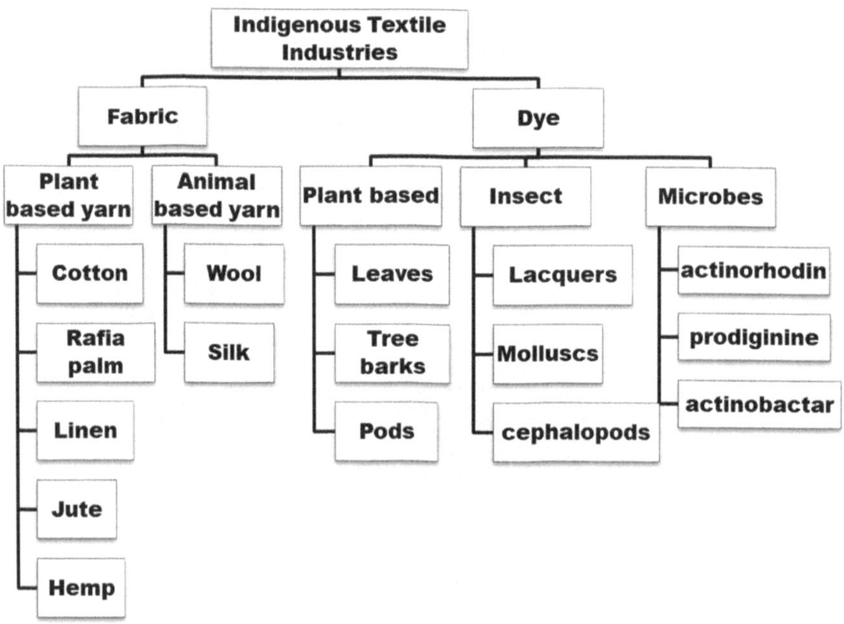

Fig. 2 Overview of indigenous textile industries (Dutsenwai, 2010; Ling et al., 2022; Shahid et al., 2013)

hand woven fabrics, and indigenous ready-mades garment were the items being dyed (Dutsenwai, 2010).

Textile production by indigenous practices requires similar processes as the recent industrial dyeing and textile processing are more complex and laborious. The preparation of yarn from plant materials, the dyeing of the yarn into required colourful threads, the acquisition of highly-technical skills for cloth weaving and the fabrication of tools and equipment such as looms, motor, spreaders, rollers and peddlers needed, are usually done manually.

In the textile wet process operations, water is used for three main purposes such as, a solvent for chemicals and dyes, as a medium for transferring dyes and chemicals to fabric, and as a washing and rinsing medium. Aside from the above mentioned processes, some parts including steam drying, ion exchange, and cleaning part of the process, consume a considerable amount of water depending on the type of product or textile processes and the specific process and equipment used (Raja et al., 2019). The four major stages involved in textile production operation unique to the wastewater produced from each stage are presented in Fig. 3.

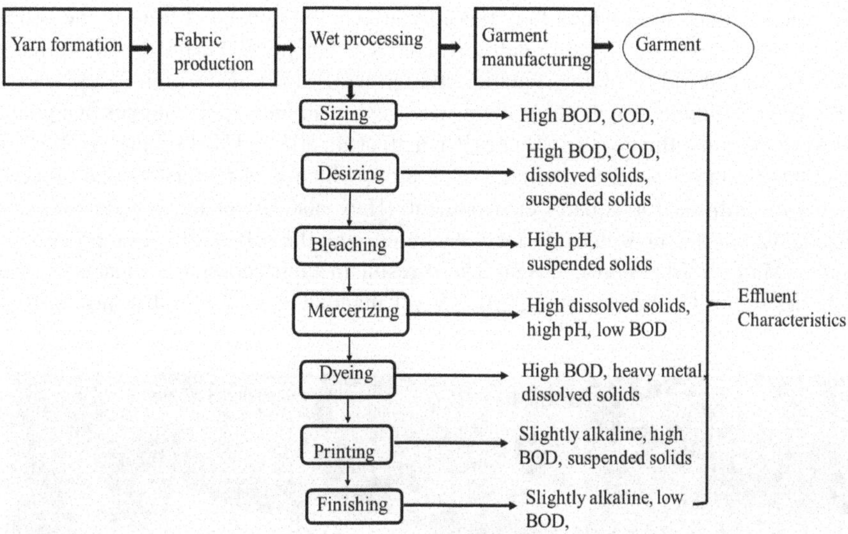

Fig. 3 Diagrammatic presentation of the four major stages in textile industry and the effluent characteristics in the each step involved in textile wet processing (Hossain et al., 2018; Madhav et al., 2018)

3 Characteristics and Impact of Wastewater from Dyeing and Textile Industries

It has been reported that textile effluents are mostly discharged into wetlands or nearby rivers without proper treatment. This action causes the alteration of chemical, physical and biological properties of the aquatic environment and poses danger to the environment and humans. Recently, environmental control of dyes has increased due to their toxic and genotoxic effects on living organisms, since dyes contain sulfur, naphthol, acetic acid, soaps, nitrates, caustic soda, chromium compounds, and heavy metals (e.g., copper, lead, arsenic, cadmium, mercury, nickel, and cobalt Holkar et al., 2016; Uddin, 2021). The discharge of these dyes into surface water (rivers and other fresh water bodies) impacts negatively on the aesthetics of the water (Fig. 4).

The presence of undesirable chemicals and heavy metals in textile dye effluent makes the wastewater highly toxic. It has a characteristic wide range of pH values, high temperature, high biochemical oxygen demand (BOD) and chemical oxygen demand (COD), high concentration of suspended solids, heavy metals and strong pigment. The discharge of high temperature wastewater into the river could increase the water temperature, which may possibly affects the fauna and flora in the water (Joshi & Santani, 2012). Additionally, the colors and oil present in the textile effluents increase the turbidity of the aquatic environment, resulting in high 'grey water' footprints, which in turn increase the water stress, instigate quick changes in aquatic ecosystems and influence the climate (Parshetti et al., 2011). The discharged effluents with color and oil prevent sunlight penetration, which is necessary for the process of photosynthesis for aquatic environment. They also affect the oxygen transfer mechanism at the air-water interface, thus hindering the self-purification process of water (Mani & Bharagava, 2018). These result in high economic impacts on the fisheries sector (Joshi & Santani, 2012; Vivekanandan et al., 2016). It is noteworthy

Fig. 4 Water pollution caused by the release of dyeing and printing liquids (Parvin et al., 2020)

that due to the release of highly polluted effluents from the textile industry, the quality of both the groundwater and surface water in the vicinity of textile facilities is degraded over time and chemicals evaporating from these effluents are mostly adsorbed through our skin, creating allergic reactions, and could cause harm to fetus (Kumar et al., 2016; Prabha et al., 2013).

4 Indigenous Management of Wastewaters from Dyeing and Textile Industries

The indigenous approaches to wastewater management are traditional and sustainable practices used by local indigenous dyers and cloth producers for the treatment of wastewater generated before it is discharge into freshwater, to minimize negative environmental impact. Like the different indigenous textile industries and their characteristic wastewater, the different indigenous wastewater management approaches and their effectiveness are also dependent on local conditions specific to communities. Though there is a dearth of report on the wastewater management in indigenous dyeing and dyeing processes, however, the wastewater discharged by the indigenous dyers is sometimes de-compartmentalized into the different parts of the environment through natural processes such as natural filtration, sunlight and microbial degradation and uptake by plants.

A common indigenous approach for the management of wastewater (especially domestic wastewater, sewage and grey water) is their reuse for agricultural irrigation and as liquid fertilizers (Oguntade et al., 2015). However, the direct reuse/recycle of wastewater from indigenous dyeing and textile industries without prior treatment is quite difficult because of the poor aesthetic of the wastewater and the possible negative impact on the properties of soil (Pattnaik et al., 2018). For instance, studies have shown that irrigation with dye wastewater could lead to heavy metal accumulation in plants (Oguntade et al., 2015). Hence, the use of untreated dye effluent for irrigation is discouraged due to the potential health risks on human and general contamination of the environment. The textile wastewater could be useful for dyeing lighter shades of fabrics, dyeing art pieces (potteries), and artefacts (Shahid et al., 2013) and in construction for concrete mixing (Patel & Pandey, 2012). The creative use of the highly coloured wastewater could provide some aesthetics and natural colouring for blocks and concretes for construction purposes. However, most of the indigenous dyeing and textile industries discharge their wastewater directly into the environment without treatment especially in the underdeveloped nations.

5 Emerging Technologies for the Management of Wastewaters from Dyeing and Textile Industries

5.1 Coagulation and Flocculation

Coagulation is recognized as a highly efficient primary treatment method for eliminating colloids and suspended particles (Liu et al., 2013; Sonal & Mishra, 2021). It plays a crucial role in textile wastewater treatment by reducing colloid stability, causing smaller particles to eventually aggregate into larger particles. These larger particles settle under gravity, effectively removing pollutants from the wastewater (Iwuozor, 2019).

For many years, aluminum sulfate (alum) has been the most commonly used coagulant. Alum forms aluminum ions when dissolved in water, neutralizing the negative charges on colloidal particles and thereby destabilizing them (Jagaba et al., 2018; Sillanpää et al., 2018). In recent years, continuous improvement has led to the development of various coagulants tailored to meet diverse wastewater qualities. Coagulants can be broadly categorized into two main types; synthetic and natural (Bhatia & Bhatia, 2016). Synthetic coagulants include inorganic and organic polymer coagulants. Inorganic polymer coagulants, such as Poly aluminum Chloride (PAC), poly ferric(III)-magnesium(II) sulfate (PFMS), and poly ferric sulfate (PFS), are widely employed due to their cost-effectiveness, versatility, low price, wide application range, and high COD removal efficiency (Liu et al., 2013; Sonal & Mishra, 2021), particularly in the treatment of industrial wastewater like printing and dyeing wastewater. A study revealed that in the treatment of textile wastewater using PFS, PAC, and poly aluminum ferric chloride (PAFC), the COD removal rate followed order: PFS > PAC > PAFC (Tianzhi et al., 2021). Organic polymer coagulants, like polyacrylamide (PAM), are also often used in combination with PAC or other coagulants to accelerate floc sedimentation (Zhao et al., 2021).

However, due to concerns about the negative effects of synthetic coagulants, there is a growing focus on natural coagulants. These materials are readily available, cost-effective, biodegradable, non-toxic, generate minimal residual sludge, and are eco-friendly (Owodunni & Ismail, 2021). In the last decades, the use of natural coagulants for textile wastewater treatment has significantly expanded. Based on their source, natural coagulants can be categorized into three primary groups: plant-based, microorganism-based, and animal-based (Freitas et al., 2018).

Specific plants like *Cactus latifaria, Moringa oleifera, Nephelium lappaceum* seeds, *Carica papaya* seeds, and *Ocimum tenuiflorum* leaves have been effectively employed for color and turbidity removal (Choy et al., 2015; Freitas et al., 2018; Owodunni & Ismail, 2021). Water-soluble coagulants sourced from Moringa seeds have shown the potential to remove color from textile wastewater (Verma et al., 2012). In another study, a combination of bentonite-based coagulant and cactus-based flocculant successfully eliminated methylene blue, a common textile dye, achieving 98.9% removal efficiency (Ihaddaden et al., 2022). *Ocimum basilicum* has been assessed as an efficient natural coagulant for the removal of dye from model textile wastewater containing Congo red, resulting in a high color and COD removal efficiency, even at low coagulant dosage. The mucilage of *O. basilicum* was found to be highly effective in treating real textile wastewater, both as a sole coagulant and with alum (Shamsnejati et al., 2015).

Another alternative natural coagulant similar to powder activated carbon (PAC), was obtained from seed gums of *Ipomoea dasysperma* and guar gum, which have been shown to be effective in decolorizing direct dyes. The combination of *Ipomoea dasysperma* and guar gumcoagulant with PAC, enhanced their coagulation for reactive and acid dyes (Sanghi et al., 2006). Additionally, papaya seed powder has been explored for treating synthetic textile wastewater of drimarene dark red (DDR) (Khee et al., 2023). Other natural coagulants, including *Plantago major L.* (Chaibakhsh et al., 2014), *Moringa stenopetala* seed extract (Dalvand et al., 2016), and *Plantago ovate* (Ramavandi, 2014) have demonstrated promising potentials for removing color and turbidity from textile wastewater.

Furthermore, microorganisms, such as algae, have found application as a source of natural coagulants (Ang et al., 2020). For example, extracted algal alginate from red algae has been harnessed for the removal of textile dye, like Congo red. In one study, algal alginate from *Sargassum sp,* a marine brown alga, was employed to treat colored wastewater primarily containing crystal violet dye from a local dyeing unit. It was observed that at optimal conditions, about 97.3% color removal was achieved, demonstrating the efficacy of algal alginate as a coagulants for dye removal (Vijayaraghavana & Shanthakumarb, 2019). This approach has also been effective in removing Acid black 1 dye (Vijayaraghavan & Shanthakumar, 2018). Similarly, algal alginate extracted from red algae was employed for removal of Congo red dye from wastewater, with the investigations revealing its high efficiency, achieving over 95% maximum color removal (Sivalingam & Gopal, 2023).

Natural coagulants offer a range of advantages over synthetic organic polymeric compounds or inorganic material because they can be sourced from renewable material and also minimizing the risk of secondary pollution (Freitas et al., 2018). However, the effectiveness of these coagulants is dependent on various hydrochemistry conditions, such as the presence of electrolytes, organic and inorganic compounds. Electrolytes, such as salts, significantly impact the coagulation process (Owodunni & Ismail, 2021). High concentrations of electrolytes enhance the coagulation efficiency of natural coagulants by neutralizing the surface charges of suspended particles, promoting floc formation. However, excessive electrolyte levels may lead to coagulation inhibition due to the stabilization of colloidal particles (Yin, 2010).

Organic compounds, including dyes and surfactants, present a complex challenge. The molecular structure and charge properties of dyes can affect the binding and aggregation efficiency of natural coagulants. For instance, dyes with high molecular weights or those that are highly anionic or cationic may interact more effectively with natural coagulants, resulting in better removal efficiency (Chinedu, 2022). Conversely, the presence of surfactants can hinder coagulation by forming protective layers around particles, thereby reducing their tendency to aggregate (Zaharia et al., 2024). Inorganic contaminants, such as heavy metals, also play a critical role in the coagulation process. Natural coagulants often exhibit chelating properties, allowing them to bind with metal ions and facilitate their removal from wastewater. The efficiency of metal removal is influenced by the specific type of coagulant and the chemistry of the wastewater. For example, natural coagulants derived from *Moringa oleifera* seeds have shown high efficacy in removing heavy metals from textile effluents (Agarwal et al., 2019). However, the presence of competing ions in the wastewater can reduce the availability of binding sites on the coagulant, thereby decreasing its overall effectiveness.

In summary, while natural coagulants offer a sustainable alternative for wastewater treatment, their performance is intricately linked to the composition of the wastewater. Understanding the interactions between natural coagulants and the various electrolytes, organic, and inorganic substances present in textile dye wastewater is essential for optimizing treatment processes and achieving the desired environmental outcomes.

5.2 *Adsorption*

In the quest for a cost-effective and environmentally sustainable approach to treating textile effluents, the focus has shifted towards adsorption techniques (Behera et al., 2021). Adsorption is a surface-oriented phenomenon wherein a substance accumulates on the surface of a solid material from its surrounding liquid or gaseous environment (Jain et al., 2020). In comparison to alternative methods for effluent treatment, adsorption proves to be both economical and eco-friendly. The adsorption of dyes is primarily influenced by the characteristics of the dyes and the surface attributes of the adsorbent.

In the twenty-first century, commercial activated carbon (CAC) gained recognition for its effectiveness in removing dyes from wastewater. Nevertheless, its relatively high cost sometimes limits its practical application (Patel, 2018; Shokry et al., 2019). In recent years, attention has been shifted toward the utilization of low cost adsorbent materials such as those derived from natural wastes, industrial and agricultural residues for water and wastewater treatment. This is due to their relative availability and abundance in nature and their high capacity for adsorption following minimal activation treatments. Different biomass wastes, including potato peels, corn straw, wheat straw, rice husk, sawdust, corncob, bagasse, coconut husk, modified ginger wastes, etc., have been investigated for their potential in the treatment of textile wastewater (Onu et al., 2022; Pandit et al., 2021; Sapci-Ayas, 2021).

A recent study conducted using biosorbent from the agricultural waste of potato peels (PP) investigates their potential for the removal of organic pollutant (chemical oxygen demand (COD)) and inorganic (Fe^{2+}, Ni^{2+}, Cu^{2+} and Cd^{2+}) from original textile wastewater (Sapci-Ayas, 2021). The PP modified with physical modification process presented an attractive solution for COD removal efficiency (69.50%) and removal efficiencies for four divalent metal ions; 78.6% for Cu^{2+}, 63.6% for Ni^{2+}, 40% for Fe^{2+}, and 34.6% for Cd^{2+}. Similarly, removal of Methylene blue and malachite green was achieved using adsorbents derived from Potato plant waste (Gupta et al., 2016). Low-cost waste materials, such as oil palm waste and bamboo have been used for textile wastewater treatment (Chan et al., 2008; Karthikeyan et al., 2008; Pang & Abdullah, 2013). Various adsorbents, including almond and hazelnut shells, as well as walnut sawdust have been investigated, and the adsorption capacities followed the order: walnut > almond > hazelnut, for Acid green 25; and almond > walnut > poplar > hazelnut for Acid red 183 (Sanghi & Verma, 2013).

Clay, a versatile material with a wide range of applications have been identified as a promising candidate for the development of clay-based adsorbents for the treatment of water and wastewater. They have been reported to exhibit both ease of processing and high efficiency (Shahadat & Isamil, 2018; Uddin, 2018). Over the past few decades, there has been a growing interest in the application of natural montmorillonite clay for the decolorization of textile wastewater (Uddin, 2018). The potential of montmorillonite nanoparticles as adsorbents in the removal of Reactive yellow 15 (RY15) and Reactive Yellow 42 (RY42) dyes from aqueous solutions has been reported (Kamranifar & Naghizadeh, 2017). Fly ash, generated from coal combustion, has shown potential in the removal of methylene blue and Congo Red from aqueous solution (Gupta et al., 2009).

One of the primary issues associated with this technology pertains to the disposal of used adsorbents, which give rise to environmental concerns, stemming from the possible release of the harmful compounds into the surrounding environment (Bakar et al., 2021). Various regeneration approaches have been used to address this issue, which include chemical, thermal, oxidation, and electrochemical processes. While thermal regeneration is widely favored, it has the potential to increase the total cost of treatment and induce alterations in the characteristics of the materials (Patel, 2018; Wang et al., 2019).

A combination of different physicochemical techniques has been reported to enhance the quality of treated textile wastewater. Arafat (2007) reported the use of sedimentation, coagulation, and adsorption for the treatment of textile wastewater samples from local textile factories in Palestinian territories. The study revealed that sedimentation helped to reduce the total suspended solids (TSS) of the wastewater, while coagulation with four different coagulants (i.e., ferric chloride, ferrous sulfate plus lime, aluminum sulfate, and aluminum sulfate plus lime) had the additional advantages of lowering the chemical oxygen demand (COD) and achieving higher filtration rates. Three of the coagulants used showed a fair removal efficiency, while ferric chloride failed to perform effectively as a coagulant. In order to further lower the COD of the treated water, adsorption using activated carbon was studied. It was found that 6 g carbon/L achieved up to 98% COD reduction of the wastewater (Arafat, 2007).

5.3 *Biological Methods*

In comparison with the physicochemical and photochemical methods, biological methods, (i.e., degradation of dyes with biological agents), such as bioremediation is a low-cost and environmentally friendly process of textile wastewater treatment (Ahmad et al., 2015). Recent studies have investigated the application of biological materials, such as algae (Gül et al., 2019), fungi (Ferri et al., 2019), bacteria (Rathour et al., 2019), yeasts (Tan et al., 2019) and enzymes (Darwesh et al., 2019) in their abilities to degrade, as well as absorb varieties of dyes. The cell wall components (lipids and heteropolysaccharides) of microorganisms, comprising of different functional groups, which include amino, hydroxyl, carboxyl, phosphate, and other charged groups promote attractive forces between the cell wall and the dye molecules (Jafari et al., 2014).

5.3.1 Microbial Degradation

Studies have reported treatment of dyeing and textile wastewater using microbial consortium isolated from textile unit effluents. The gene encoded for microbial dye decolourisation is located in the plasmid of microbial isolates (Adebajo et al., 2023). Samuchiwal et al. (2021) investigated the decolorization of textile effluent using microbial consortium enriched from the textile effluents. Microbial consortium was developed from pretreatment range (PTR) effluent, under anaerobic conditions at 35 ± 2 °C, and further enhanced via subsequent sub-culturing with steady increase in the concentration of dye effluent, with yeast extract (1 g/L) as a supplement. The process achieved 70–74% colour removal from dyeing unit wastewater with PTR effluent as carbon source and 0.5 g/L yeast extract. The maximum colour removal (74%) occurred at 35 ± 2 °C, while there was no significant reduction in colour removal till 55 ± 2 °C (less than 70%). There was a decline in colour removal efficiency (42.6%) at lower temperature of 25 ± 2 °C, and the pH of the wastewater impacted the colour removal efficiency. At low pH of 4 and 6, decreased decolorization of 22.6 and 54% were obtained, respectively. When the pH value was increased to 8, there was an increase in colour removal efficiency up to 72.3%, and no significant reduction was observed at higher pH of 10 and 12. The reason for the efficiency of the microbial consortium at high temperature and pH was attributed to the adaptation of the inoculum to extreme conditions. The applicability of the microbial consortium was investigated for the continuous treatment of the textile wastewater effluent, and an average colour removal of 61–71% was recorded with removal rate of 1236 hazen/day during the sixty (60) days of continuous treatment. The study demonstrated easy applicability of this indigenously developed process for the remediation of both local and industrial textile wastewater.

In the remediation of vat dyes used by indigenous dyers in Itoku, Abeokuta, Nigeria, four bacterial isolates, namely: *Klebsiella oxytoca, Bacillus firmus, Staphylococcus aureus* as well as *Bacillus macerans* were used for the wastewater containing 100 mg/L of vat dyes (vat red 15, vat brown 1 and vat black 27) in mineral salt medium for 5 days (Adebajo et al., 2023). The study was repeated with bacterial isolates that were cured of their plasmid using sodium dodecylsulphate (0.06, 0.08, 0.11, 0.22, 5 and 10 g) and 10 μL of ethidium bromide (EtBr) at 40 °C. The plasmid weight of the bacterial isolates was 1500 branch-point sequence (bps) and their plasmids were not cured until when sodium dodecylsulphate and ethidium bromide were combined and used. High decolourisation occurred before plasmid curing, with the highest value of 96% by *Bacillus macerans* for vat black 27 dye and vat red dye, while the decolourization activity of *Staphylococcus aureus* was the lowest at 40%. After plasmid curing, the decolourization activity of the isolate reduced tremendously to as low as 7% for *Klebsiella oxytoca.*

Generally, there are two ways in which dye decolorization is achieved using the biological treatment method: (i) adsorption on microbial biomass, and (ii) biodegradation of dye by microbial cells (Popli & Patel, 2015). Microbial treatments can be performed in aerobic, anaerobic or combined anaerobic/aerobic conditions. Reports have shown that bioremediation of azo dyes occurs in a two-step process viz: the first step is the anaerobic reduction of the azo bond, which results in the production of two aromatic amines; and the second step entails the aerobic digestion of the resulting aromatic amines by a mixed bacterial population, which lead to their complete mineralization (Şen & Demirer, 2003).

5.3.2 Aerobic Process

Aerobic treatment process purifies wastewater by decomposing the wastes and reducing unpleasant odors via the help of facultative aerobic bacteria (Mani & Bharagava, 2018). Some of aerobic bacteria that decolorize dyes are listed in Table 1. In principle, many of these strains require organic carbon as a growth substrate, because they cannot use the dye molecule for such. Hence, aerobic treatment with activated sludge is often used in large-scale textile effluent treatment.

Biological aerated filter (BAF) is an aerobic system that involves the growth of an organism on inert media, held in fixed position and exposed to air. The BAF allows up to 34–44% color removal (Carmen & Daniela, 2012) in the treatment of textile wastewater. Aside bacteria, white-rot fungi, mixed microbial cultures from a variety of habitats are also found to degrade dyes using enzymes such as lignin peroxidases (LiP), manganese dependent peroxidases (MnP), H_2O_2 producing enzymes like glucose-1-oxidase and glucose-2-oxidase, laccase and phenoloxidase (Carmen & Daniela, 2012).

A wide range of bacterial strains have been proposed for the decolorization of textile dye, and their effectiveness depends on their adaptability and activities under certain environmental conditions (Aassila et al., 2018). The mechanisms of azo dye decolorization by bacteria involve a series of complex biochemical processes that are essential for the successful remediation of textile wastewater. Understanding these mechanisms is crucial for developing sustainable and effective bioremediation strategies that contribute to environmental protection and sustainability efforts. The proposed mechanism of azo dye degradation by bacteria are presented in Fig. 5 (Khan et al., 2013; Pearce et al., 2006). Typically, this degradation occurs in a two-step process (Pearce et al., 2006; Pinheiro et al., 2022). The initial step involves a reductive cleavage of the azo bonds, leading to the formation of potentially toxic aromatic amines (Pearce et al., 2006). Subsequently, the second step focuses on the degradation

Table 1 Decolorization of different dyes using aerobic bacteria

Textile dye	Microorganisms	References
Congo red	*Acinetobacter baumannii* YNWH 226	Li et al. (2015)
Acid blue 25	*Klebsiella sp.*	Prasad (2015)
Reactive orange 4	*Lysinibacillus sp.*	Saratale et al. (2015)
Reactive blue 5	*Streptococcus thermophilus*	Zhang et al. (2015)
Reactive black 5	*Pseudomonas entomophila BS1*	Khan and Malik (2016)
Acid yellow	*Pseudomonas fluorescens*	Pandey et al. (2007)
Remazol black B	*Halomonas sp.*	Asad et al. (2007)
Dispersive dye	*Bacillus sp.*	Maier et al. (2004)
Acid orange 7	*Enterobacter, Pseudomonas, morganella*	Barragán et al. (2007)
Methyl red and sulfonated azo dyes	*Enterococcus faecalis*	Chen et al. (2003)
Remazol orange	*Brevibacillus latrosporus*	Sandhya et al. (2008)
Acid red 88, reactive black 5, direct red 81, and disperse orange 3	*Shewanella putrefaciens*	Khalid et al. (2008)

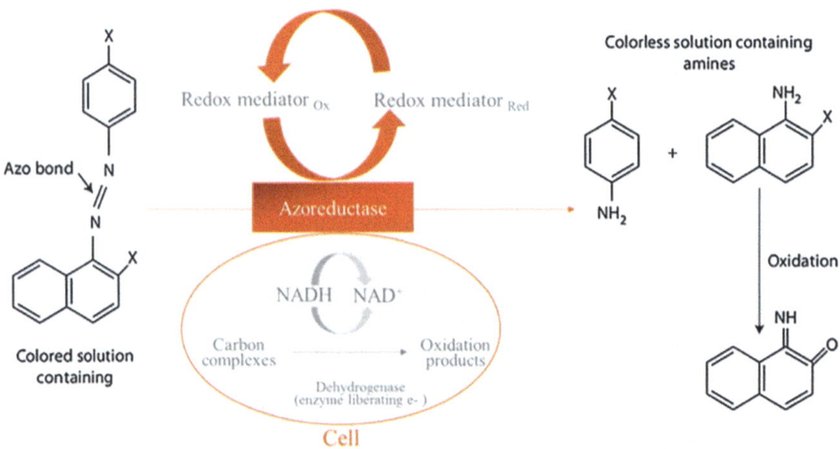

Fig. 5 Schematic illustration of the proposed mechanism of azo dyes degradation by bacteria (Khan et al., 2013)

of these aromatic amines. It is noteworthy that while the bacterial degradation process can occur in the presence or absence of oxygen, the biodegradation of these amines predominantly happens through aerobic processes.

Moreover, the ideal method for treating industrial waste contaminated by azo compounds often involves a combination of anaerobic and aerobic steps. This combined approach ensures the safe and efficient removal of environmental and human risk factors associated with these compounds (Pinheiro et al., 2022). The presence of toxic aromatic amines resulting from dye breakdown poses challenges as they are resistant to traditional effluent treatment methods and can persist longer than the dyes themselves. This persistence can hinder the dye degradation process due to their toxicity to bacterial life.

5.3.3 Anaerobic Biological Treatment

This method employs anaerobic bacteria to break down organic matter under anaerobic conditions (Mostafa, 2015). This technique has found great applicability in the textile industry, particularly for treating highly concentrated organic wastewater from processes such as wool washing, textile printing, and dyeing effluents, which often contain organic matter exceeding 1000 mg/L (Mani et al., 2019; Mostafa, 2015).

The primary anaerobic treatment process is the hydrolysis acidification process, encompassing the first two stages of anaerobic treatment (Xing et al., 2018). This process harnesses anaerobic and facultative bacteria to break down complex macromolecules, heterocyclic organic matter, and other biodegradable organic substances into smaller molecules (Mostafa, 2015). This transformation enhances the wastewater's biodegradability and disrupts the colored groups of dye molecules, resulting in the removal of a portion of color from the wastewater. Importantly, the structural alterations made by anaerobic bacteria make these organic and colored materials become more susceptible to decomposition and decolorization under aerobic conditions, thereby improving the overall decolorization efficiency (Khan et al., 2013; Supaka et al., 2004).

Other processes have also been combined with anaerobic processes in the treatment of textile dye wastewater treatment. A typical example is the combination of anaerobic hydrolysis and acidification, with subsequent aerobic biological treatment and flotation (Mostafa, 2015). In this technique, the wastewater undergoes a series of steps, starting with a bar-screen, to remove larger particles, before it is regulated in a tank. It then undergoes an anaerobic hydrolysis reaction, where insoluble organic matter is degraded into soluble organic matter. Concurrently, acid-producing bacteria convert complex macromolecules into biodegradable small molecules. Following this, the wastewater proceeds to a biological contact oxidation tank (de Lemos Chernicharo, 2007), before it finally undergoes flotation treatment via a pressurized dissolved air flotation process. During this phase, polymer flocculants are added into the flotation tank, causing the particulates to aggregate. Meanwhile, a sufficient quantity of air is dissolved in the wastewater, and when pressure is rapidly released, uniformly fine bubbles formed and adhered to the small particles, enabling them to float and facilitate the separation of solids from the liquid phase.

Anaerobic technique has become a well-established and widely employed method to break down poorly biodegradable polymer materials and insoluble substances in textile wastewater into smaller, soluble components. This enhances the wastewater's biodegradability and desirable BOD_5/COD_{Cr} ratio, setting the stage for effective aerobic biological treatment (Mostafa, 2015).

6 Roadmap for Sustainable Wastewater Management

Sustainable management of wastewater from indigenous dyeing and textile industries requires a well-defined roadmap to guide stakeholders (the industry players, such as the weavers and artisans, relevant government agencies, and local communities). Figure 6 presents a schematic for enhanced resource utilization and waste minimization for indigenous dyeing and textile industries as proposed by Shahid et al. (2013). A comprehensive roadmap takes cognizance and are tailored to address the issue, taking into account the social and environmental aspects. Government and its agencies should identify the contributions of the indigenous industry, formalize their processes and provide adequate support in providing solutions to the challenges facing them. The solution to the challenges emanating from the industry must involve a multi-stakeholder collaboration and knowledge sharing to foster sustainability and efficiency.

To implement a sustainable management of wastewater from any source, it is important to first understand the characteristics of the wastewater in order to properly manage them. Hence, it is imperative to provide data from a baseline assessment of the state of wastewater discharge from indigenous dyeing and textile industries. This baseline assessment would include the understanding of the water use or requirement of the industry, volume of wastewater generated as well as the physicochemical properties and possible pollutants present. In addition, and as a part of providing baseline assessment of the wastewater generated from the textile industry, identification of the key sources of the pollution recognized in the wastewater would assist in finding a plausible way to reduce the pollution from source. The baseline assessment will then inform the selection of appropriate approaches and technologies for their remediation prior to discharge or reuse. This is important because indigenous dyeing materials differs with region and culture based on materials use, and there is potential for suitable management technologies to also differ across these regions and culture.

The challenge of the facilities to assess the wastewater from the indigenous industries can be mitigated through technical support from the government and other non-governmental organizations (NGOs), and scientists in academic communities. Government can provide support to scientists for routine monitoring of the environmental impact of the wastewater from these artisanal indigenous textile dyeing industries while the research output is communicated and used for the sensitization of the practitioners (dyers) on the potential hazards of their processes and practice. Understanding the potential hazards and the impact of their processes and wastes will aid the cooperation of the dyers and relevant stakeholders. In addition, collaboration between the dyers and the research community could also provide more sustainable improvement to their practices, a higher output and lesser waste generated.

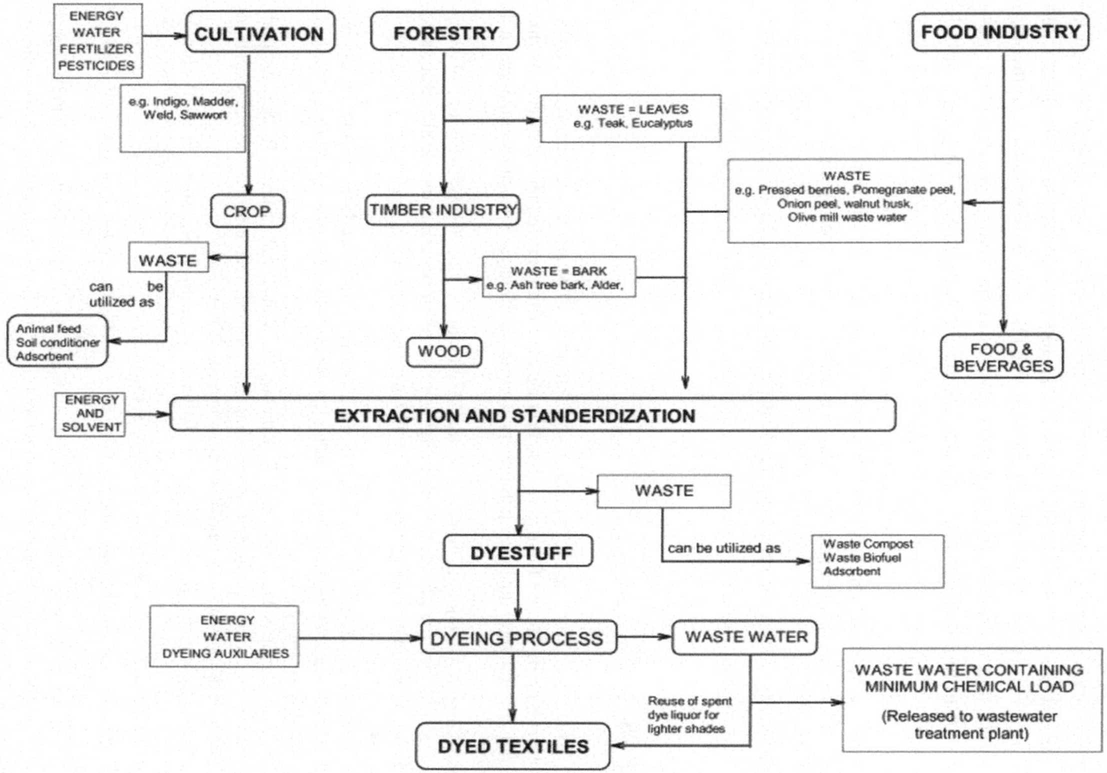

Fig. 6 A schematic representation of enhanced resource utilization and waste minimization in a natural textile dyeing (Shahid et al., 2013)

7 Conclusion

Indigenous dyeing and textile industries are considered relatively environmentally friendly due to the use of natural materials; however, the process is not without its challenges. Textile production generally requires a large volume of water for different purposes and hence, the process generate a lot of wastewaters. It is essential for indigenous dyers to consider the characterization and treatment of the generated wastewater before discharge into the environment. This can only be achieved through a multi-stakeholder collaborative approach involving the dyers, government agencies, non-governmental agencies, and the research communities. Processes should be designed to reduce waste generation to the barest minimum, and/or in such ways that wastes (including wastewater) generated can be reused in the production. To properly treat wastewater generated from these industries, a combination of different techniques will be more efficient than the use of a singular technique.

Acknowledgements Authors acknowledge the Redeemer's University Nigeria for her support.

References

Aassila, H., Aboussabiq, F., Dari, K., Aarfane, A., Tahiri, S., El Krati, M., & Etahiri, S. (2018). Biodecolorization of Methyl Orange by bacteria isolated from textile industrial wastes: Optimization of cultural and nutritional parameters. *Journal of Material Environmental Science, 9*, 2779–2787.

Adebajo, S., Balogun, S., Ojo, A., Bankole, P., & Akintokun, A. (2023). Evaluation of plasmid-mediated decolourisation of vat dyes by indigenous bacterial isolates from local textile factories in Itoku, Abeokuta. *International Research Journal of Science, Technology, Education, Management, 3*(2), 1–12.

Agarwal, V., Dixit, D., & Bhatt, M. J. (2019). Use of Moringa oleifera seeds as a primary coagulant in textile wastewater treatment. In *Waste management and resource efficiency: Proceedings of 6th IconSWM 2016* (pp. 1231–1236).

Ahmad, A., Mohd-Setapar, S. H., Chuong, C. S., Khatoon, A., Wani, W. A., Kumar, R., & Rafatullah, M. (2015). Recent advances in new generation dye removal technologies: Novel search for approaches to reprocess wastewater. *RSC Advances, 5*, 30801–30818.

Ang, T.-H., Kiatkittipong, K., Kiatkittipong, W., Chua, S.-C., Lim, J. W., Show, P.-L., Bashir, M. J., & Ho, Y.-C. (2020). Insight on extraction and characterisation of biopolymers as the green coagulants for microalgae harvesting. *Water, 12*, 1388.

Arafat, H. A. (2007). Simple physical treatment for the reuse of wastewater from textile industry in the Middle East. *Journal of Environmental Engineering Science, 6*, 115–122.

Asad, S., Amoozegar, M., Pourbabaee, A. A., Sarbolouki, M., & Dastgheib, S. (2007). Decolorization of textile azo dyes by newly isolated halophilic and halotolerant bacteria. *Bioresource Technology, 98*, 2082–2088.

Bakar, N. A., Othman, N., Yunus, Z. M., Altowayti, W. A. H., Tahir, M., Fitriani, N., & Mohd-Salleh, S. N. A. (2021). An insight review of lignocellulosic materials as activated carbon precursor for textile wastewater treatment. *Environmental Technology & Innovation, 22*, Article 101445.

Barber, E. J. W. (1991). *Prehistoric textiles: The development of cloth in the Neolithic and Bronze Ages with special reference to the Aegean*. Princeton University Press.

Barragán, B. E., Costa, C., & Marquez, M. C. (2007). Biodegradation of azo dyes by bacteria inoculated on solid media. *Dyes and Pigments, 75*, 73–81.

Behera, M., Nayak, J., Banerjee, S., Chakrabortty, S., & Tripathy, S. K. (2021). A review on the treatment of textile industry waste effluents towards the development of efficient mitigation strategy: An integrated system design approach. *Journal of Environmental Chemical Engineering, 9*, Article 105277.

Bhatia, S., & Bhatia, S. (2016). Natural polymers vs synthetic polymer. In *Natural polymer drug delivery systems: Nanoparticles, plants, and algae* (pp. 95–118).

Carmen, Z., & Daniela, S. (2012). *Textile organic dyes-characteristics, polluting effects and separation/elimination procedures from industrial effluents-a critical overview* (Vol. 3). IntechOpen Rijeka.

Chaibakhsh, N., Ahmadi, N., & Zanjanchi, M. A. (2014). Use of Plantago major L. as a natural coagulant for optimized decolorization of dye-containing wastewater. *Industrial Crops and Products, 61*, 169–175.

Chan, L. S., Cheung, W. H., & McKay, G. (2008). Adsorption of acid dyes by bamboo derived activated carbon. *Desalination, 218*, 304–312.

Chen, K.-C., Wu, J.-Y., Liou, D.-J., & Hwang, S.-C.J. (2003). Decolorization of the textile dyes by newly isolated bacterial strains. *Journal of Biotechnology, 101*, 57–68.

Chinedu, E. T. (2022). Application of agricultural waste in anionic dyes removal from wastewater. In *Textile wastewater treatment: Sustainable bio-nano materials and macromolecules* (Vol. 2, pp. 111–141). Springer.

Choy, S., Prasad, K., Wu, T., & Ramanan, R. (2015). A review on common vegetables and legumes as promising plant-based natural coagulants in water clarification. *International Journal of Environmental Science and Technology, 12*, 367–390.

Dalvand, A., Gholibegloo, E., Ganjali, M. R., Golchinpoor, N., Khazaei, M., Kamani, H., Hosseini, S. S., & Mahvi, A. H. (2016). Comparison of Moringa stenopetala seed extract as a clean coagulant with Alum and Moringa stenopetala-Alum hybrid coagulant to remove direct dye from textile wastewater. *Environmental Science and Pollution Research, 23*, 16396–16405.

Darwesh, O. M., Matter, I. A., & Eida, M. F. (2019). Development of peroxidase enzyme immobilized magnetic nanoparticles for bioremediation of textile wastewater dye. *Journal of Environmental Chemical Engineering, 7*, Article 102805.

de Lemos Chernicharo, C. A. (2007). *Anaerobic reactors*. IWA Publishing.

Dutsenwai, S. (2010). Indigenous Hausa indigo dyeing in Northern Nigeria. *Multidisciplinary Journal of Research Development, 15*, 80–85.

Ferri, M., Campisi, S., Scavini, M., Evangelisti, C., Carniti, P., & Gervasini, A. (2019). In-depth study of the mechanism of heavy metal trapping on the surface of hydroxyapatite. *Applied Surface Science, 475*, 397–409.

Freitas, T. K. F. S., Almeida, C. A., Manholer, D. D., Geraldino, H. C. L., de Souza, M. T. F., Garcia, J. C. (2018). Review of utilization plant-based coagulants as alternatives to textile wastewater treatment. In *Detox fashion: Waste water treatment* (pp. 27–79).

Gül, Ü., Taştan, B., & Bayazıt, G. (2019). Assessment of algal biomasses having different cell structures for biosorption properties of acid red P-2BX dye. *South African Journal of Botany, 127*, 147–152.

Gupta, V. K., Carrott, P., Ribeiro Carrott, M., & Suhas. (2009). Low-cost adsorbents: Growing approach to wastewater treatment—A review. *Critical Reviews in Environmental Science and Technology, 39*, 783–842.

Gupta, N., Kushwaha, A. K., & Chattopadhyaya, M. (2016). Application of potato (Solanum tuberosum) plant wastes for the removal of methylene blue and malachite green dye from aqueous solution. *Arabian Journal of Chemistry, 9*, S707–S716.

Holkar, C. R., Jadhav, A. J., Pinjari, D. V., Mahamuni, N. M., & Pandit, A. B. (2016). A critical review on textile wastewater treatments: Possible approaches. *Journal of Environmental Management, 182*, 351–366.

Hossain, L., Sarker, S. K., & Khan, M. S. (2018). Evaluation of present and future wastewater impacts of textile dyeing industries in Bangladesh. *Environmental Development, 26*, 23–33.

Ihaddaden, S., Aberkane, D., Boukerroui, A., & Robert, D. (2022). Removal of methylene blue (basic dye) by coagulation-flocculation with biomaterials (bentonite and Opuntia ficus indica). *Journal of Water Process Engineering, 49*, Article 102952.

Islam, M. M., Mahmud, K., Faruk, O., & Billah, M. (2011). Textile dyeing industries in Bangladesh for sustainable development. *International Journal of Environmental Science and Development, 2*, 428.

Iwuozor, K. O. (2019). Prospects and challenges of using coagulation-flocculation method in the treatment of effluents. *Advanced Journal of Chemistry-Section A, 2*, 105–127.

Jafari, N., Soudi, M. R., & Kasra-Kermanshahi, R. (2014). Biodegradation perspectives of azo dyes by yeasts. *Microbiology, 83*, 484–497.

Jagaba, A. H., Abubakar, S., Lawal, I. M., Latiff, A. A. A., & Umaru, I. (2018). Wastewater treatment using alum, the combinations of alum-ferric chloride, alum-chitosan, alum-zeolite and alum-moringa oleifera as adsorbent and coagulant. *International Journal of Engineering Management, 2*, 67–75.

Jain, M., Mudhoo, A., Ramasamy, D. L., Najafi, M., Usman, M., Zhu, R., Kumar, G., Shobana, S., Garg, V. K., & Sillanpää, M. (2020). Adsorption, degradation, and mineralization of emerging pollutants (pharmaceuticals and agrochemicals) by nanostructures: A comprehensive review. *Environmental Science and Pollution Research, 27*, 34862–34905.

Joshi, V., & Santani, D. (2012). Physicochemical characterization and heavy metal concentration in effluent of textile industry. *Universal Journal of Environmental Research & Technology, 2*(2), 39.

Kamranifar, M., & Naghizadeh, A. (2017). Montmorillonite nanoparticles in removal of textile dyes from aqueous solutions: Study of kinetics and thermodynamics. *Iranian Journal of Chemistry and Chemical Engineering (IJCCE), 36*, 127–137.

Karthikeyan, S., Sivakumar, P., & Palanisamy, P. (2008). Novel activated carbons from agricultural wastes and their characterization. *Journal of Chemistry, 5*, 409–426.

Khalid, A., Arshad, M., & Crowley, D. E. (2008). Decolorization of azo dyes by Shewanella sp. under saline conditions. *Applied Microbiology and Biotechnology, 79*, 1053–1059.

Khan, S., & Malik, A. (2014). Environmental and health effects of textile industry wastewater. In: *Environmental deterioration and human health* (pp. 55–71). Springer.

Khan, R., Bhawana, P., & Fulekar, M. (2013). Microbial decolorization and degradation of synthetic dyes: A review. *Reviews in Environmental Science and Bio/technology, 12*, 75–97.

Khan, S., & Malik, A. (2016). Degradation of Reactive Black 5 dye by a newly isolated bacterium Pseudomonas entomophila BS1. *Canadian Journal of Microbiology, 62*, 220–232.

Khee, Y., Kiew, P., & Chung, Y. (2023). Valorizing papaya seed waste for wastewater treatment: A review. *International Journal of Environmental Science Technology, 20*, 2327–2346.

Kumar, S., Chaurasia, P., & Kumar, A. (2016). Isolation and characterization of microbial strains from textile industry effluents of Bhilwara, India: Analysis with bioremediation. *Journal of Chemical and Pharmaceutical Research, 8*, 143–150.

Li, R., Ning, X.-A., Sun, J., Wang, Y., Liang, J., Lin, M., & Zhang, Y. (2015). Decolorization and biodegradation of the Congo red by Acinetobacter baumannii YNWH 226 and its polymer production's flocculation and dewatering potential. *Bioresource Technology, 194*, 233–239.

Ling, T. C., Inta, A., Armstrong, K. E., Little, D. P., Tiansawat, P., Yang, Y.-P., Phokasem, P., Tuang, Z. K., Sinpoo, C., & Disayathanoowat, T. (2022). Traditional knowledge of textile dyeing plants: A case study in the Chin Ethnic Group of Western Myanmar. *Diversity, 14*, 1065.

Liu, Z., Li, S., Zhang, H., Nie, F., & Wang, Q. (2013). Preparation, characterization and coagulation behaviour of a novel inorganic coagulant-polyferric (III)-magnesium (II) sulfate. *Environment Protection Engineering, 39*(3).

Madhav, S., Ahamad, A., Singh, P., & Mishra, P. K. (2018). A review of textile industry: Wet processing, environmental impacts, and effluent treatment methods. *Environmental Quality Management, 27*, 31–41.

Maier, J. R., Kandelbauer, A., Erlacher, A., Cavaco-Paulo, A., & Gübitz, G. M. (2004). A new alkali-thermostable azoreductase from Bacillus sp. strain SF. *Applied and Environmental Microbiology, 70*, 837–844.

Mani, S., & Bharagava, R.N. (2018). Textile industry wastewater: Environmental and health hazards and treatment approaches. In *Recent advances in environmental management* (pp 47–69). CRC Press.

Mani, S., Chowdhary, P., & Bharagava, R. N. (2019). Textile wastewater dyes: Toxicity profile and treatment approaches. In *Emerging and eco-friendly approaches for waste management* (pp. 219–244).

Mostafa, M. (2015). Waste water treatment in textile Industries-the concept and current removal technologies. *Journal of Biodiversity and Environmental Sciences, 7*, 501–525.

Oguntade, O. A., Adetunji, M. T., & Azeez, J. O. (2015). Uptake of manganese, iron, copper, zinc and chromium by Amaranthus cruentus L. irrigated with untreated dye industrial effluent in low land field. *Journal of Environmental Chemical Engineering, 3*, 2875–2881.

Onu, C. E., Asadu, C. O., Ohale, P. E., Ojukwu, E. V., Ogunaobi, N. C., Onu, C. P., & Akaeme, F. C. (2022). Recent trends in textile wastewater treatment using agricultural waste. In *Textile wastewater treatment: Sustainable bio-nano materials and macromolecules* (Vol. 2, pp. 89–110). Springer.

Owen, M., Ogunleye, C., & Orekoya, E. (2016). The Nigerian textile industry: An overview. *Nigerian Journal of Polymer Science and Technology, 11*, 99–108.

Owodunni, A. A., & Ismail, S. (2021). Revolutionary technique for sustainable plant-based green coagulants in industrial wastewater treatment—A review. *Journal of Water Process Engineering, 42*, Article 102096.

Pandey, A., Singh, P., & Iyengar, L. (2007). Bacterial decolorization and degradation of azo dyes. *International Biodeterioration & Biodegradation, 59*, 73–84.

Pandit, P., Singha, K., Maity, S., Maiti, S., & Kane, P. (2021). Treatment of textile wastewater by agricultural waste biomasses. In *Sustainable technologies for textile wastewater treatments* (pp. 137–156). Elsevier.

Pang, Y. L., & Abdullah, A. Z. (2013). Current status of textile industry wastewater management and research progress in Malaysia: A review. *Clean-Soil, Air, Water, 41*, 751–764.

Parshetti, G., Parshetti, S., Telke, A., Kalyani, D., Doong, R., & Govindwar, S. P. (2011). Biodegradation of crystal violet by Agrobacterium radiobacter. *Journal of Environmental Sciences, 23*, 1384–1393.

Parvin, F., Islam, S., Akm, S. I., Urmy, Z., & Ahmed, S. J. (2020). A study on the solutions of environment pollutions and worker's health problems caused by textile manufacturing operations. *Biomedical Journal of Scientific & Technical Research, 28*, 21831–21844.

Patel, H. (2018). Charcoal as an adsorbent for textile wastewater treatment. *Separation Science and Technology, 53*, 2797–2812.

Patel, H., & Pandey, S. (2012). Evaluation of physical stability and leachability of Portland Pozzolona Cement (PPC) solidified chemical sludge generated from textile wastewater treatment plants. *Journal of Hazardous Materials, 207*, 56–64.

Pattnaik, P., Dangayach, G., & Bhardwaj, A. K. (2018). A review on the sustainability of textile industries wastewater with and without treatment methodologies. *Reviews on Environmental Health, 33*, 163–203.

Pearce, C. I., Christie, R., Boothman, C., Von Canstein, H., Guthrie, J. T., & Lloyd, J. R. (2006). Reactive azo dye reduction by Shewanella strain J18 143. *Biotechnology and Bioengineering, 95*, 692–703.

Pinheiro, L. R. S., Gradíssimo, D. G., Xavier, L. P., & Santos, A. V. (2022). Degradation of azo dyes: Bacterial potential for bioremediation. *Sustainability, 14*, 1510.

Popli, S., & Patel, U. D. (2015). Destruction of azo dyes by anaerobic–aerobic sequential biological treatment: A review. *International Journal of Environmental Science and Technology, 12*, 405–420.

Prabha, S., Kumar, M., Kumar, A., Das, P., & Ramanathan, A. (2013). Impact assessment of textile effluent on groundwater quality in the vicinity of Tirupur industrial area, southern India. *Environmental Earth Sciences, 70*, 3015–3022.
Prasad, D. (2015). Biodecolorization of anthraquinone textile (Acid Blue 25) dye by Klebsiella sp. *International Journal of Recent Scientific Research, 6*, 3216–3222.
Raja, A., Arputharaj, A., Saxena, S., & Patil, P. (2019). Water requirement and sustainability of textile processing industries. In *Water in textiles and fashion* (pp. 155–173).
Rajagopalan, V. (2016). A new synergetic nanocomposite for dye degradation in dark and light. *Scientific Reports, 6*, 1–10.
Ramavandi, B. (2014). Treatment of water turbidity and bacteria by using a coagulant extracted from Plantago ovata. *Water Resources and Industry, 6*, 36–50.
Rathour, R., Jain, K., Madamwar, D., & Desai, C. (2019). Microaerophilic biodegradation of raw textile effluent by synergistic activity of bacterial community DR4. *Journal of Environmental Management, 250*, Article 109549.
Samuchiwal, S., Gola, D., & Malik, A. (2021). Decolourization of textile effluent using native microbial consortium enriched from textile industry effluent. *Journal of Hazardous Materials, 402*, Article 123835.
Sandhya, S., Sarayu, K., & Swaminathan, K. (2008). Determination of kinetic constants of hybrid textile wastewater treatment system. *Bioresource Technology, 99*, 5793–5797.
Sanghi, R., Bhattacharya, B., Dixit, A., & Singh, V. (2006). Ipomoea dasysperma seed gum: An effective natural coagulant for the decolorization of textile dye solutions. *Journal of Environmental Management, 81*, 36–41.
Sanghi, R., & Verma, P. (2013). Decolorisation of aqueous dye solutions by low-cost adsorbents: A review. *Coloration Technology, 129*, 85–108.
Sapci-Ayas, Z. (2021). Investigation of direct applicability of modified agricultural waste for contaminant removal from real textile wastewater. *Water Policy, 13*, 1354.
Saratale, R. G., Saratale, G. D., Govindwar, S. P., & Kim, D. S. (2015). Exploiting the efficacy of Lysinibacillus sp. RGS for decolorization and detoxification of industrial dyes, textile effluent and bioreactor studies. *Journal of Environmental Science and Health, Part A, 50*, 176–192.
Saxena, S., Raja, A., & Arputharaj, A. (2017). Challenges in sustainable wet processing of textiles. In *Textiles and clothing sustainability: Sustainable textile chemical processes* (pp. 43–79).
Şen, S., & Demirer, G. (2003). Anaerobic treatment of real textile wastewater with a fluidized bed reactor. *Water Research, 37*, 1868–1878.
Shahadat, M., & Isamil, S. (2018). Regeneration performance of clay-based adsorbents for the removal of industrial dyes: A review. *RSC Advances, 8*, 24571–24587.
Shahid, M., Shahid ul, I., & Mohammad, F. (2013). Recent advancements in natural dye applications: A review. *Journal of Cleaner Production, 53*, 310–331.
Shamsnejati, S., Chaibakhsh, N., Pendashteh, A. R., & Hayeripour, S. (2015). Mucilaginous seed of Ocimum basilicum as a natural coagulant for textile wastewater treatment. *Industrial Crops and Products, 69*, 40–47.
Shokry, H., Elkady, M., & Hamad, H. (2019). Nano activated carbon from industrial mine coal as adsorbents for removal of dye from simulated textile wastewater: Operational parameters and mechanism study. *Journal of Materials Research and Technology, 8*, 4477–4488.
Sillanpää, M., Ncibi, M. C., Matilainen, A., & Vepsäläinen, M. (2018). Removal of natural organic matter in drinking water treatment by coagulation: A comprehensive review. *Chemosphere, 190*, 54–71.
Sivalingam, S., & Gopal, V. (2023). Coagulation study on extracted algal alginate from red algae as natural coagulant for remediation of textile dye Congo red. *Environment and Natural Resources Journal, 21*, 90–101.
Sonal, S., & Mishra, B. K. (2021). Role of coagulation/flocculation technology for the treatment of dye wastewater: Trend and future aspects. In *Water pollution and management practices* (pp. 303–331).

Supaka, N., Juntongjin, K., Damronglerd, S., Delia, M.-L., & Strehaiano, P. (2004). Microbial decolorization of reactive azo dyes in a sequential anaerobic–aerobic system. *Chemical Engineering Journal, 99*, 169–176.

Tan, L., Xu, B., Hao, J., Wang, J., Shao, Y., & Mu, G. (2019). Biodegradation and detoxification of azo dyes by a newly isolated halotolerant yeast Candida tropicalis SYF-1. *Environmental Engineering Science, 36*, 999–1010.

Tianzhi, W., Weijie, W., Hongying, H., & Khu, S.-T. (2021). Effect of coagulation on bio-treatment of textile wastewater: Quantitative evaluation and application. *Journal of Cleaner Production, 312*, Article 127798.

Uddin, F. (2018). *Montmorillonite: An introduction to properties and utilization* (Vol. 817). IntechOpen.

Uddin, F. (2021). Environmental hazard in textile dyeing wastewater from local textile industry. *Cellulose, 28*, 10715–10739.

Verma, A. K., Dash, R. R., & Bhunia, P. (2012). A review on chemical coagulation/flocculation technologies for removal of colour from textile wastewaters. *Journal of Environmental Management, 93*, 154–168.

Vijayaraghavan, G., & Shanthakumar, S. (2018). Effective removal of acid black 1 dye in textile effluent using alginate from brown algae as a coagulant. *Iranian Journal of Chemistry and Chemical Engineering, 37*(3).

Vijayaraghavana, G., & Shanthakumarb, S. (2019). Removal of crystal violet dye in textile effluent by coagulation using algal alginate from brown algae Sargassum sp. In *Presented at the First International Conference on Recent Trends in Clean Technologies for Sustainable Environment (CTSE-2019)* (p. 27).

Vivekanandan, E., Hermes, R., & O'Brien, C. (2016). Climate change effects in the Bay of Bengal large marine ecosystem. *Environmental Development, 17*, 46–56.

Wang, W.-L., Hu, H.-Y., Liu, X., Shi, H.-X., Zhou, T.-H., Wang, C., Huo, Z.-Y., & Wu, Q.-Y. (2019). Combination of catalytic ozonation by regenerated granular activated carbon (rGAC) and biological activated carbon in the advanced treatment of textile wastewater for reclamation. *Chemosphere, 231*, 369–377.

Xing, X., Yu, Y., Xi, H., Song, G., Wang, Y., Zuo, J., & Zhou, Y. (2018). Reduction of wastewater toxicity and change of microbial community in a hydrolysis acidification reactor pre-treating trimethylolpropane wastewater. *Frontiers of Environmental Science & Engineering, 12*, 1–13.

Yin, C.-Y. (2010). Emerging usage of plant-based coagulants for water and wastewater treatment. *Process Biochemistry, 45*, 1437–1444.

Zaharia, C., Musteret, C.-P., & Afrasinei, M.-A. (2024). The use of coagulation-flocculation for industrial colored wastewater treatment—(I) the application of hybrid materials. *Applied Sciences, 14*, 2184.

Zhang, C., Xin, Y., Wang, Y., Guo, T., Lu, S., & Kong, J. (2015). Identification of a novel dye-decolorizing peroxidase, EfeB, translocated by a twin-arginine translocation system in Streptococcus thermophilus CGMCC 7.179. *Applied and Environmental Microbiology, 81*, 6108–6119.

Zhao, C., Zhou, J., Yan, Y., Yang, L., Xing, G., Li, H., Wu, P., Wang, M., & Zheng, H. (2021). Application of coagulation/flocculation in oily wastewater treatment: A review. *Science of the Total Environment, 765*, Article 142795.

Management of Wastewater from Abattoirs: Sustainable Practices for a Cleaner Environment

Eduardo Alberto López-Maldonado and Christian L. Castro-Riquelme

Abstract This chapter discusses the challenges and opportunities of dealing with wastewater generated from abattoir operations. It emphasizes the importance of sustainable wastewater management and the need to comply with regulations. An insight into the characteristics of abattoir wastewater and its potential environmental implications is provided. Wastewater treatment technologies, ranging from primary and secondary treatment methods to emerging innovations are explored. The importance of resource recovery, water recycling, and energy generation from wastewater treatment processes is emphasized as essential components of responsible abattoir wastewater management. Real-world case studies that illustrate the successful implementation of sustainable practices in abattoirs, offering valuable lessons for industry professionals and policymakers alike are showcased. Challenges and barriers, including economic, technical, and social aspects, along with strategies to overcome them, are discussed. Future trends and technologies for better wastewater management in abattoirs were discussed.

1 Introduction

Abattoirs (slaughterhouses) are at the heart of the global food supply chain. They hold a vital position, ensuring that the world's growing population has access to a consistent and safe supply of meat products (Bustillo-Lecompte & Mehrvar, 2015). However, this indispensable role comes at a cost which often remains hidden from the public view: the generation of copious amounts of wastewater (Yildirim et al., 2019). This wastewater, rich in organic matter, blood, fats, and chemicals from animal processing, poses significant environmental and public health challenges (Yildirim

E. A. López-Maldonado (✉) · C. L. Castro-Riquelme
Faculty of Chemical Sciences and Engineering, Autonomous University of Baja California, Calzada Universidad, Parque Internacional Industrial, Tijuana, Baja California, Mexico
e-mail: elopez92@uabc.edu.mx

C. L. Castro-Riquelme
e-mail: riquelme31139@uabc.edu.mx

N. A. Oladoja and E. I. Unuabonah (eds.), *Strategic Management of Wastewater from Intensive Rural Industries*, Springer Water,
https://doi.org/10.1007/978-3-031-90314-4_5

et al., 2019). The improper management of those waters can lead to dire consequences, including soil and water contamination, damage to aquatic ecosystems, and the spread of diseases (Ng et al., 2022). This challenge cannot be underestimated and thus, require urgent and sustainable solutions, especially in the context of a changing climate and heightened environmental consciousness (Bazrafshan et al., 2012; Matheyarasu et al., 2016). As our global population continues to expand, so does the demand for meat products, prompting the establishment of more abattoirs worldwide (Bustillo-Lecompte & Mehrvar, 2015). This chapter seeks to raise awareness, educate, and inform a broad audience, ranging from industry professionals, policymakers, and researchers to concerned citizens. Its objectives are to shed light on the environmental challenges of abattoir wastewater, inform readers about its sources, characteristics, and legal framework, and inspire action by showcasing best practices and innovative technologies for sustainable wastewater management in abattoirs of all sizes. In this chapter, we will embark on a journey through the world of abattoir wastewater, exploring and unveiling strategies to transform this challenge into an opportunity for a cleaner and more sustainable future. Through case studies, expert insights, and practical recommendations, we aim to empower individuals, organizations, and governments to take meaningful steps toward reducing the environmental footprint of abattoir operations and ensuring a more responsible and sustainable meat industry for generations to come.

2 The Abattoirs

Abattoirs are essential components of the food supply chain, ensuring that meat products are safe, accessible, and affordable for consumers worldwide (Al-Mutairi et al., 2004). They serve as the destination for livestock, where animals are processed into various cuts of meat for human consumption (Mofijur et al., 2021). This process, however, generates a substantial volume of wastewater, which is loaded with organic matter, blood, fats, and chemicals used in the slaughter and processing of animals. If not appropriately managed, this wastewater can have devastating environmental consequences, including soil and water pollution, contamination of aquatic ecosystems, and the spread of diseases (Rajpal et al., 2022). Different management approaches are needed depending on the composition of the wastewater. Doing so not only safeguards our natural resources but also contributes to the long-term viability of the meat industry itself.

The operations within an abattoir involve the humane slaughtering of animals, followed by the dressing, butchering, and processing of carcasses into cuts of meat. Abattoirs are responsible for separating edible meat from inedible parts, such as bones, skin, and organs, as well as processing by-products like hides and offal (Bustillo-Lecompte & Mehrvar, 2015; Ng et al., 2022; Rajpal et al., 2022). This division of labour within the facility ensures that every part of the animal is used efficiently, minimizing waste and maximizing the yield of meat products (Mofijur et al., 2021).

Abattoirs play a pivotal role in the meat industry, serving several essential functions:

Food Supply: Abattoirs are central to meeting the global demand for meat products. They ensure a consistent and reliable source of meat that adheres to safety and quality standards.

Hygiene and Safety: Abattoirs are equipped with state-of-the-art facilities and processes to ensure the sanitary and humane treatment of animals. This includes rigorous inspection procedures to identify and eliminate any health risks associated with the meat.

Economic Contribution: Abattoirs contribute significantly to the economy by providing employment opportunities, supporting livestock farmers, and generating revenue through the sale of meat and by-products.

Value Addition: They add value to the raw product (livestock) by processing it into various cuts of meat, making it suitable for diverse culinary purposes and consumer preferences.

Quality Control: Abattoirs implement strict quality control measures, from animal welfare standards to meat processing practices, to guarantee that the final product is safe and of high quality.

2.1 Types of Abattoirs and Their Operations

Abattoirs come in various types, catering for different needs within the meat industry (Table 1). Abattoirs of all sizes and specializations must follow regulatory standards and best practices to ensure food safety, animal welfare, and environmental responsibility. However, each type has unique procedures, equipment, and challenges (Roupas et al., 2007).

3 Wastewater Generation in Abattoirs

Abattoirs are classified as small (up to 20 m^3/day), medium (up to 500 m^3/day) and large (more than 500 m^3/day) depending on the volume of wastewater discharge they generate (Best Available Techniques in the Slaughterhouses and Animal By-Products Industries, 2005).

Table 1 Functionality of the types of abattoirs

Type	Description
Commercial abattoirs	These large-scale facilities are designed to process a high volume of livestock daily. They often serve as primary suppliers to retail markets, supermarkets, and the food service industry
Custom or small-scale abattoirs	These are smaller facilities that cater to local or niche markets. They provide custom processing services for small-scale farmers and speciality meat products, ensuring traceability and unique processing methods (Nastasijevic et al., 2022; Roupas et al., 2007)
Mobile abattoirs	They can be transported to different locations to process livestock on-site, benefiting small-scale and remote farming operations
Integrated abattoirs	Some abattoirs are part of vertically integrated agribusinesses that oversee the entire meat production process, from raising livestock to processing and distribution. This allows for greater control over quality and supply chain efficiency

3.1 Sources of Abattoir Wastewater

1. **Slaughter Processes**: Abattoirs' primary source of wastewater is the slaughtering process, which involves bleeding, scalding, dehairing, and evisceration of animals. Blood and organic matter are released during these processes, contributing significantly to the wastewater volume and composition (Mofijur et al., 2021; Nastasijevic et al., 2022; Roupas et al., 2007; Yildirim et al., 2019).
2. **Processing and Cleaning**: Following slaughtering, carcasses undergo further processing, including skinning, dressing, and butchering. Additionally, the equipment and facilities must be regularly cleaned to maintain hygiene standards. These activities produce wastewater containing fats, proteins, and detergents (Ng et al., 2022).
3. **Animal By-Products**: Abattoirs generate by-products such as offal (internal organs), bones, hides, and feathers. These by-products are often processed separately, resulting in wastewater streams with distinct characteristics (Mittal, 2006).
4. **Carcass Washing**: To remove contaminants and ensure meat quality, carcasses may be washed before processing, which generates additional wastewater (Johns et al., 1995).
5. **Sanitizers and Disinfectants**: Abattoirs employ various chemicals for cleaning and disinfection purposes. Residues from these chemicals can find their way into the wastewater, posing environmental risks (Fig. 1).

3.2 Characteristics of Abattoir Wastewater

Abattoir wastewater is characterized by its unique composition, which sets it apart from municipal or industrial wastewater (Johns et al., 1995; Ng et al., 2022; Roupas

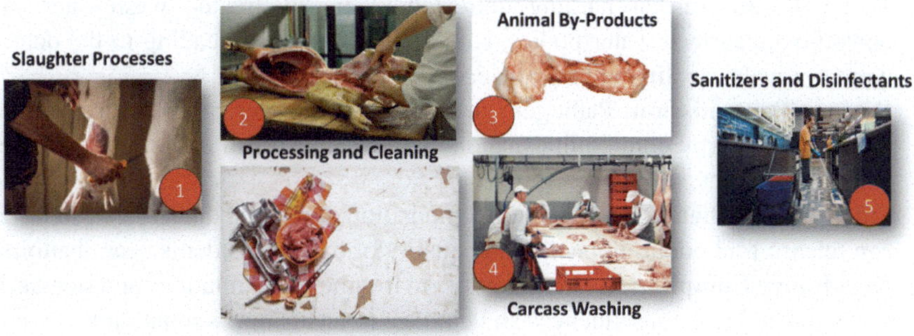

Fig. 1 Slaughterhouse wastewater generation points

et al., 2007, Bustillo-Lecompte and Mehrvar, 2017). The key characteristics of abattoir wastewater include:

1. **High Organic Content**: Abattoir wastewater contains a significant amount of organic matter, primarily in the form of blood, animal tissue, and fats. This organic load can lead to oxygen depletion and the release of foul odours if not properly treated. These impacts have been reported with a value of Chemical Oxygen Demand from 4000 mg/L to 11,000 mg/L, depending on the operating characteristics of the slaughterhouse (Hailu et al., 2015; Matheyarasu et al., 2016).
2. **Nutrient Enrichment**: The wastewater is rich in nutrients like nitrogen and phosphorus, which, when discharged into water bodies, can contribute to eutrophication, harmful algal blooms, and aquatic ecosystem disruptions (Mittal, 2006).
3. **High Strength**: Abattoir wastewater is considered "high-strength" due to its elevated levels of organic pollutants. Treating such wastewater is more challenging and energy-intensive compared to typical municipal sewage (Shengquan et al., 2008).
4. **Pathogens**: Pathogenic microorganisms, including bacteria and viruses, can be present in abattoir wastewater. This poses a risk to public health and necessitates thorough disinfection (Nastasijevic et al., 2022; Yildirim et al., 2019).
5. **Chemical Residues**: Chemicals from sanitizers, disinfectants, and veterinary drugs may be present in the wastewater, requiring specialized treatment to remove or neutralize them (Alfonso-Muniozguren et al., 2018).

3.3 Environmental and Public Health Implications

The improper management of abattoir wastewater can have significant environmental and public health implications:

1. **Water Pollution:** Discharging untreated or poorly treated wastewater can contaminate surface water bodies, soil, and groundwater, leading to the degradation of water quality and aquatic ecosystems.
2. **Disease Transmission:** Pathogens in abattoir wastewater can pose health risks to workers, nearby communities, and downstream users of contaminated water sources.
3. **Odor and Aesthetic Issues:** The decomposition of organic matter in wastewater can release foul odour, impacting the quality of life for residents near abattoirs.
4. **Regulatory Compliance:** Violations of environmental regulations and standards can result in legal consequences and damage to an abattoir's reputation.

Given these challenges, it is imperative to implement effective wastewater management practices in abattoirs to mitigate the environmental and public health risks associated with wastewater generation. In the following sections, we will explore various strategies and technologies for the treatment and sustainable management of abattoir wastewater, highlighting best practices and innovations aimed at minimizing its impact on the environment.

The proper management of wastewater from abattoirs is not only an environmental imperative but also a legal requirement. In this chapter, we will explore the legal and regulatory framework governing abattoir wastewater management, both at the national and international levels. We will delve into the significance of compliance and the potential consequences of non-compliance, as well as examine case studies of regulatory success stories that highlight the positive outcomes of stringent wastewater regulations.

3.4 National and International Regulations on Wastewater Management

3.4.1 National Regulations

The regulation of abattoir wastewater management primarily falls under national jurisdiction. Governments around the world have established specific laws, regulations, and standards to ensure that abattoirs handle their wastewater responsibly. These regulations encompass various aspects, including the quality of discharged wastewater, the treatment technologies to be employed, and reporting requirements.

- **Water Quality Standards:** Many nations have set water quality standards that limit the permissible levels of pollutants in discharged wastewater. These standards typically cover parameters such as biochemical oxygen demand (BOD), total suspended solids (TSS), pH, and nutrient concentrations.
- **Permitting and Reporting:** Abattoirs are often required to obtain permits for wastewater discharge and adhere to reporting requirements. These permits specify the conditions under which wastewater can be released into the environment and often include monitoring and reporting obligations.

- **Treatment Requirements:** Regulations may prescribe specific wastewater treatment technologies and methodologies that abattoirs must implement to meet water quality standards. Common treatments include primary and secondary biological treatment, chemical precipitation, and disinfection.

3.4.2 International Agreements

While national regulations govern the bulk of abattoir wastewater management, some international agreements and guidelines complement some of these efforts. Organizations like the World Health Organization (WHO), the Food and Agriculture Organization (FAO), and the United Nations Environment Programme (UNEP) play crucial roles in promoting best practices for wastewater management in the context of food production. The Codex Alimentarius Commission, jointly established by the WHO and FAO, sets international food standards and guidelines. These standards encompass aspects of food safety and quality, including the handling and processing of meat products, which indirectly influence wastewater management practices. The United Nations' Sustainable Development Goals, particularly Goal 6 (Clean Water and Sanitation) and Goal 12 (Responsible Consumption and Production), emphasize the importance of sustainable water use and environmental responsibility in food production, indirectly encouraging the adoption of wastewater management best practices.

3.5 Compliance and Consequences

Compliance with wastewater regulations is of paramount importance for abattoirs. Failure to adhere to these regulations can lead to a range of consequences:

1. **Legal Penalties:** Non-compliance with wastewater regulations can result in fines, legal action, and even the temporary or permanent closure of abattoirs. Such penalties are designed to deter negligent behavior and protect the environment and public health. In fact, many countries have strict regulations in place to ensure that abattoirs are not polluting the environment or endangering public health with their wastewater. For example, in the United States, the Clean Water Act regulates the discharge of pollutants into surface waters, and abattoirs must obtain permits and meet specific discharge standards.
2. **Reputation Damage:** Public perception of abattoirs can be significantly affected by environmental violations. Negative publicity can harm a company's reputation and customer trust, potentially leading to reduced business opportunities and profitability.
3. **Environmental Impact:** The most significant consequence of non-compliance is the environmental harm caused by untreated or poorly treated wastewater. This can lead to pollution of water bodies, damage to ecosystems, and harm to aquatic and human lives.

4. **Public Health Risks:** Inadequate wastewater management can pose health risks to nearby communities through the spread of waterborne diseases and contamination of local water sources.

3.6 Case Studies of Regulatory Success Stories

To highlight the positive impact of stringent wastewater regulations, let's explore a few case studies of abattoirs that have successfully complied with and even exceeded regulatory requirements:

1. **Danish Crown:** The Danish Crown Group, one of the world's largest meat processing companies, has implemented advanced wastewater treatment technologies in its abattoirs. They have not only met Danish regulatory standards but have also reduced water consumption and environmental impact, contributing to their sustainability goals (*Home—Danish Crown*, n.d.).
2. **Smithfield Foods:** Smithfield Foods, a major pork producer in the United States, faced regulatory pressure due to the environmental impact of its activities, especially from its wastewater. In response, they invested in innovative wastewater treatment processes and renewable energy solutions, significantly reducing their carbon footprint and improving their compliance record.
3. **Brazilian Beef Industry:** In Brazil, the beef industry has worked closely with the government and environmental organizations to improve wastewater management practices. This collaborative approach has resulted in the implementation of wastewater treatment systems that have helped to reduce the sector's environmental impact (Morais et al., 2023).

These case studies demonstrate that compliance with wastewater regulations not only avoids penalties but also can drive innovation, reduce environmental impact, and enhance a company's reputation. They serve as inspiration for abattoirs worldwide to prioritize sustainable wastewater management practices.

4 Wastewater Treatment Technologies

Effective wastewater treatment is critical for abattoir management, as it ensures that wastewater generated is purified before discharge into the environment. This chapter provides a comprehensive overview of the various wastewater treatment processes, including primary, secondary, and tertiary treatment methods. We also will delve into the emerging technologies that hold great promise for more efficient and sustainable abattoir wastewater treatment.

4.1 Overview of Abattoir Wastewater Treatment Processes

Abattoir wastewater is treated to remove pollutants through physical, chemical, and biological processes. The process typically consists of the following stages:

4.1.1 Preliminary Treatment

This initial stage involves the removal of large debris, such as bones, hair, and solid particles, through screens and settling tanks. It prevents these materials from interfering with downstream treatment processes. The initial filtering usually removes particles between 25 and 6 mm, while the finer one removes smaller particles. Filtration processes can be static, rotary and vibrational, at this stage 25 to 40% of the biological oxygen demand (BOD) and 50 to 70% of the soluble solids (SS) are usually eliminated (Masse & Masse, 2000).

4.1.2 Primary Treatment

In primary treatment, suspended solids and settleable organic matter are removed through sedimentation (López-Maldonado and Oropeza-Guzmán, 2019). Wastewater flows into large settling tanks, allowing heavier particles to settle to the bottom, and the clarified water is then separated from the solids.

Dissolved Air Flotation (DAF)

It is one of the most used primary methods for the clarification of abattoir wastewater, by separating solids, fats and oils. This method is based on the agglomeration of impurities in the microbubbles generated, and the size distribution and concentration of the air bubbles are key parameters that influence the process efficiency. To achieve this, air is bubbled to the bottom of the tank (0.25–0.30 MPa) containing the abattoir wastewater, which displaces small solids, grease, and oils. The DAF is a faster method compared to the gravity method (Manjunath et al., 2000). The DAF is usually combined with polyacrylamide polymers, as coagulants, to obtain a synergistic effect that accelerates the separation of the constituents of the wastewater.

This method is limited by the low or null removal of dissolved solids, nitrogen and soluble proteins, but specific ways to improve the process have been provided. One of these improvements is through the use of thin-layer cells and the improvement of the size of the flotation aggregates, through the secondary saturation of the flotation cell with air bubbles of 0.1 to 0.3 mm in diameter that float in a regime of laminar flow and therefore create slight turbulence (Andreev et al., 2023; Schmideder et al., 2022).

Coagulation-Flocculation

The process is typically used in slaughterhouse wastewater treatment for the removal of Chemical Oxygen Demand (COD), Total Suspended Solids (TSS), and Total Phosphorus (TP). The most used materials for this process are alum, ferric chloride and ferric sulfate. With these materials, TSS and TP removal efficiencies of 65% and 98% have been achieved, respectively. Alum showed greater efficiency in the removal of TSS and TP, while ferric sulfate was more effective in COD reduction (Amuda & Alade, 2006; Okey-Onyesolu et al., 2023). The elimination of TP is dependent on the concentration of alum, obtaining the maximum removal with a dose of 1 g/L with a confidence interval of 95%.

Abattoir water treatment using electrocoagulation with iron electrodes has been successfully carried out by analyzing variables such as pH, current intensity, electrolysis time, sedimentation time and temperature. The pH was analyzed in a range from 1 to 7, the current intensity from 0.5 to 2.5 A, the electrolysis time from 10 to 30 min, the separation time from 35 to 55 min and the temperature from 35 to 55 °C. The best results produced clarification percentages of up to 97.42%.

4.1.3 Secondary Treatment

Secondary treatment focuses on the biological removal of dissolved and colloidal organic matter. It typically involves the use of aerobic (with oxygen) or anaerobic (without oxygen) biological processes, such as activated sludge, trickling filters, or lagoons. Microorganisms break down organic pollutants, transforming them into stable substances (Borja et al., 1998; Gannoun et al., 2009; Manjunath et al., 2000).

Biological treatments can remove up to 90% of contaminants in water and decrease the biological oxygen demand (BOD) and total soluble solids (TSS). The microorganisms are kept in suspension within the liquid by mixing with some static medium, such as ceramic or plastic materials. The detergents used in abattoirs must be compatible with the biological process else they will harm the microorganisms. Both processes (flocculation and coagulation) require an optimal contact time between the wastewater and the microorganisms.

Anaerobic Treatment

Abattoir water is suitable for anaerobic treatments due to its high organic matter content. The anaerobic treatment transforms the organic compounds to methane and CO_2 mediated by bacteria in the absence of oxygen, reducing the BOD and allowing the use of methane, however, these processes are limited by the presence of particles and fats, so a pretreatment stage is necessary. The main anaerobic treatments are anaerobic contact ponds (AC), up-flow anaerobic sludge blankets (UASB), anaerobic sequence batch reactors (ASBR), and anaerobic filter (AF) processes, which are capable of reducing the physicochemical characteristics (e.g., 97% BOD, 95% DS and 96% COD, with high levels of organic N_2 (70–250 g/m^3).

Anaerobic Contact Ponds (AC)

The focus of anaerobic ponds has shifted from the removal of solids to optimizing biogas production for their reuse in the production of alternative energy. In a comparative study between an anaerobic pond with horizontal and vertical baffles, clear differences were observed in COD (Chemical Oxygen Demand), with particulate COD removal of 78% and 32%, respectively. Methane production at the end of the study was 6.09 and 9.04 L/m^3 of treated wastewater volume for the pond with horizontal baffles, used as control, and staged pond, respectively (Cruddas et al., 2014; Ruiz et al., 1997).

The effect of temperature in the anaerobic process for the treatment of contaminated water has also been studied as a sludge management strategy since these processes are limited to places with very low temperatures during most of the year. During the study, the effluent temperature oscillated between 0.1 and 21.1 °C. The production of soluble COD was observed at temperatures higher than 10 °C and coincided with an increase in the concentration of volatile fatty acids (VFA) in the effluent, indicating anaerobic degradation. Biogas generation is observed predominantly in the initial periods and increases throughout the anaerobic process. An increase in the average biogas production of more than an order of magnitude was obtained between winters. The maximum methane recorded was 2630 mg PE^{-1} d^{-1}, approximately fifty times higher than that estimated from storage sludge (57 mg PE^{-1} d^{-1}), which could provide a viable opportunity for power generation (Cruddas et al., 2014).

Upflow Anaerobic Sludge Blankets (UASB)

In UASB reactors, the effluent enters at the bottom of the digester, flows up through a compacted layer of bacteria (the sludge blanket), and exits at the top of the reactor. It consists of three phases: liquid, solid (sludge or biomass), and gas (gases formed during the digestion process, predominantly CO_2 and CH_4) (Manjunath et al., 2000). A successful operation depends on the formation of bacterial flocs or granules that accumulate and easily settle to the bottom of the digester. One of the limitations of anaerobic processes is the high content of fat particles and proteins. To address this challenge, pretreatments have been used. One of the processes that have been used as pretreatment is hydrolysis in water from slaughterhouses with a high content of organic solids (fat and protein) without aeration. The pretreatment was applied in static and stirred reactors, with a microbial population in the storage tank. The microbial analysis showed that the main populations contributing to the pretreatment belong to the gender Bacillus and the stirring favours hydrolysis and solubilizes 80% of the suspended matter. The abattoir water pretreated in the stirred reactor was given 2 days retention time and fed to an up-flow anaerobic filter (UAF) under mesophilic conditions (37 °C). Mesophilic conditions achieved COD removal efficiencies of 70 to 72%, while efficiencies of 80 to 90% were obtained under thermophilic conditions (55 °C). The biogas yield was approximately 0.32 to 0.45 L of biogas per gram of COD removed. Mesophilic anaerobic digestion has been shown to partially destroy pathogens, while the thermophilic process was more efficient in removing indicator microorganisms and pathogenic bacteria at different organic loading rates (Manjunath et al., 2000; Ng et al., 2022).

Processes that combine the use of a sludge blanket and a filter arrangement made up of an anaerobic reactor in which the lower two-thirds were occupied by a layer of sludge and the upper third by small submerged cubes of polyurethane foam have also been used. The system was evaluated for the treatment of abattoir wastewater at 35 °C. In the first study, the COD was increased from 3.74 to 10.41 gL^{-1} and the hydraulic retention time (HRT) was 1.5 days. In the second experiment, the HRT was reduced from 1.5 to 0.50 days, maintaining a constant COD of 10.41 gL^{-1}. The results showed that this type of reactor is suitable for the anaerobic treatment of slaughterhouse wastewater and demonstrated a high removal of organic load and COD between 90.2 and 93.4% (Borja et al., 1998).

Aerobic Treatments

Aerobic treatment involves the degradation of organic compounds by microorganisms in the presence of oxygen, these systems require daily maintenance by a trained technician and daily drainage of accumulated sludge. Microorganisms need free dissolved oxygen to reduce biomass in wastewater and the biological sludge must be treated before disposal, these treatments are very good at treating odour and pathogens. The main treatments are aerobic lagoon, activated sludge processes, extended aeration, oxidation ditches, batch sequencing, reactors (SBR), trickling filtration and rotary biological contactors (RBC).

4.1.4 Tertiary Treatment

Tertiary treatment further polishes the effluent to meet specific water quality standards. This includes advanced filtration, chemical coagulation, and disinfection processes, to remove remaining suspended solids, nutrients, and pathogens (Alfonso-Muniozguren et al., 2018; Shengquan et al., 2008).

Membrane Sequencing Batch Reactor (MSBR)

Abattoir wastewater is highly suitable for MSBR treatment using nitrifying and denitrifying bacteria because it has a high content of organic compounds (Le-Clech, 2010). This system is very convenient in the long term since it can be self-controlled and does not require the addition of chemicals or bacteria for its operation. Through this system, typical reductions of up to 98% BOD, 93% SS, 98% COD, 95% NH_3-N and 95% lipids have been obtained (Shengquan et al., 2008). Wastewater in slaughterhouses can be effectively treated with the membrane filtration process to produce recyclable water. Total COD can be reduced below 100 mg/L. It has been shown that the MSBR process has a better treatment effect and is easier to operate and manage compared with the SBR process (Shengquan et al., 2008).

Combination of Processes Sludge-Filtration-Ozonation

A combination of methods has been used for the treatment of wastewater for possible reuse and allowing compliance with the standards of different regions in the world. A combination of filtration, ozonation and activated sludge systems have been used to treat abattoir wastewater. An activated sludge process was used with a hydraulic retention time of 24 h and a solid retention time of 13 days, followed by the filtration process (mesh size of 4–7 µm) and subsequently ozonation as a tertiary treatment. Average reductions of 93% and 98% of COD and BOD respectively with final values of 128 mg/L of COD and 12 mg/L of BOD were obtained. The average reduction of total suspended solids (TSS) reached 99% in the same system, reducing the final TSS value to 3 mg/L. In addition, a 98% reduction in phosphorus (P) and a complete inactivation of total coliforms (TC) were obtained after 17 min of ozonation with an ozone dose of 71 mg/L. The combined process was sufficient to meet the discharge requirements without additional treatment for COD, BOD, TSS, P, TC and TVC parameters (Alfonso-Muniozguren et al., 2018).

Disinfection

The final step in the treatment process involves the disinfection of effluent to eliminate harmful microorganisms. Common disinfection methods include chlorination, ultraviolet (UV) irradiation, and ozonation.

4.2 Emerging Technologies for Abattoir Wastewater Treatment

The meat industry is continually exploring innovative wastewater treatment technologies to enhance efficiency and sustainability. Some of these emerging technologies include:

1. **Membrane Bioreactors (MBRs):** MBRs can be combined with activated sludge treatment as well as with membrane filtration to produce high-quality wastewater effluent in a compact system. As a compact and low-maintenance process, MBRs presented some advantages in resorts, hotels, ships and small office buildings. The

membranes are generally mounted in modules or cassettes, which include aeration ports, permeate flow connections and supporting frames. The aeration stage provides extra oxygen and keeps the biomass in suspension, but also produces a turbulent flow velocity (0.2–0.4 m/s), hereby limiting fouling deposition.

2. **Advanced Oxidation Processes (AOPs):** AOPs use chemical reactions involving ozone, hydrogen peroxide, or UV light to break down recalcitrant organic compounds and disinfect wastewater (Miklos et al., 2018).
3. **Constructed Wetlands:** Natural or engineered wetlands are used for polishing wastewater, and the operation includes wetland vegetation, soil, and the associated microbial assemblages, to assist in treating wastewater through biological and physical processes. Suspended solids are effectively removed via settling and filtration through the dense vegetation. Nitrogen is removed primarily through nitrification (in the water column) and subsequent denitrification (in the litter layer), and ammonia volatilization under higher pH values, caused by algal photosynthesis. This process has been used to treat animal wastes, dairy pasture runoff, agriculture drainage, refinery process water, sugar factories, shrimp aquaculture and a wide variety of wastewater (Vymazal, 2010).
4. **Nutrient Recovery:** Technologies that capture and recover nutrients like phosphorus and nitrogen from wastewater for use as fertilizers or other applications. This is motivated by unregulated discharge of wastewater, poor access to fertilizers in developing countries and high fertilizer costs. This process is usually used to treat wastewater from agriculture, domestic and industrial operations. Plant growth and yield potentials with recovered nutrients are either similar or better than that of the conventional fertilizer (Saliu & Oladoja, 2021)
5. **Electrocoagulation:** An electrochemical process that destabilizes and coagulates suspended particles and contaminants for easy removal (Obi et al., 2022). It has a wide range of applications in wastewater, due to its versatility, ease of setup, low footprint and eco-friendly nature.

The choice of wastewater treatment technology depends on factors such as the volume and composition of wastewater, local regulations, available infrastructure, and the sustainability goals of the abattoir. By adopting these emerging technologies and tailoring treatment systems to their specific needs, abattoirs can not only achieve compliance but also reduce their environmental footprint and potentially recover valuable resources from their wastewater streams.

5 Best Practices in Abattoir Wastewater Management

Effective wastewater management in abattoirs goes beyond compliance with regulations; it involves adopting sustainable practices that minimize environmental impact, reduce resource consumption, and enhance overall operational efficiency. In this section, we explore the best practices in abattoir wastewater management, which include pollution prevention strategies, recycling and reuse of wastewater, and the potential for energy recovery from wastewater treatment processes.

5.1 Pollution Prevention Strategies

1. **Source Control:** Abattoirs can implement source control measures to reduce the generation of wastewater at its origin. This includes optimizing processes to minimize blood and organic matter release during slaughter and processing. For example, they can optimize their processes to reduce the release of blood and organic matter during slaughter and processing. This can be achieved by using sharp knives, efficient processing machines, and proper handling techniques. Additionally, abattoirs can implement waste segregation and recycling programs to reduce the amount of waste generated. Some abattoirs also use waterless or low-water processing methods to reduce water consumption and minimize wastewater generation. These source control measures not only reduce the environmental impact of abattoirs but also contribute to their sustainability goals.
2. **Efficient Water Use:** Implementing water-efficient practices, such as low-flow fixtures and water recycling systems, can significantly reduce water consumption and wastewater generation.
3. **Chemical Management:** Proper management of chemicals used in cleaning and disinfection is essential. Minimizing the use of harmful chemicals and ensuring their proper disposal can prevent their release into wastewater. Some examples of proper chemical management in abattoirs can include using eco-friendly cleaning products that are less harmful to the environment and reducing the usage of harsh chemicals. Proper disposal of cleaning chemicals is also important to prevent contamination of wastewater. Abattoirs can also implement measures such as regular monitoring of chemical use and storage to prevent leaks and spills that can pollute the environment. Additionally, training employees on chemical handling and management can help prevent accidents and ensure the safe and responsible use of chemicals in the abattoir.
4. **Training and Education:** Training employees on wastewater management best practices, spill prevention, and the importance of responsible resource use can foster a culture of environmental stewardship within the abattoir.

5.2 Recycling and Reuse of Wastewater

1. **On-Site Treatment:** Abattoirs can invest in on-site treatment systems that purify wastewater to a quality suitable for reuse within the facility. Treated water can be used for cleaning, cooling, or even in animal watering systems.

2. **Irrigation:** In some cases, treated wastewater can be used for irrigation of non-food crops or land application after meeting regulatory requirements and ensuring it poses no health risks.
3. **Closed-Loop Systems:** Closed-loop systems can be implemented in certain processes to minimize water consumption and reduce wastewater generation. For example, water used for carcass cooling can be recirculated and reused.
4. **Collaboration:** Collaborating with neighbouring businesses or farms can lead to innovative solutions for wastewater reuse. Abattoirs can explore partnerships for jointly managing wastewater and sharing the benefits of treated water.

5.3 Energy Recovery from Wastewater Treatment

1. **Biogas Production:** Anaerobic digestion of wastewater can produce biogas, primarily methane, which can be captured and used as an energy source for heating or electricity generation within the abattoir.
2. **Heat Recovery:** Heat generated during wastewater treatment processes can be harnessed for various purposes, such as preheating incoming water or providing space heating within the facility.
3. **Combined Heat and Power (CHP):** Abattoirs can consider combined heat and power systems that utilize biogas or other recovered energy sources to simultaneously produce electricity and heat, enhancing energy efficiency.
4. **Energy-Efficient Technologies:** Implementing energy-efficient technologies and equipment in wastewater treatment processes can reduce overall energy consumption and operational costs.

By adopting these pollution prevention strategies, recycling and reuse practices, and energy recovery measures, abattoirs can significantly reduce their environmental footprint, lower operating costs, and contribute to a more sustainable and responsible meat processing industry.

6 Case Studies

There are successful case studies of abattoirs that have demonstrated commitment to implementing sustainable wastewater management practices. These examples showcase how responsible wastewater management not only ensures compliance with regulations but also leads to environmental stewardship, cost savings, and positive impacts on the surrounding communities. Additionally, valuable lessons from these case studies are meant to inspire and guide other abattoirs on their journey toward sustainable wastewater management.

Case Study 1: Danish Crown Group—Reducing Environmental Impact

Overview: Danish Crown Group, one of the world's largest meat processing companies, operates numerous abattoirs in Denmark and beyond. They have made significant strides in reducing their environmental impact through sustainable wastewater management.

Actions Taken:

- Investment in advanced wastewater treatment technologies, including anaerobic digestion for biogas production.
- Reuse of treated wastewater within the facility for cleaning and cooling purposes.
- Implementation of water-efficient practices and source control measures to minimize wastewater generation.
- Collaboration with local farmers to provide them with nutrient-rich treated wastewater for crop irrigation.

Results and Lessons:

Results

- Significant reduction in freshwater consumption and wastewater discharge.
- Production of biogas, which is used for on-site energy generation.
- Positive community and public relations impact by reducing environmental footprint.

Lessons: Proactive investment in sustainable wastewater treatment can lead to substantial resource savings and improve a company's environmental image.

Case Study 2: Smithfield Foods—Enhancing Resource Efficiency

Overview: Smithfield Foods, a leading pork producer in the United States, has faced regulatory pressure and public scrutiny over its environmental impact. They have responded by adopting innovative wastewater management practices.

Actions Taken:

- Implementation of advanced treatment technologies, including membrane bioreactors (MBRs) for enhanced wastewater purification.
- Energy recovery from wastewater treatment processes through heat exchangers.
- Water recycling systems to minimize freshwater use.
- Collaboration with environmental organizations to develop sustainability initiatives.

Results and Lessons:

- Improved compliance with environmental regulations.
- Significant reductions in freshwater consumption and wastewater discharge.
- Recovery of energy from wastewater treatment processes for cost savings.
- Recognition in the public domain for efforts made towards environmental stewardship.

Lessons: Collaborating with environmental organizations can result in inventive solutions and enhanced sustainability (*Good Is What We Do*, n.d.; *Home—Danish Crown*, n.d.; Morais et al., 2023).

Case Study 3: Brazilian Beef Industry—Collaborative Approach

Overview: The Brazilian beef industry is one of the largest in the world. However, it has faced environmental challenges related to wastewater management. To tackle this issue, the industry has collaborated with the government and environmental organizations. This approach has been successful in addressing the problem.

They have taken specific actions to improve the process's sustainability. These include implementing wastewater treatment systems in abattoirs to meet regulatory standards, collaborating with government agencies to develop and enforce wastewater regulations, and running public awareness campaigns to educate consumers and stakeholders on sustainability. These efforts have helped to ensure the attainment of our sustainability goals and contribution to a cleaner environment.

Results and Lessons:

- Improved wastewater management practices and compliance with regulations.
- Reduced environmental impact and positive contributions to local ecosystems.
- Strengthened reputation for environmental responsibility.

Lessons: Collaboration is key for sustainable wastewater management and addressing environmental challenges (Morais et al., 2023).

These case studies highlight the diverse approaches that abattoirs can take to implement sustainable wastewater management practices. Lessons learned from these examples include the importance of proactive investment, collaboration with stakeholders, and the potential for resource recovery from wastewater treatment. By following these examples and tailoring strategies to their specific contexts, abattoirs can contribute to a more environmentally responsible and sustainable meat processing industry.

7 Challenges and Barriers

Sustainable wastewater management is crucial for abattoirs, but it can be difficult to achieve due to various challenges and barriers. These challenges can be of an economic, technical, or social nature. To overcome these obstacles and encourage responsible wastewater management practices, potential strategies are analyzed and presented.

7.1 Common Challenges Faced by Abattoirs in Wastewater Management

Abattoir wastewater has a high organic load, which can increase treatment costs and overwhelm the system. Advanced treatment technologies are expensive, energy-intensive, and require skilled labour. Regulatory compliance can be challenging, and negative public perception can harm an abattoir's reputation. Also, in regions with water scarcity, obtaining a reliable water supply for processing and wastewater treatment can be difficult.

7.2 Economic, Technical, and Social Barriers

1. **Economic Barriers**

Abattoirs incur significant capital and operational expenses when upgrading their wastewater treatment systems, which can put a strain on their financial resources. Additionally, these facilities often have to balance various competing priorities when allocating capital investments, which can make it difficult to set aside funds for wastewater management.

2. **Technical Barriers**

Advanced treatment technologies require technical expertise and may involve complex operational procedures. Some treatments are energy-intensive, resulting in higher operational costs.

3. **Social Barriers**

Negative perceptions of the abattoir's environmental practices can lead to community opposition, hindering expansion. A lack of public awareness of wastewater management can also slow progress.

7.3 Strategies to Overcome Challenges and Barriers

1. **Investment in Efficiency:** Abattoirs must invest in water-efficient technologies to reduce consumption, waste, and operational costs.
2. **Resource Recovery:** Operational costs can be offset by recovering resources like biogas from anaerobic digestion or heat from wastewater treatment.
3. **Collaboration:** Work with government agencies, environmental groups, and businesses to exchange information, resources, and solutions for managing wastewater.
4. **Training and Education:** Train employees to operate and maintain advanced treatment systems.

5. **Public Relations:** Inform and address community concerns about wastewater management efforts to improve public perception.
6. **Regulatory Compliance:** Meet wastewater treatment standards and reporting requirements by working closely with regulatory authorities. Address compliance issues proactively.
7. **Sustainability Integration:** Align the abattoir's wastewater management with sustainable business goals by incorporating environmental and social objectives.

By addressing these challenges and barriers through a combination of investment, collaboration, education, and sustainability integration, abattoirs can overcome obstacles to sustainable wastewater management. These efforts not only ensure compliance with regulations but also contribute to environmental responsibility, operational efficiency, and enhanced stakeholder relationships, paving the way for a more sustainable and responsible meat processing industry.

8 Future Directions

The future of abattoir wastewater management hinges on promising innovations and opportunities for sustainable practices. The evolving landscape of abattoir wastewater management, including emerging technologies, trends, and the critical role of public awareness and education in shaping a more environmentally responsible meat processing industry are presented herein.

8.1 Innovations in Abattoir Wastewater Management

1. **Smart Monitoring and Automation:** The integration of sensors, real-time monitoring, and automation in wastewater treatment systems can enhance operational efficiency and reduce the risk of non-compliance (*Transforming Wastewater Treatment with IoT Technologies*, n.d.).
2. **Decentralized Treatment Systems:** Smaller-scale, decentralized treatment systems, including modular and mobile units, can provide cost-effective solutions for abattoirs in remote or under-served areas (Swan et al., 2023).
3. **Circular Economy Practices:** Abattoirs are exploring circular economy approaches, such as nutrient recovery from wastewater for use in agriculture, as a means to reduce waste and promote resource efficiency (Wainaina et al., 2020).
4. **Water Reuse Technologies:** Advanced water purification and desalination technologies are being adapted to treat wastewater for direct reuse within the abattoir, reducing the demand for freshwater (Implementing Circular Economy for Nutrient Recovery in Urban Wastewater Treatment Plants—European Commission (europa.eu), n.d.; Wainaina et al., 2020)

5. **Green Infrastructure:** Natural and engineered green infrastructure, such as constructed wetlands and vegetative buffer zones, are being integrated into wastewater management systems for enhanced treatment and environmental benefits (Implementing Circular Economy for Nutrient Recovery in Urban Wastewater Treatment Plants—European Commission (europa.eu)).

8.2 Trends and Technologies on the Horizon

1. **Advanced Membrane Technologies:** Membrane technologies, including forward osmosis and electro-dialysis, hold promise for more efficient separation and treatment of abattoir wastewater components.
2. **Artificial Intelligence (AI) and Machine Learning:** AI and machine learning algorithms are being employed to optimize wastewater treatment processes, predict maintenance needs, and improve overall system performance.
3. **Resource Recovery:** The focus on recovering valuable resources, such as biogas, nutrients, and clean water, from wastewater is expected to continue as part of the circular economy approach.
4. **Green Energy Integration:** Abattoirs may increasingly integrate renewable energy sources, such as solar and wind power, into their wastewater treatment systems to reduce energy consumption.
5. **Zero Liquid Discharge (ZLD):** ZLD technologies, which aim to recover all water from wastewater, will likely gain prominence in regions facing water scarcity.

8.3 The Role of Public Awareness and Education

1. **Consumer Demand for Sustainability:** Growing public awareness of environmental issues and ethical considerations is driving consumer demand for sustainably produced meat products. Abattoirs that prioritize responsible wastewater management can meet this demand and enhance their market position.
2. **Educational Initiatives:** Industry associations, environmental organizations, and government agencies play a crucial role in educating abattoirs, stakeholders, and the public about the importance of sustainable wastewater management.
3. **Stakeholder Engagement:** Engaging with local communities, regulators, and NGOs through transparent communication can foster support for abattoir wastewater management efforts.
4. **Regulatory Influence:** Public pressure can influence regulatory changes, pushing for stricter wastewater management standards and incentivizing abattoirs to adopt sustainable practices.
5. **Knowledge Sharing:** Abattoirs can benefit from sharing best practices and success stories, both within the industry and across sectors, to collectively drive innovation and sustainability.

The future of abattoir wastewater management is marked by innovation, sustainability, and heightened public awareness. By embracing emerging technologies, trends, and educational initiatives, abattoirs can not only meet regulatory requirements but also contribute to a more environmentally responsible and socially conscious meat processing industry. This holistic approach will ensure that abattoirs play a vital role in safeguarding the environment and public health while meeting the growing global demand for meat products.

9 Conclusion

Here we discussed the role of abattoirs in providing meat products while maintaining the health of the environment. The critical importance of sustainable wastewater management in abattoirs was emphasized. The meat processing industry, with its significant contributions to the global food supply, faces a critical imperative in the responsible management of wastewater. Sustainable wastewater management in abattoirs is not just a regulatory requirement but an ethical obligation and an opportunity for the industry to:

- Protect the Environment: By reducing pollution, conserving water resources, and safeguarding ecosystems.
- Ensure Public Health: By preventing the spread of diseases and contaminants.
- Enhance Reputation: By demonstrating commitment to sustainability and social responsibility.
- Achieve Efficiency: By optimizing resource use, reducing costs, and harnessing the potential for resource recovery.

Sustainable wastewater management in abattoirs is not just a responsibility, but a catalyst for positive change. It is a testament to our commitment to safeguarding the environment, public health, and the future of the meat processing industry. By working together, abattoirs, policymakers, and the public can ensure a cleaner, healthier, and more sustainable world for generations to come.

Acknowledgements Authors gratefully acknowledge Consejo Nacional de Humanidades, Ciencia y Tecnología (CONAHCyT) for the support to Grant No. A1-S-38139.

References

Alfonso-Muniozguren, P., Lee, J., Bussemaker, M., Chadeesingh, R., Jones, C., Oakley, D., & Saroj, D. (2018). A combined activated sludge-filtration-ozonation process for abattoir wastewater treatment. *Journal of Water Process Engineering, 25*, 157–163. https://doi.org/10.1016/J.JWPE.2018.07.009

Al-Mutairi, N. Z., Hamoda, M. F., & Al-Ghusain, I. (2004). Coagulant selection and sludge conditioning in a slaughterhouse wastewater treatment plant. *Bioresource Technology, 95*(2), 115–119. https://doi.org/10.1016/J.BIORTECH.2004.02.017

Amuda, O. S., & Alade, A. (2006). Coagulation/flocculation process in the treatment of abattoir wastewater. *Desalination, 196*(1–3), 22–31. https://doi.org/10.1016/J.DESAL.2005.10.039

Andreev, S. Y., Lebedinskiy, K. V., & Stepanov, S. V. (2023). A novel technology for optimizing dissolved air flotation unit efficiency via secondary saturation of the flotation cell with air bubbles and thin-layer settling. *Chemical Engineering and Processing—Process Intensification, 184*, Article 109292. https://doi.org/10.1016/J.CEP.2023.109292

Bazrafshan, E., Kord Mostafapour, F., Farzadkia, M., Ownagh, K. A., & Mahvi, A. H. (2012). Slaughterhouse wastewater treatment by combined chemical coagulation and electrocoagulation process. *PLoS ONE, 7*(6), 40108. https://doi.org/10.1371/JOURNAL.PONE.0040108

Borja, R., Banks, C. J., Wang, Z., & Mancha, A. (1998). Anaerobic digestion of slaughterhouse wastewater using a combination sludge blanket and filter arrangement in a single reactor. *Bioresource Technology, 65*(1–2), 125–133. https://doi.org/10.1016/S0960-8524(98)00004-2

Bustillo-Lecompte, C. F., & Mehrvar, M. (2015). Slaughterhouse wastewater characteristics, treatment, and management in the meat processing industry: A review on trends and advances. *Journal of Environmental Management, 161*, 287–302. https://doi.org/10.1016/J.JENVMAN.2015.07.008

Bustillo-Lecompte, C. F., & Mehrvar, M. (2017). Treatment of actual slaughterhouse wastewater by combined anaerobic–aerobic processes for biogas generation and removal of organics and nutrients: An optimization study towards a cleaner production in the meat processing industry. *Journal of Cleaner Production, 141*, 278–289. https://doi.org/10.1016/J.JCLEPRO.2016.09.060

Cruddas, P. H., Wang, K., Best, D., Jefferson, B., Cartmell, E., Parker, A., & McAdam, E. J. (2014). Diagnosis of an anaerobic pond treating temperate domestic wastewater: An alternative sludge strategy for small works. *Ecological Engineering, 63*, 64–71. https://doi.org/10.1016/J.ECOLENG.2013.12.011

Gannoun, H., Bouallagui, H., Okbi, A., Sayadi, S., & Hamdi, M. (2009). Mesophilic and thermophilic anaerobic digestion of biologically pretreated abattoir wastewaters in an upflow anaerobic filter. *Journal of Hazardous Materials, 170*(1), 263–271. https://doi.org/10.1016/J.JHAZMAT.2009.04.111

Good Is What We Do. (n.d.). Retrieved Mar 1, 2024, from https://www.smithfieldfoods.com/good-is-what-we-do/?gad_source=1&gclid=CjwKCAiAuYuvBhApEiwAzq_YiZfnOPaH9xzmH2PcFBSlth4vWv7IaE4RfGhd6DCZb7GY7KzfF7PvjBoCNWAQAvD_BwE

Hailu, A. M., Mulu, A., & Ayenew, T. (2015). Characterization of abattoir wastewater and evaluation of the effectiveness of the wastewater treatment systems in Luna and Kera abattoirs in Central Ethiopia. *International Journal of Scientific & Engineering Research, 6*(4). http://www.ijser.org

Home—Danish Crown. (n.d.). Retrieved Mar 1, 2024, from https://www.danishcrown.com/global/

Implementing Circular Economy for Nutrient Recovery in Urban Wastewater Treatment Plants—European Commission. (n.d.). Retrieved March 1, 2024, from https://environment.ec.europa.eu/topics/urban-environment/european-green-capital-award/implementing-circular-economy-nutrient-recovery-urban-wastewater-treatment-plants_en

Integrated Pollution Prevention and Control Reference Document on Best Available Techniques in the Slaughterhouses and Animal By-products Industries (2005).

Johns, M. R., Harrison, M. L., Hutchinson, P. H., & Beswick, P. (1995). Sources of nutrients in wastewater from integrated cattle slaughterhouses. *Water Science and Technology, 32*(12), 53–58. https://doi.org/10.1016/0273-1223(96)00138-2

Le-Clech, P. (2010). Membrane bioreactors and their uses in wastewater treatments. *Applied Microbiology and Biotechnology, 88*(6), 1253–1260. https://doi.org/10.1007/S00253-010-2885-8/FIGURES/1

López-Maldonado, E.A., & Oropeza-Guzmán, M. T. (2019). Chapter 8: Evaluating new biopolyelectrolytes for the meat processing wastewater treatment via coagulation-flocculation. In E. Vollan (Ed.), *Flocculation: Processes and applications* (pp. 187–196). Nova Science Publishers, Inc.

Manjunath, N. T., Mehrotra, I., & Mathur, R. P. (2000). Treatment of wastewater from slaughterhouse by DAF-UASB system. *Water Research, 34*(6), 1930–1936. https://doi.org/10.1016/S0043-1354(99)00337-1

Masse, D. I., & Masse, L. (2000). Characterization of wastewater from hog slaughterhouses in eastern Canada and evaluation of their in-plant wastewater treatment systems. *Canadian Agricultural Engineering, 42.*

Matheyarasu, R., Bolan, N. S., & Naidu, R. (2016). Abattoir wastewater irrigation increases the availability of nutrients and influences on plant growth and development. *Water, Air, and Soil Pollution, 227*(8), 1–16. https://doi.org/10.1007/S11270-016-2947-3/TABLES/6

Mittal, G. S. (2006). Treatment of wastewater from abattoirs before land application—A review. *Bioresource Technology, 97*(9), 1119–1135. https://doi.org/10.1016/J.BIORTECH.2004.11.021

Miklos, D. B., Remy, C., Jekel, M., Linden, K. G., Drewes, J. E., & Hübner, U. (2018). Evaluation of advanced oxidation processes for water and wastewater treatment—A critical review. *Water Research, 139*, 118–131. https://doi.org/10.1016/J.WATRES.2018.03.042

Mofijur, M., Fattah, I. M. R., Kumar, P. S., Siddiki, S. Y. A., Rahman, S. M. A., Ahmed, S. F., Ong, H. C., Lam, S. S., Badruddin, I. A., Khan, T. M. Y., & Mahlia, T. M. I. (2021). Bioenergy recovery potential through the treatment of the meat processing industry waste in Australia. *Journal of Environmental Chemical Engineering, 9*(4), Article 105657. https://doi.org/10.1016/J.JECE.2021.105657

Morais, H. B., Chardulo, L. A. L., Baldassini, W. A., Lippi, I. C. de C., Orsi, G. B., & Ruviaro, C. F. (2023). Environmental impacts of high-quality Brazilian beef production: A comparative life cycle assessment of premium and super-premium beef. *Animals, 13*(22), 3578. https://doi.org/10.3390/ANI13223578/S1

Nastasijevic, I., Boskovic, M., & Glisic, M. (2022). Abattoir hygiene. *Present knowledge in food safety: A risk-based approach through the food chain* (pp. 412–438). https://doi.org/10.1016/B978-0-12-819470-6.00002-0

Ng, M., Dalhatou, S., Wilson, J., Kamdem, B. P., Temitope, M. B., Paumo, H. K., Djelal, H., Assadi, A. A., Nguyen-tri, P., & Kane, A. (2022). Characterization of slaughterhouse wastewater and development of treatment techniques: A review. *Processes, 10*(7), 1300. https://doi.org/10.3390/PR10071300

Obi, C. C., Nwabanne, J. T., Igwegbe, C. A., Ohale, P. E., & Okpala, C. O. R. (2022). Multi-characteristic optimization and modeling analysis of electrocoagulation treatment of abattoir wastewater using iron electrode pairs. *Journal of Water Process Engineering, 49*, Article 103136. https://doi.org/10.1016/J.JWPE.2022.103136

Okey-Onyesolu, C. F., Chukwuma, E. C., Okoye, C. C., & Umobi, C. O. (2023). Application of fish bone chitosan-protein bio-coagulant for abattoir wastewater treatment: Comparative process optimization and evaluation. *Waste Management Bulletin, 1*(1), 49–59. https://doi.org/10.1016/J.WMB.2023.03.003

Rajpal, A., Ali, M., Choudhury, M., Almohana, A. I., Alali, A. F., Munshi, F. M. A., Khursheed, A., & Kazmi, A. A. (2022). Abattoir wastewater treatment plants in India: Understanding and performance evaluation. *Frontiers in Environmental Science, 10*, Article 881623. https://doi.org/10.3389/FENVS.2022.881623/BIBTEX

Roupas, P., De Silva, K., Smithers, G. W., & Ferguson, A. (2007). Waste management and co-product recovery in red and white meat processing. *Handbook of Waste Management and Co-Product Recovery in Food Processing, 1*, 305–331. https://doi.org/10.1533/9781845692520.4.305

Ruiz, I., Veiga, M. C., De Santiago, P., & Blázquez, R. (1997). Treatment of slaughterhouse wastewater in a UASB reactor and an anaerobic filter. *Bioresource Technology, 60*(3), 251–258. https://doi.org/10.1016/S0960-8524(97)00020-5

Schmideder, S., Thurin, L., Kaur, G., & Briesen, H. (2022). Inline imaging reveals evolution of the size distribution and the concentration of microbubbles in dissolved air flotation. *Water Research, 224*, Article 119027. https://doi.org/10.1016/J.WATRES.2022.119027

Shengquan, Y., Siyuan, G., & Hui, W. (2008). High effective to remove nitrogen process in abattoir wastewater treatment. *Desalination, 222*(1–3), 146–150. https://doi.org/10.1016/J.DESAL.2007.01.140

Saliu, T. D., & Oladoja, N. A. (2021). Nutrient recovery from wastewater and reuse in agriculture: A review. *Environmental Chemistry Letters, 19*(3), 2299–2316. https://doi.org/10.1007/S10311-020-01159-7

Swan, K. H., Surinkul, N., & Visvanathan, C. (2023). Investigation of decentralized wastewater treatment system in urban wastewater management: Case study in Yangon, Myanmar. *Sustainability, 15*(24), 16756. https://doi.org/10.3390/SU152416756

Tahreen, A., Jami, M. S., & Ali, F. (2020). Role of electrocoagulation in wastewater treatment: A developmental review. *Journal of Water Process Engineering, 37*, Article 101440. https://doi.org/10.1016/J.JWPE.2020.101440

Transforming Wastewater Treatment with IoT Technologies. (n.d.). Retrieved March 1, 2024, from https://www.teamtweaks.com/blog/wastewater-treatment/

Yildirim, N. C., Tanyol, M., Serdar, O., & Yildirim, N. (2019). Gammarus pulex as a model organism to assess the residual toxicity of slaughterhouse wastewater treated by electrocoagulation process. *Bulletin of Environmental Contamination and Toxicology, 103*(3), 447–452. https://doi.org/10.1007/S00128-019-02666-2

Wainaina, S., Awasthi, M. K., Sarsaiya, S., Chen, H., Singh, E., Kumar, A., Ravindran, B., Awasthi, S. K., Liu, T., Duan, Y., Kumar, S., Zhang, Z., & Taherzadeh, M. J. (2020). Resource recovery and circular economy from organic solid waste using aerobic and anaerobic digestion technologies. *Bioresource Technology, 301*, Article 122778. https://doi.org/10.1016/J.BIORTECH.2020.122778

Vymazal, J. (2010). Constructed wetlands for wastewater treatment. *Water*, *2*(3), 530–549. https://doi.org/10.3390/W2030530

Management of Wastewater from Tannery and Leather Processing Industries

Eric T. Anthony, Olayinka S. Adanlawo, and Nurudeen A. Oladoja

Abstract Wastewater effluent from tannery and leather processing industries is characterized by significant amounts of salts, dyes, chemicals, and solids, thereby raising environmental concerns. The environmental challenges created by these wastewaters are more exacerbated in the rural areas due to the limited wastewater treatment infrastructures. In treating wastewater from tannery and leather industries in rural areas, the identified treatment technologies include adsorption, coagulation, and membrane filtration. Other technologies that have been investigated but may not be applicable to the rural settings include heterogeneous photocatalysis and biological treatment. Integrated technologies have also been studied as they combine the use of multiple technologies to achieve the treatment.

Keywords Tannery wastewater · Leather wastewater · Wastewater management · Adsorption · Coagulation · Membrane filtration · Heterogenous catalysis

1 Introduction

The constant and indiscriminate disposal of industrial, domestic and nuclear waste, septic/sewage leakages and emission of particulate matter close to water sources has exacerbated the impact of wastewater on the ecosystem (Karri et al., 2021). Thus, the burden exerted by the wastewater remains a significant barrier to ecological balance. This barrier is even more significant in rural communities, due to the rapid development of rural industries and poor wastewater treatment infrastructure in rural Centres (Deng & Wheatley, 2016; Pan et al., 2007). While household-generated

E. T. Anthony · O. S. Adanlawo · N. A. Oladoja (✉)
Hydrochemistry Research Laboratory, Department of Chemical Sciences, Adekunle Ajasin University, Akungba-Akoko, Nigeria
e-mail: bioladoja@yahoo.com; nurudeen.oladoja@aaua.edu.ng

E. T. Anthony
Department of Chemistry and Biochemistry, Faculty of Mathematics and Natural Sciences, Institute for Inorganic Chemistry, University of Cologne, Cologne, Germany

N. A. Oladoja and E. I. Unuabonah (eds.), *Strategic Management of Wastewater from Intensive Rural Industries*, Springer Water,
https://doi.org/10.1007/978-3-031-90314-4_6

wastewater (grey and black water) is the most common source in rural areas (Cheng et al., 2021), wastewater from rural industries presents the most significant ecological danger (Afrad et al., 2020).

Due to the large amount of wastewater generated, the tannery and leather processing industries are typical industries that pose a significant threat to the environment (Sivaram & Barik, 2019). The unit processes in tanning include soaking, degreasing, unhairing and liming of animal hide, and deliming, softening, pickling, tanning, neutralization, re-tanning, dyeing and fat-liquoring (Zhao & Chen, 2019). The aqueous waste from the tannery and leather processing industries accounts for about 90% of the pollution from these industries (Chidambaram et al., 2012; Appiah-Brempong et al., 2022; Ghorab et al., 2022). The effluent from the tannery and leather processing industry is rich in heavy metals, salts, phenols, chromophores and particulates (Chowdhury et al., 2015). Thus, treating the effluent before discharging into water bodies is important. The different methods, which can be fitted into either centralized or decentralized wastewater treatment approach (Oladoja, 2017; Pasciucco et al., 2022), have been employed in the treatment of wastewaters from these industries. These methods include flotation, electrochemical treatment, sedimentation, coagulation, filtration and advanced oxidation processes.

Considering the process efficiency, high-technology wastewater treatment plants are more attractive for the treatment of wastewater from these industries. For example, the Advanced Oxidation Processes (AOPs) have been found to be effective in the management of tannery wastewater (Sivagami et al., 2018; Vijayalakshmi et al., 2011), and for the promotion of sustainable development (Zhao & Chen, 2019). For instance, the electro-fenton process has been reported to consume less energy and is capable of reducing the Chemical Oxygen Demand (COD) of the wastewater from 41 to 99% (Kurt et al., 2007). The snag with the high-end wastewater treatment options is that they present a high operational and maintenance cost (Garcia & Bécares, 1997). For example, the performance of an electro-oxidation system depends on the type of anodic material, which contributes to operational cost (Suman et al., 2021).

Rural industries are vital for economic development (Suman et al., 2021), but establishing industries in rural areas has led to a significant increase in rural water pollution (Wang et al., 2008). Therefore, indigenous methods of water and wastewater treatment are vital in combating pollution. Adsorption (Chaudhary et al., 2022), coagulation (Oladoja et al., 2017), membrane filtration (Arhin et al., 2019) and constructed wetland (Hendy et al., 2023) are robust water and wastewater treatment options, and have been reported to be effective in eliminating the major and minor constituent of tannery and leather wastewater (Aber et al., 2010; Degache & Hans, 2006; Moges et al., 2022; Nur-E-Alam et al., 2020; Zapana et al., 2020). This chapter is thus, focused on the methods employed in the management and treatment of wastewater effluent from tannery and leather processing industries situated in rural areas.

2 Wastewater Sources in Rural Areas

Wastewater in rural areas originates from point and nonpoint sources. The point source water pollution refers to water contamination that can be linked to certain, identified discharge locations. These sources include industrial discharges, wastewater treatment plants, sewer overflows, oil and chemical spills, landfill leachate etc. The non-point source water pollution refers to the contamination of water bodies by dispersed sources, as opposed to discrete, recognizable discharge locations (Jamwal et al., 2011; Yadav & Pandey, 2017). These sources include agricultural runoff, mining activities, atmospheric deposition etc. The source of rural wastewater can also be categorized into: domestic wastewater; Agricultural wastewater, and Industrial wastewater.

3 Wastewater Generation from Tannery and Leather Processing

The raw materials used in leather industries are obtained from goat and sheep skins and hides from cows and buffalo, and the animal skins are converted to end-use via chemical treatments. The chemicals include phenolic compounds, surfactants, sulfonated oil, dyes, biocides, acrylic resins, organic acids and salts (Degache & Hans, 2006). Thus, the effluent from tannery is a complex mixture. The effluent is characterized by high chemical/biological oxygen demand, total dissolved solids, total suspended solids, Cr^{3+} and phenolics with high pH, strong odour and dark brown. The composition of tannery effluent is dictated by the source of the hides (Durai & Rajasimman, 2011) and the tanning conditions (Zhou et al., 2012).

The introduction of chemicals occurs initially, during the curing of animal skin, and salts such as NaCl (3 kg per square meter area) are added since they are required for the preservation of the quality of the animal skin before processing. The first wastewater is generated from the soaking of the cured animal skin in water, to remove salt and other impurities (blood, dirt, dung, and proteins) (Tünay et al., 1995). Calcium hydroxide, sodium sulfide, or a mixture of both are added to dehair the animal skin and remove the epidermal and fleshy layer from the skin. The second wastewater is generated during deliming, by washing and soaking the hide and skin in freshwater or acidic solution, boric acid, or a salt (ammonium chloride or ammonium sulfate) (Nazer et al., 2006) (Fig. 1). Proteolytic enzymes (biodates, pancolin, pancreol or orapan) are added to the hide and skin to breakdown degraded-proteins (Hasan et al., 2022). The bate skins and hides are treated with sulfuric acid and common salt (pickling), to preserve the stock for tanning process (Sutherland, 1947). The bating and pickling process generates additional wastewater. The stock is then tanned using the bark of trees (Das et al., 2022) e.g., avaram, konnam, babul and wattle bark (vegetable tanning), chromic acid solution (chrome tanning) or alum solution (Hansen et al., 2020). The chemical composition of wastewater generated during tanning include chromium, salts, sulfate, aluminium and carbonates. The final wastewater is generated during dyeing to impart colour, and fat liquoring, to improve leather softness, flexibility, and finishing (Lofrano et al., 2013).

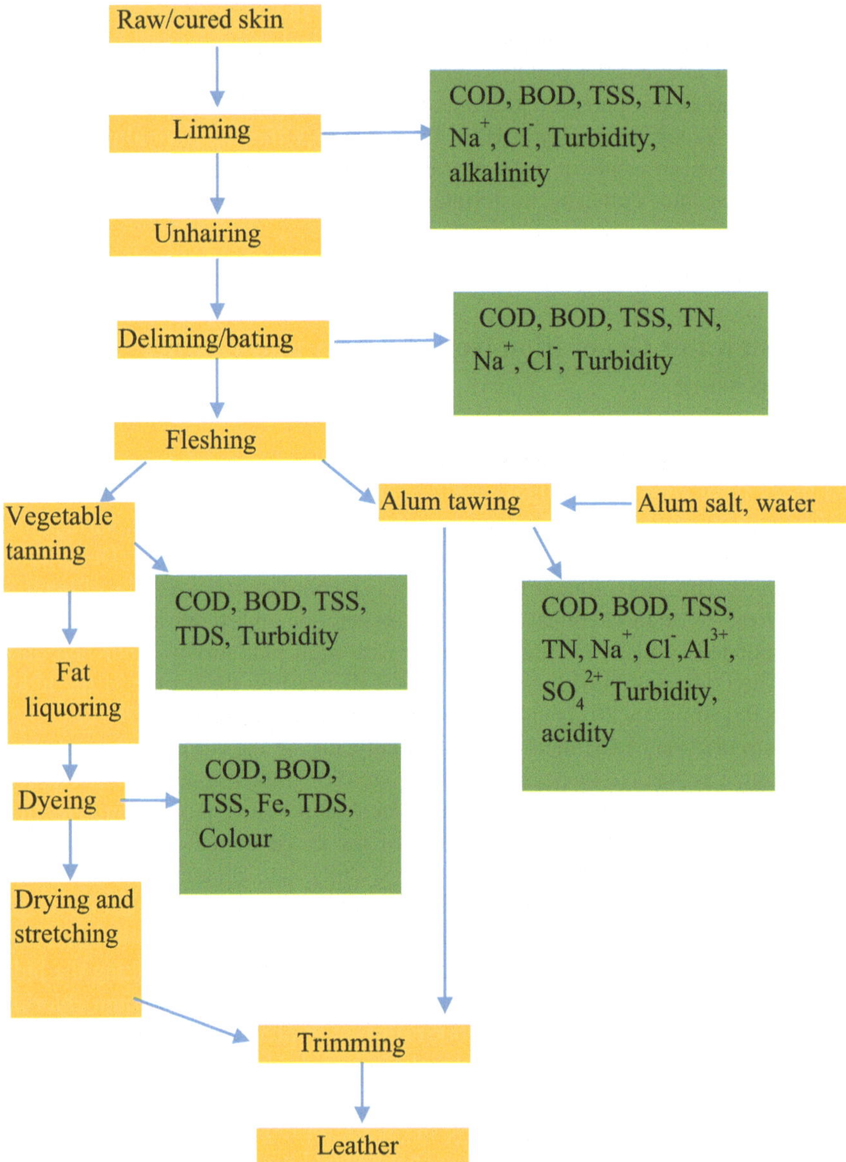

Fig. 1 Unit processes involved in a leather processing industry. Adopted from Appiah-Brempong et al. (2022)

4 Composition of Wastewater from Tannery and Leather Processing

The composition of wastewater from the tannery and leather industry is important for the design of wastewater treatment plants. Among these compositions, COD and Biochemical Oxygen Demand (BOD) are the most important parameters to consider (Table 1) (Cotman et al., 2004; Leta et al., 2004). Tannery wastewaters are characterized by high COD and BOD, due to the high amount of organic compounds and dichromate (Aboulhassan et al., 2018) used during hide processing. The impact of COD and BOD of tannery wastewater is crucial due to the presence of low biodegradable recalcitrant organic compounds and metal ions susceptible to oxidation e.g., $Cr_2O_7^{2-}$ (AY & Ahmed, 2016). $Al_2(SO_4)_3$ and $FeCl_3$ [45], and MgO (Oladipo et al., 2017) have been employed as coagulants in reducing the amount of the COD and BOD in tannery wastewater.

Chromium salts are used extensively in the tanning and leather industries. Thus, chromium ions are important constituents of tannery wastewater. $Cr_2O_7^{2-}$ contributes significantly to the level of COD (Sungur & Özkan, 2017) and Cr^{3+} is the source of the chrome. Traditionally, hexavalent chromium ions species contained in the wastewater are removed via the reduction of $Cr_2O_7^{2-}$ to Cr^{3+} (Liu et al., 2016), followed by precipitation of Cr^{3+} (Fahim et al., 2006). Microbial degradation and adsorption have been investigated as sustainable methods for the removal of $Cr_2O_7^{2-}$ and Cr^{3+} (Balaji et al., 2016; Durairaj, 2016; Vankar & Bajpai, 2008) from the wastewater. Constructed wetlands have also been found to be efficient in removing $Cr_2O_7^{2-}$ (Makos & Hrncir, 1995), however, the effectiveness of the technology greatly depends on the wetland plants and growth (Younas et al., 2022).

Sodium sulfide (Na_2S) is also an important chemical that is used in the leather industries, independently or with hydrated lime ($Ca(OH)_2$), as depilation agent. However, $Ca(OH)_2$ is also used solely as an de-hairing agent. The high amount of nitrogen in tannery wastewater is due to the addition of ammonium salt, which is used in removing Ca^{2+} ions. The use of Na_2S and $Ca(OH)_2$ contribute 84%, 75% and 92% to the BOD, COD and suspended solids, respectively (Sawalha et al., 2020).

Total suspended solids (TSS) are undissolved dry matter and particulates, contributing to the organic load and temperature of tannery wastewater. Suspended solids are removed by size exclusion or coagulation. *Moringa oleifera* been employed as a biogenic coagulant to remove suspended solids from tannery wastewater (Marichamy et al., 2021). Excluding TSS could help improve the turbidity of wastewater.

Table 1 Composition of effluent from tannery industries

Total chromium (mg/L)	Ammonia-N (mg/L)	Total suspended solid (mg/L)	Total solid (mg/L)	COD (mg/L)	BOD	Chloride (mg/L)	Conductivity (mScm^{-1})	pH	References
49.8				–		787	6.10	4.0	Costa et al. (2008)
		1500	29,000	5000	1500			9.0	Song et al. (2000)
		976					9.60	6.9	Munz et al. (2008)
62	130	1520		2810	910	6400	19,950	7.2	Kurt et al. (2007)
		4011.60	26,859.80	2825.90	1911.20		2045.50	6.95	Oladipo et al. (2017)
100		1500	29,000	5000	1500			9.0	Song et al. (2000)
21	460	1374	14,250	3350		72		8.1	Marichamy et al. (2021)
200		2100	3000	3800	5000			5.3	Mohammed and Sahu (2019)
		1398	1080	203			2190	8.2	Fouda et al. (2021c)
69.25			3195	5950	1900				Kamaraj et al. (2020)
382.4			19,946.10				35,303.20	3.7	Kaseva and Mbuligwe (2010)
		18,436		5503			9652	4.1	Yadav et al. (2021)
		6800	20,800	3980	920		18.65	7.5	Chowdhury et al. (2013)
6853		2384	85,340	16,231	8875	23,125	5600	12.8	Das et al. (2006b)
136.36	260	1550	6140	3280	463.25	-	12,250	9.04	Ayoub et al. (2011)
2007.08				5485	90			4.13	Reyes-Serrano et al. (2020)

(continued)

Table 1 (continued)

Total chromium (mg/L)	Ammonia-N (mg/L)	Total suspended solid (mg/L)	Total solid (mg/L)	COD (mg/L)	BOD	Chloride (mg/L)	Conductivity (mScm^{-1})	pH	References
5,200				23,350	1475		12		GilPavas et al. (2011)
		20,230	57,400	25,000			9.5	4	Borchate et al. (2012)
29				5280		3438		3.8	Rezgui et al. (2022)

5 Treating Tannery and Leather Wastewater in Rural Area

Rural area constitutes almost half the world's population and in developing nations, this accounts for about 70% (Dasgupta et al., 2015). Rural areas are characterized by the abundance of natural capital with limited infrastructure (Wiggins & Protor, 2001), leading to indiscriminate discharge of wastewater (Sawalha et al., 2019). Therefore, decentralized water and wastewater treatment has been suggested as an ideal method for treating aqueous effluents in rural areas. Some of the methods includes coagulation and flocculation, adsorption, advanced oxidation process, membrane separation, biological treatment and constructed wetland.

5.1 Adsorption

Among the different treatment methods for the remediation of tannery wastewater, adsorption has proven effective. Adsorption is distinguished by its cost-effectiveness, ease of use, and high pollutant removal efficiency, making it a preferred choice for wastewater treatment (Hansen et al., 2021; Nur-E-Alam et al., 2020). Typically, cattle hair and shavings from chrome-tanned leather serve as the most common adsorbents in post-tanning wastewater treatment (Mella et al., 2017, 2019; Tsamo et al., 2023). However, various other materials, including zeolite, green macro-algae, modified kaolin, and commercial activated carbon, have also found applications (Dabai & Mohammed, 2020; Devi et al., 2023; Hevira et al., 2023; Saira & Shanthakumar, 2023).

Simple metal oxides such as MgO, CaO, Al_2O_3, and SiO_2 are among the most studied adsorbents due to their abundance and their low levels of toxicity. In treating tannery wastewater, MgO nanoparticles achieved a removal efficiency of 99%, 97%, 71%, 69% and 21% for Fe^{2+}, Ni^{2+}, TOC, COD and sulfate ion, respectively (Masindi et al., 2023). The underlying removal mechanisms for treating tannery wastewater were attributed to the complexation and dissolution character of MgO nanoparticles at alkaline pH. Increasing pH towards the alkaline region, results in the deprotonation of the adsorbent active sites (Nur-E-Alam et al., 2018), thus, increasing adsorption efficiency (Masindi et al., 2023). Acidic conditions is also beneficial for the removal of hexavalent chromium from tannery wastewater. The removal of hexavalent chromium using hydrogen-bond rich alginate from tannery wastewater was 30% at neutral pH, 50% at pH 5 and 95% at pH 2 (Kubra et al., 2024). The higher removal efficiency at acidic pH is attributed to the electrostatic attraction between positively charged adsorbent (alginate) and negatively charged chromate ($HCrO_4^-$) (Kubra et al., 2024). Activated carbon also relies on the electrostatic force of attraction at acidic pH in removing negatively charged chromate ions from tannery wastewater (Palani et al., 2017).

SiO_2 synthesized by calcination of rice husk, followed by acidification, achieved an adsorption capacity of 185 mg/g for the removal of Cr^{3+} ions from tannery wastewater (Amirhandeh et al., 2023). The high adsorption efficiency at acidic pH (pH = 2) was attributed to the diffusion of Cr^{3+} into the pore of silica nanoparticles. Mixed oxides and double layered Oxides/Hydroxides (LDO/LDH) have also shown tremendous potentials in removing chrome, dye, COD and suspended solids from tannery wastewater (Ma et al., 2022). LDO exhibits high charge density caused by mixed positively-charged metal ions, which can attract negatively charged undesirable chemicals in tannery wastewater (Ma et al., 2022). Similarly, the high removal efficiency in LDH is due to their anionic exchange capacity and complexing characteristics (De Martino et al., 2013).

Bare Fe_3O_4 (Li et al., 2021) and modified Fe_3O_4 (Moges et al., 2022) (Farahat et al., 2021) have been used for removing chromium ions from tannery wastewater. In this process, initial solution pH is one of the most influential parameters in removing Cr^{3+} ions from tannery wastewater, as alkaline pH favoured the removal of Cr^{3+} ions in the tannery wastewater (Fig. 2). Ferrites are another class of magnetic materials that have gained attention, due to sustainability concerns. $CoFe_2O_4$ was used in removing chromium ions and reducing the concentration of TDS, BOD and COD from tannery wastewater (Albalah et al., 2020). Activated carbon modified with $CoFe_2O_4$ via impregnation method has also been employed in removing hexavalent chromate ions and Pb^{2+} from tannery wastewater (Yahya et al., 2020b).

The functional groups on the surface of an adsorbent also play an important role in treating tannery wastewater, especially, in removing charged contaminants. Therefore, traditional adsorbents have been functionalized with organic molecules (e.g. chelates) for removing charged contaminants (e.g., Cr^{3+} and $Cr_2O_7^{2-}$)) in tannery wastewater (Hamza et al., 2022). Acetate and imidazole functionalized chitosan/Fe_3O_4 nanoparticles were employed in removing $Cr_2O_7^{2-}$) from tannery wastewater. The imidazole functionalized chitosan/Fe_3O_4 nanoparticles were selective for $Cr_2O_7^{2-}$ compared to the acetate functionalized chitosan/Fe_3O_4 nanoparticles [94].

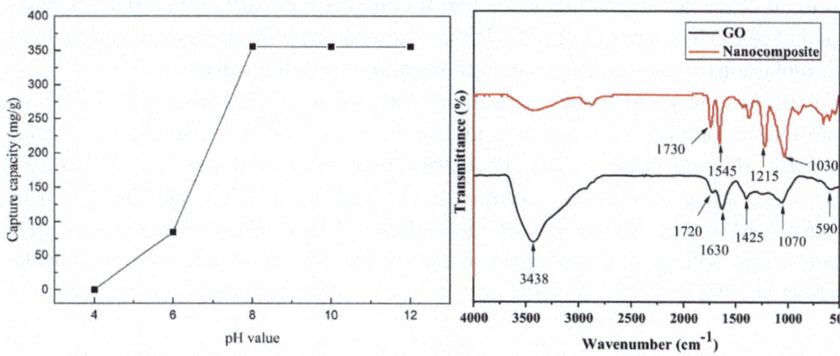

Fig. 2 Effect of pH on the removal of Cr^{3+} from tannery wastewater and the FTIR of pristine GO (Gaber et al., 2017; Roy et al., 2023a, 2023b)

The authors suggested that the removal of $Cr_2O_7^{2-}$ was attributed to its attraction to protonated amine (Hamza et al., 2022). The adsorption efficiency for the removal of Cr^{3+} were also dependent on the SiO_2 surface functional groups. Carboxylate functionalized SiO_2 were preferred for removing Cr^{3+} in tannery wastewater compared with cyano-functionalized SiO_2 since the former is hard acid (Cr^{3+}) vs hard base (OH^-), while the latter (CN^-) is a soft base (Gervas et al., 2016).

Graphene oxides contain hydroxyl, carbonyl, and carboxylic functional groups (Fig. 2) (Gaber et al., 2017). These functional groups contribute to their high adsorption capacity for removal of Cr^{3+}, COD, TDS, chloride and sulfide from tannery wastewater (Pal et al., 2020). Blending graphene oxide with nanoparticles can also be employed to improve the adsorption capacity of the resulting adsorbent in treating tannery wastewater. The nanoparticle serves as core or support (Roy et al., 2023a), while the graphene oxides serves as the binding /adsorption sites for removing tannery wastewater contaminants (Roy et al., 2023b) (Fig. 2). Clays (Espantaleón et al., 2003) are also sustainable materials used in the treatment of tannery wastewater. Due to ionic interactions, negatively charged bentonite have been used in the recovery of Cr^{3+} ions from tannery wastewater (Tahir & Naseem, 2007). Kaolin (Mustapha et al., 2019, 2020) and montmorillonite (Tao et al., 2019; Hao et al., 2022) have also been used in treating tannery wastewater.

Another sustainability approach to treat tannery and leather wastewaters involves the use of solid waste, as adsorbent. Tannery sludge, converted to biochar (Hashem et al., 2022; Li et al., 2024) was used in removing ($Cr_2O_7^{2-}$) from tannery wastewater. The FTIR analysis of the biochar revealed the presence of OH^- functional groups, chiefly responsible for removing $Cr_2O_7^{2-}$ from the tannery wastewater (Li et al., 2024), since both ions are hard acid and hard base, respectively. The removal efficiency has also attributed to biochar's high specific surface area (Hashem et al., 2022). Biochar prepared from surface water pollutants (Water hyacinth) has been employed in removing Cr^{3+}, chloride, BOD and COD from tannery wastewater (Hashem et al., 2020). Ashes from various sources have also been used in treating wastewater from tannery and leather industries (El mouhri et al. 2021; Elkarrach et al., 2023). The most important characteristic of the ash is that it contains both soft and hard acids/bases (Fauzi et al., 2016; Rees et al., 2007) that showed different degrees of interactions with metal ions (Pearson, 1963) and other pollutants in the wastewater.

Treating tannery and leather wastewater using waste-derived adsorbents is gaining a lot of attention (Table 2), and agricultural waste has shown tremendous potential. Tea waste (Nigam et al., 2019) and *Carica papaya* (Hashem et al., 2024b) trees have been used as adsorbents in removing chromium ions (Cr^{3+} and $Cr_2O_7^{2-}$) from tannery wastewater. Activated carbon synthesized from *Rumex Abyssinicus* were characterized with high surface area was used in decolorization of tannery wastewater (Nure et al., 2023).

Table 2 Overview of the treatment of tannery wastewater via adsorption technology

Adsorbent	Initial concentration	Contaminant	Adsorption capacity (mg/g)	Adsorbent dosage (mg)	Solution pH	References
Egg shell	3210	Cr^{3+}	200.25	20,000	3.81	Elabbas et al. (2016a)
Fly ash	1000.3	Cr^{3+}	20.32	5000	2.5	Rahaman et al. (2016)
Fly ash	1291	Cr^{3+}	16.32	15,000	2.5	Rahaman et al. (2016)
Fly ash	1524	Cr^{3+}	16.5	5000	2.5	Rahaman et al. (2016)
Avocado kernel seeds	5	Cr^{6+}	10.08	500	1	Mekonnen et al. (2015)
Sunflower	100	Cr^{3+}	7.85	4000	2	Jain et al. (2013)
Cucurbita moschata (pumpkin) peel	3178.6	Cr^{3+}	3164.46	2000	7.2	Hashem et al. (2024a)
Bio-char	5.17	Cr^{6+}	0.6593	500	2	Katenta et al. (2020)
Almond shell	67.5	Cr^{3+}/Cu^{2+}	21.92	16,100	3.7	Yahya et al. (2020a)
Juniperus procera sawdust	5	Cr^{6+}	16.06	500	1	Mekonnen et al. (2015)
Rice husk	1414	Cr^{3+}	95.23	5000	4	Sa ad et al. (2022)
Papaya peels	5	Cr^{6+}	7.16	500	1	Mekonnen et al. (2015)
Leather waste activated carbon (LIW)	150	Acid-red (dye)	14.40	250	2.5	Carpenter et al. (2013)
Activated carbon	600	Cr^{3+}/Cr^{6+}	202	30	–	Jimenez-Paz et al. (2023)
Powdered marble	3210	Cr^{3+}	434.82	12,000	3.81	Elabbas et al. (2016a)
Fleshing waste activated carbon	100	Cr^{3+}	60.8	500	3	Palani et al. (2017)
Spent tea leaves	616.8	Cr^{3+}	10.64	1400	10	Nur-E-Alam et al. (2018)

5.2 Coagulation/Flocculation

Coagulation and Flocculation have been used in pre-treatment and post-treatment of tannery wastewater (Ranjan et al., 2020). Common electrolytes and hydro/oxides (e.g., $MgCl_2.7H_2O$, $Al_2(SO_4)_3$, $FeCl_3$, CaO and $Ca(OH)_2$) have been investigated for the removal of COD (Amin, 2015; Garrote et al., 1995; Reyes-Serrano et al., 2020), chromium and TSS from tannery wastewater. $FeCl_3$ has shown to be suitable in treating tannery effluent from PICUSA tanning plant in Spain, producing colourless and odourless water (Garrote et al., 1995).

$FeCl_3$ has also been shown to be a better coagulants in the removal of COD and chromium concentration than $Al_2(SO_4)_3$ from tannery wastewater (Song et al., 2004). The wastewater is characterized by a pH value of 9.2, COD 3300 mg/L, 15,000 mg/L TS, 260 mg/L SS, and 16.8 mg/L chromium. The removal efficiency is significant at low initial chromium concentration (Song et al., 2004). The addition of the coagulants alters the initial pH of the solution because the coagulants are Lewis acids (Fe^{3+} and Al^{3+}), which react with Lewis base (OH^-) to precipitate ($Fe(OH)_3$ and $Al(OH)_3$. The addition of $FeCl_3$ results in a lower solution pH value (5.72) as compared to the pH value of (6.57) when $Al_2(SO_4)_3$ is added (Song et al., 2004). Therefore, $FeCl_3$ was more effective in removing COD from the tannery wastewater. In removing chromium ions, alkaline pH was more beneficial due to competition between H^+ and Cr^{3+} under acid conditions (Song et al., 2004).

A combination of alum and ferric chloride has shown to be more effective in removing COD, BOD, suspended solids, total dissolved solids and conductivity from tannery wastewater (Barman et al., 2017). Some of the challenges associated with coagulation include slow settling of sludge and high sludge volume. MgO is a superior coagulant to NaOH, $NaHCO_3$, Na_2CO_3 and CaO with respect to sludge volume, while CaO is a superior coagulant to NaOH, MgO, $NaHCO_3$ and Na_2CO_3 with respect to sludge purity and Cr^{3+} recovery (Guo et al., 2006). Therefore, a more effective coagulant was obtained by blending CaO and MgO in the ratio 4:1. The CaO/MgO coagulant was effective in removing Cr^{3+}, COD, Suspended Solid (SS), and colour at lower sludge volume and higher sludge purity (Guo et al., 2006).

Polyaluminum chloride (PAC) and $AlCl_3$ are among the most studied coagulants, and have been employed for the removal of BOD, COD and TDS in wastewater (Sabur, 2012; Solanki et al., 2013). PAC is more advantageous because the flocs settle faster, are larger, and thus, are more stable and also generate lower volume of sludge (Fig. 3) (Solanki et al., 2013) Other coagulants that have been used in the treatment of tannery and leather wastewater includes polyferric sulfate and polyaluminum ferric chloride (Song et al., 2004; Tianzhi et al., 2021). Polyaluminum ferric chloride outperformed the conventional coagulants (ferrous chloride and aluminium sulfate) in removing pollutants from leather tanning wastewater effluents (Lofrano et al., 2018).

Coagulant aids have been used in support of coagulants during the treatment of tannery wastewater. Coagulant aid is added to improve the settling characteristics0 of flocs during the coagulation process to produce bigger flocs. Coagulant aid could vary

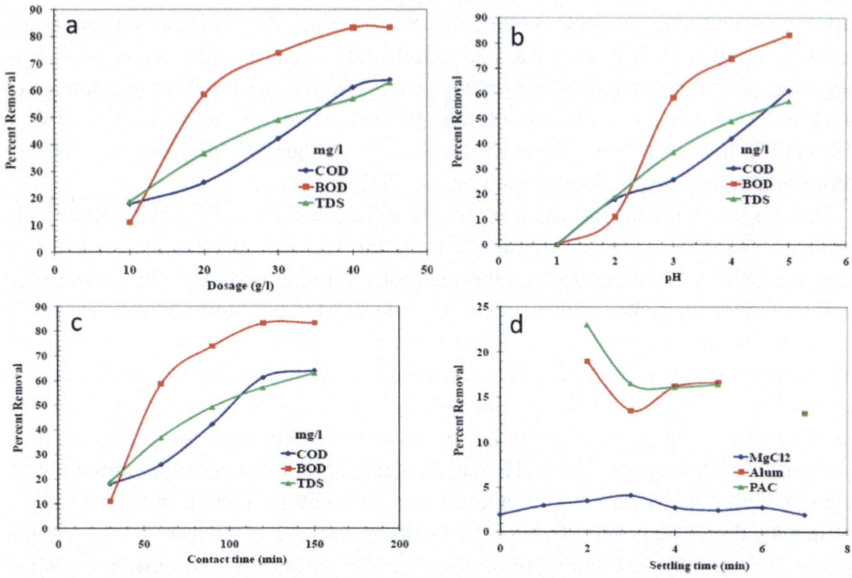

Fig. 3 Removal of COD, BOD and TDS as a function of PAC: dosage (**a**); pH (**b**); contact time (**c**). Sludge settling time of PAC, alum and $MgCl_2$ (**d**) (Elabbas et al., 2016a, 2016b)

from organic, inorganic, polymer, electrolytes and to plant-based (Brandt et al., 2017). Polyaluminium chloride is used as a coagulant and polyelectrolyte as a coagulant aid for removing COD, TS, TSS, TDS and colour from vegetable tannery wastewater (Borchate et al., 2012). Different inorganic coagulants ($Al_2(SO_4)_3$, $FeSO_4$, $FeCl_3$ and poly aluminium chloride, and inorganic coagulant aids (CaO, $CaCO_3$, Na_2SiO_3, and Na_2CO_3) were compared for treating tannery wastewater (Aber et al., 2010). $FeCl_3$ was the most effective coagulant in removing Cr^{3+} at pH between 7.5 and 9 and CaO was the most suitable coagulant aid. The coagulant aid (CaO) have a significant impact on the solution pH by precipitating $Cr(OH)_3$, thus, producing stronger and stable flocs at a reduce coagulant dosage (Aber et al., 2010).

Biogenic-based coagulants e.g., plants and animal extracts and agricultural waste have also been used as coagulants in treating tannery and leather wastewater. Biogenic coagulants are used because of their environmental friendliness e.g., *Moringa oleifera* and *Cicer aretinum* (Kazi & Virupakshi, 2013), algae (Paul et al., 2018) and *Strychnos potatorum, Azadirachta indica* (Sankaranarayannan & Gowdam, 2020). In a study, *Moringa oleifera* seeds were used in treating tannery wastewater effluent (Kumar et al., 2016). The coagulation process follows the charge neutralization mechanism at low coagulant dosage and sweep process at high coagulant dosage (Kumar et al., 2016). In some cases, biogenic-based coagulants have shown more efficiency in treating tannery and leather wastewater than the synthetic inorganic coagulants (Dotto et al., 2019). Egg shell, as coagulant, was more efficient in removing turbidity, as compared with alum, and produced sludge with better settling time, when compared with Ferric acid/alum coagulant (Precious Sibiya et al., 2021). However, biogenic-based coagulants such as cassava starch and orange peel powder were inferior to inorganic coagulant (Appiah-Brempong et al., 2023). The removal efficiency for COD and TSS were 37.25% and 73.95% for cassava starch, respectively; 17.97% and 66.08% for organ peel, respectively; and 38.51% and 76.09%, respectively for aluminium sulfate (Appiah-Brempong et al., 2023).

The coagulation and flocculation process is valued for its capacity to effectively eliminate a wide spectrum of organic and inorganic recalcitrant chemicals in solution (Yargeau, 2012). However, this method exhibits some drawbacks. Due to the poor quality of the tannery and leather processing wastewater, it requires a huge amount of coagulant and flocculant for effective floc formation (Zheng et al., 2014). Other disadvantages include the formation of large quantities of sludge and the inability to handle large-volume of wastewater for treatment (Wei et al., 2018; Nayeri & Mousavi, 2022), which results in sub-optimal pollutant removal and increased energy consumption (Vadasarukkai & Gagnon, 2017). Hence, the sludge produced during treatment often contains high concentrations of contaminants and may pose environmental risks if not managed and disposed of properly. Furthermore, the effectiveness of coagulation and flocculation is contingent on the characteristics of the wastewater, making them less effective for certain types of pollutants, such as dissolved contaminants or heavy metals (Turunen et al., 2019). These drawbacks can be overcome by optimizing the operational parameters e.g., pH, coagulant dosage, initial concentration, ionic strength, contact and settling time (Khayet et al., 2011).

5.3 *Electrocoagulation*

Electrocoagulation is a subdivision of the coagulation process that has recently gained prominence in the treatment of tannery wastewater. This is because the conventional coagulation and flocculation processes struggle with the handling of high concentrations of chemicals and toxic sludge (Barrera-Díaz et al., 2018). Electrocoagulation has emerged as an effective method for the removal of suspended solids, oil, grease, etc., (Cañizares et al., 2008). This technique consists of metal plates (anode and cathode), which are immersed in the wastewater. It relies on the dissolution of the electrode material, employed as an anode (often referred to as a sacrificial anode), which produces metal ions that function as coagulating agents within the tannery effluent (Deghles & Kurt, 2016). Aluminum (Zouboulis et al., 2008), iron or stainless steel (Apaydin et al., 2009) is the most commonly used metal electrodes because they are readily available, effective and inexpensive (Aber et al., 2010). Iron (Aboulhassan et al., 2018; Espinoza-Quiñones et al., 2009), aluminium (Benhadji et al., 2011), and steel (Feng et al., 2007; Varank et al., 2014) electrodes have been reported to be effective in removing Cr^{3+}, NH_3-N, TOC, and sulfide. Steel electrodes showed a high affinity for sulfide (Caliari et al., 2021; Moradi & Moussavi, 2019). The removal of Cr^{3+}, NH_4^+ and COD was negligible using titanium anode and stainless steel cathode even at high current intensity and different resident times

(Elsheikh, 2009). Pure aluminium and aluminium alloy (duralumin) electrodes were compared as electrocoagulants for removing COD and Cr^{3+} (Elabbas et al., 2016b). At lower current density (200 A/m^2), duralumin and aluminum Cr^{3+} removal efficiency was 93% and 75%, respectively. Duralumin electrodes also showed higher removal efficiency for COD compared to pure aluminium electrodes (Elabbas et al., 2016b).

The efficiency of electrocoagulation depends on the interplay of various factors viz: (i) the physicochemical properties of the water, such as pH, conductivity, total solids, and more; (ii) the magnitude of the applied current density; (iii) the properties of the electrodes, their separation distance, and their arrangement; and (iv) the hydrodynamic conditions (Kuokkanen et al., 2013). Other parameters that are peculiar to the performance of electrocoagulation includes electrode type, distance between electrodes, current density, stirring velocity and initial concentration (GilPavas et al., 2011). Optimization of these operation parameters, 99.76%, 64% and 61% removal efficiency for Cr^{3+}, TOC and COD was achieved, respectively (GilPavas et al., 2011). Mild steel and aluminium electrodes were used in the removal of COD, TOC, sulfide, colour and ammonia from tannery wastewater effluent (Feng et al., 2007). High cell current was beneficial for reducing the COD concentration (Fig. 4). Mild steel electrodes were more favourable for removing sulfide while aluminium electrodes were favourable for removing ammonia and TOC, respectively (Feng et al., 2007). Iron plate electrodes have also shown affinity for sulfides in tannery wastewater (Apaydin et al., 2009). Increasing electrolysis time has also been beneficial for removing chromium (Deghles & Kurt, 2016; Espinoza-Quiñones et al., 2009) and parallel monopolar configuration of electrodes was more suitable in reducing COD and chromium concentration compared to monopolar electrodes in series and dipolar electrodes in parallel (Kongjao et al., 2008).

One of the major advantages of electrocoagulation is that it combines electrochemical oxidation with coagulation. Therefore, in an electrocoagulation system, simultaneous oxidation of the organic content (Elabbas et al., 2016b) and reduction of the high oxidation state metal ions occur.

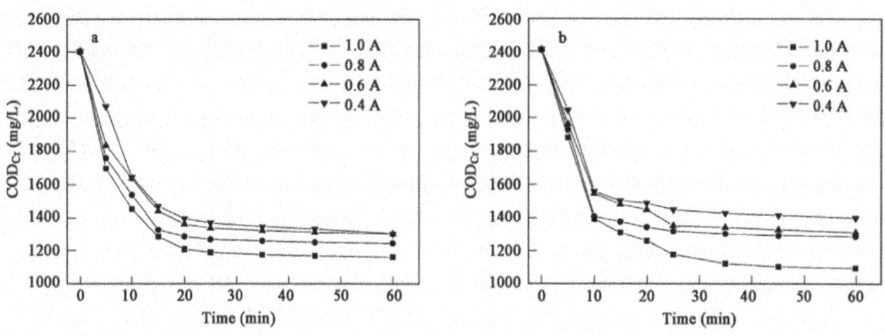

Fig. 4 Effect of cell current on COD removal using aluminium (**a**) and mild steel (**b**) electrodes (Kazi & Virupakshi, et al., 2013)

5.4 Membrane Separation Process

Membrane technologies (i.e., ultrafiltration, microfiltration, nanofiltration, reverse and forward osmosis, electrodialysis and membrane bioreactors) have received considerable attention in recent years, because of the continuous cost reductions and an expanding range of applications (Mohammed & Sahu, 2019). Membrane technology has made significant progress in treating wastewater effluent from the leather and tanning industry (Yang et al., 2023), particularly in chromium recovery (Mohammed & Sahu, 2019; Nur-E-Alam et al., 2020). Polyethersulfone (PES) is a polymeric compound that have been employed in the design of membranes. The performance of PES as an ultrafiltration membrane has been investigated based on its Molecular Weight Cut-offs (MWCO) in removing colour from tannery wastewater (Brites Alves & Norberta de Pinho, 2000). Membranes with higher MWCO (i.e. 40 and 10 kDa) are susceptible to fouling and concentration polarization, while membrane with lower MWCO (i.e., 5 and 3 kDa) showed higher colour rejection (95%) (Brites Alves & Norberta de Pinho, 2000).

To enhance the performance efficiency of the PES, it has been appropriately modified via different routes. PES was prepared as an ultrafiltration membrane via the phase inversion method with a blend of $CaCl_2$. The $CaCl_2$ enhanced the mechanical stability, surface hydrophilicity, water flux and hydraulic resistance (Fig. 5) (Rambabu & Velu, 2016). The PES/$CaCl_2$ membrane showed affinity for sulphides, and removed COD, BOD and TDS (Rambabu & Velu, 2016). Commercially available PES and Polyvinylidene fluoride (PVDF) have also showed similar filtration capacity and fouling characteristics and efficiency depending on operating conditions (Yang et al., 2021). Unfortunately, ultrafiltration suffers setbacks due to severe membrane fouling (Mendoza-Roca et al., 2010). Other polymeric membranes that have been reported include OT050 and GR60PP, for the recovery of tannin from vegetable tanning wastewater (Romero-Dondiz et al., 2015).

Other studies have investigated and compared the performance of ultrafiltration membrane with nanofiltration membrane. Two commercial ultrafiltration membranes (OT050 and GR60PP) and four nanofiltration membranes (DK, CK, TFC-SR3 and MPF-34) were tested for removing tannin from vegetable tanning wastewater (Romero-Dondiz et al., 2016). The nanofiltration membranes were characterized by lower MWCO (300–150 Da) compared to the ultrafiltration with high MWCO (50,000–25,000 Da) (Romero-Dondiz et al., 2016). The nanofiltration membranes were more effective in treating the tannery wastewater, especially, MPF-34 (Fig. 5) which showed the highest retention rate for tannin, because of the significant decline in the permeate flux (Romero-Dondiz et al., 2016). Four different commercial nanofiltration membranes (DS_5, CA, DS_5* and BQ_01) were examined for treating tannery wastewater (Ortega et al., 2005), and the success of the nanofiltration membranes were found to depend on the membrane type and initial Cr^{3+} concentration (Ortega et al., 2005). The nanofiltration membranes showed preference for multivalent species (Cr^{3+}) rather than monovalent specie (Na^+), and membrane DS_

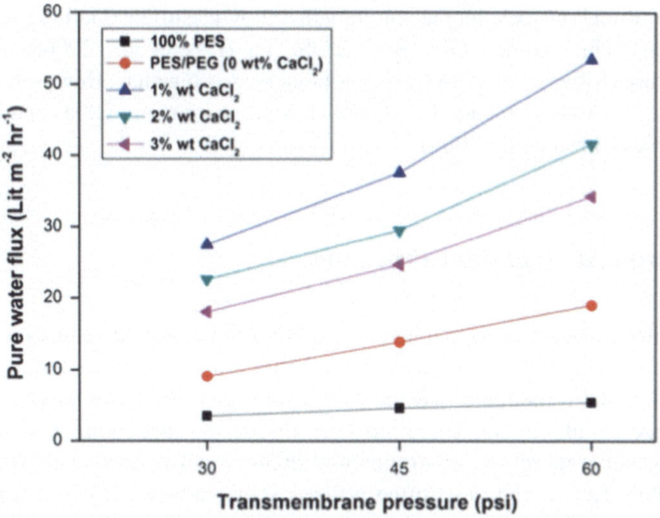

Fig. 5 Pure water flux for $CaCl_2$ modified Polyethersulfone ultrafiltration membrane used in treating tannery wastewater (Religa et al., 2011)

5 was the most effective (99%) in removing (Cr^{3+}). However, the operations of the DS_5* and BQ_01 were characterized by severe fouling (Ortega et al., 2005) (Fig. 6).

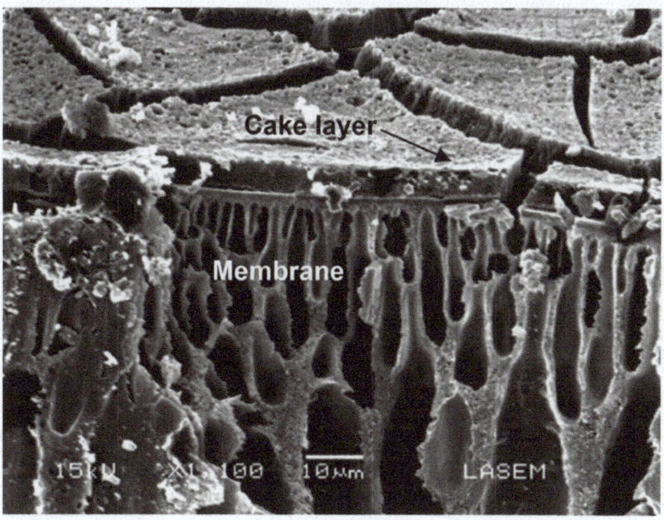

Fig. 6 Scanning Electron image of the cross-section of MPF-34 nanofiltration membrane (Romero-Dondiz et al., 2016)

Nanofiltration is an excellent technology in removing sulfate (Galiana-Aleixandre et al., 2005), chromium, COD, Bod, TDS, TS (Das et al., 2006a) and chloride (Cuartas-Uribe et al., 2006) from tannery wastewater. However, it suffer setbacks such as low permeate flux (Galiana-Aleixandre et al., 2005) and hydraulic permeabilities (Streit et al., 2009).

5.5 *Advanced Oxidation Processes*

Owing to the capacity to trigger photocatalytic oxidation and reduction reaction, when exposed to light illumination, semiconductors have become materials of choice for environmental remediation (Khodabandeloo et al., 2023; Kumari et al., 2023; Urbina-Suarez et al., 2021). There has been discussion that semiconductor photocatalysis can solve energy and environmental challenge (Kuspanov et al., 2023; Wang et al., 2019c). TiO_2 is the most often utilized semiconductor because it possesses a valence band with adequate positive potential to induce oxidation, low toxicity, chemical stability, suitability for thin-film coating on substrates, and eco-friendliness (Monteiro et al., 2023; Tripathi & Narayanan, 2018). TiO_2 has been employed for the degradation and mineralization of toxic organic compounds in wastewater (El Sharkawy et al., 2023), including effluents from tannery and leather processing industries (Oller et al., 2011; Urbina-Suarez et al., 2021). The redox activities of TiO_2 (or simple semiconductors) in treating tannery wastewater under visible (Wahyuni et al., 2023) or UV light (Shahbazi & Pedram, 2021) have been improved by introducing foreign elements or compounds into the lattice of the semiconductor (Elmehasseb et al., 2020). In the case of TiO_2, it has been doped with N (Vaiano et al., 2014), Co (Renugadevi et al., 2016), Zn, N and S (Elmehasseb et al., 2020) to extend the photocatalytic activities to visible light region by narrowing the bandgap, for treating tannery wastewater. The narrowing of TiO_2 (semiconductor) bandgap has also been achieved by precipitating TiO_2 (semiconductor) in a solution of leave extract, so that the ions/compounds in the leave extracts can be trapped within the lattice of the semiconductor. TiO_2 was modified with *Jatropha curcas* extract (Goutam et al., 2018) and clay (Mustapha et al., 2022) for the treatment of tannery wastewater. *Jatropha curcas* modified TiO_2 introduced C into the lattice of TiO_2 while clay introduced Si, Al, and Fe into the lattice of the TiO_2. The modified TiO_2 nanoparticles were more effective in removing COD from tannery wastewater.

Electron–hole recombination (Wang et al., 2019a) is one of the most critical factors impeding the large-scale application of semiconductor photocatalysis in treating wastewater from tannery and leather industry (Li et al., 2022). Doping photocatalyst with noble metals significantly improved electron–hole recombination time (Guo et al., 2016). However, in practice, non-precious elements are preferable, and carbon quantum dots have been found as alternatives due to cost, low toxicity, stability and biocompatibility (Wang & Hu, 2014). Therefore, carbon quantum dots loaded onto SnS_2 was used in the reduction of $Cr_2O_7^{2-}$ in tannery wastewater (Wang et al., 2019b).

Other metal oxides such as ZnO and MgO have also gained attention in treating tannery wastewater because of their ultraviolent light activation, cost and biocompatibility (Vaiano & Iervolino, 2018; Fouda et al., 2021b). ZnO activated using UV light was used in removing COD, BOD, total solids, total organic carbon and turbidity (Hasegawa et al., 2014). The photo-oxidation depended on the pH (Hasegawa et al., 2014). Magnetic nanoparticles such as ferrite and iron oxides have also found application in treating tannery wastewater (Zhang et al., 2017; Fouda et al. 2021a; Riyat et al., 2022).

The photo-oxidation of the organic and inorganic contents of tannery wastewater is achieved by hydroxyl radical (·OH). Hydroxyl radical is generated by the decomposition of water or/and the reaction of positive hole with hydroxyl ions. The generation of hydroxyl radical is affected by solution pH, thus, an alkaline pH is more favorable (Akyol et al., 2004; Chakrabarti & Dutta, 2004). The photo-reduction of reducible parts of tannery wastewater is achieved by the electron transfer (Alharbi et al., 2023).

Another form of AOPs that has gained attention in the treatment of tannery wastewater is irradiation (UV or O_3) or oxidants (e.g. H_2O_2, persulfates) or a combination of both (Balasubramanian and Pugalenthi, 1999). H_2O_2/UV, O_3 and UV have been reported in the mineralization and degradation of active ingredients contained in tannery wastewater (Andrade de Souza et al., 2020; Iqbal et al., 2019; Moradi & Moussavi, 2019; Schrank et al., 2004b, 2005). One of its challenges is the issue of high energy consumption (Schrank et al., 2004a), thus making it impractical in areas deficient in infrastructure.

5.6 *Biological Treatment*

Biological treatment of wastewater has been studied as one of the most effective techniques for treating industrial wastewater. The reliance on living organisms in biological treatment techniques makes this method vulnerable to toxic substances being treated within the wastewater system. Prior to the biological treatment of wastewater, the primary treatment process, which involves the physicochemical processes, includes screening, equalization, pH adjustment, and chemical treatment, are crucial to the process efficiency (Lofrano et al., 2013). The mechanism of biological treatment by microbes is complex and involves different sub-techniques, such as adsorption, degradation, coagulation, and detoxification of contaminants. In the treatment of industrial effluents, bioprocesses like activated sludge, biofilm, anaerobic hydrolysis, acidification, biological filtering, etc. are examples of processes widely utilized, due to the affordability and provision of desirable functions, such as denitrification, dephosphorization, etc. (Biswas et al., 2015; Goswami & Mazumder, 2014). A series of studies have been undertaken to remediate color and metal-containing wastewater using microalgae, fungal, activated sludge, bacterium strains, enzymes, etc. (Huang et al., 2015; Mahmood et al., 2013; Morris, 2003; Ortiz-Monsalve et al., 2019). Implementing biological treatment involves the identification and culturing of microbes that exhibit tolerance to salinity and heavy metals, because tannery wastewater often presents with high salinity with toxic substances that induce the development and physiological activity of commonly used microbes, which consequently reduce the biological treatment efficiency (Chaudhary et al., 2017).

Treatment of wastewater via biological techniques has primarily focused on assessing the overall detoxification capabilities of various microorganisms. Smiley and Piyush (Mustapha et al., 2022), investigated the detoxification potentials of four fungal strains immobilized on nylon mesh. After a period of 120 h, removal rates of 82.52% for chemical oxygen demand (COD), 86.19% for color, 99.92% for total chromium (Cr), and 95.91% for total Pb^{2+} were reported (Lofrano et al., 2006). Sivaprakasam's report highlighted an 80% reduction in COD within tannery saline wastewater, achieved through the successful isolation of four salt-tolerant bacterial strains (*Pseudomonas aeruginosa, Bacillus flexus, Exiguobacterium homiense,* and *Staphylococcus aureus*) (Sivaprakasam et al., 2008). the use of binary bacterial strains (*Acinetobacter sp.* and *Rhodococcus sp.*) that led to the total removal of phenolic compounds and a significant reduction of BOD and COD in phenol-contaminated tannery effluent has been reported (Paisio et al., 2014) Investigations have shown that during the treatment of tannery wastewater in aeration lagoons of common effluent treatment plants, significant reductions in COD, total Cr, and color was achieved, accompanied by a decrease in the biotoxicity of the treated wastewater. This showed the potentials of this treatment operations in the effective management of challenging wastewater (Appiah-Brempong et al., 2023).

Given that tannery wastewater often contains ample nutrient content, anaerobic digestion has emerged as an alternative method for pollutant removal and biogas production. Technologies like up-flow anaerobic sludge blanket reactors, membrane bioreactors, non-aerated biofilm systems, and packed bed reactors have been explored for processes such as denitrification, dephosphorization, anaerobic ammonium oxidation, detoxification, and bio-gas generation through the cultivation of resilient microbial species (Yargeau, 2012). However, tannery wastewater typically poses challenges to biological treatment due to its inhibitory effects on enzymatic reactions in anaerobic digestion (Khayet et al., 2011; Barrera-Díaz et al., 2018).

Biological treatment processes are generally considered to be sustainable, cost-effective and environmentally friendly. They rely on the action of microorganisms to break down pollutants, making them effective in treating a wide range of contaminants while reducing the toxicity of the industrial effluent (Garcia & Bécares, 1997). However, there are certain drawbacks associated with biological treatment. A prominent disadvantage is the large amount of land required, particularly in the case of constructed wetlands, which require a significant amount of space. In addition, these processes can be sensitive to variations in environmental conditions, requiring precise control measures to maintain peak performance. Furthermore, the efficiency of biological treatment processes can suffer when exposed to toxic substances or elevated levels of organic matter, resulting in reduced treatment efficiency and sludge accumulation. Despite these challenges, biological treatment processes offer several advantages. However, they require careful management and continuous monitoring to ensure optimal performance. Ongoing research and development efforts hold promise for tackling these challenges and increasing the overall effectiveness of biological treatments (Garcia & Bécares, 1997).

5.7 *Integrated Technologies*

Tannery wastewater management is a complex undertaking that requires careful evaluation of different treatment methods. In previous sections, several of such methods were presented, each with its own set of advantages and disadvantages. Despite the ongoing efforts to improve these approaches, achieving complete detoxification of persistent organic and inorganic contaminants in tannery wastewater effluent using a single method remains challenging and expensive. Combining different treatment methods may provide a more effective and cost-effective solution to the detrimental effect of tannery wastewater. By carefully weighing the strengths and weaknesses of each method, it is possible to select an appropriate combination of techniques that can deliver multiple performance improvements while reducing costs. This approach has gained popularity among tannery effluent treatment researchers, who have experimented with different techniques to achieve better results. Integrated technologies include the combination of reverse osmosis, nanofiltration, gravity settling, and coagulation/flocculation etc. (Deghles & Kurt, 2016; Garcia & Bécares, 1997; Kurt et al., 2007). The efficiencies of different tannery wastewater combined treatment strategies are presented in Table 3.

Integrated or combined technologies for tannery effluent treatment offer several advantages, such as high contaminant removal efficiencies, cost-effectiveness and the ability to address a wide range of contaminants. However, there are also associated disadvantages. For example, the complexity of the treatment process can lead to operational challenges and maintenance issues. In addition, the initial investment costs can be significant, which may raise feasibility issues for small-scale industries. Furthermore, the effectiveness of these methods depends mainly on various factors such as pH, temperature and the presence of other contaminants, requiring careful monitoring and adjustment. Hence, selecting the appropriate combination of techniques and optimizing treatment parameters can be a complex task requiring considerable expertise. While integrated technologies offer many advantages, their successful implementation and operation requires careful consideration of these potential drawbacks (Cañizares et al., 2008; Garcia & Bécares, 1997; Kurt et al., 2007).

Coagulation/flocculation and membrane filtration are attractive. Coagulation/flocculation could be used in the pretreatment of tannery wastewater (Liu et al., 2022) because coagulation/flocculation are effective in removing salts which could potentially clog membrane pores. Coagulation significantly reduced turbidity, chroma and COD by 96.5%, 53.7% and 45.8%, respectively. After the nanofiltration process, the recovery of the spent-water was 90% (Liu et al., 2022). Membrane filtration has also been used as the preferred pretreatment technique, followed by coagulation/flocculation as post-treatment option (Ahmad et al., 2016). Oxidation process has

Table 3 Integrated technologies for the treatment of tannery effluent

Methods	Efficiency	References
AOP/Microbial consortia	84% COD, 60% colour, 100% odor, 90% inorganic carbon, 82% NH_4^+-N, 100% PO_4^{3-}–P, 97% Cr and 10% TDS	Saranya and Shanthakumar (2020)
Coagulation-flocculation/ adsorption	50.04% TOC, 53.13% COD, 17.05% Na + and 61.13% color	Mella et al. (2018)
O_3 (AOP) + Aerobic sequential batch reactor	98% color and 64% COD removal	Srinivasan et al. (2012)
Electrocoagulation/ Electrooxidation	9.9% removal of COD and color	Parthasarathy and Narayanan (2014)
Electrocoagulation/Active sludge process/ Microfiltration	76% COD removal, 93% color removal	Keerthi et al. (2013)
Coagulation-Flocculation/ Ozonation	46.50% TOC, 53.25% COD, 11.10% Na + and 83.34% color	Mella et al. (2018)
Anaerobic sludge bed/ Sequencing batch reactor/ Electrochemical oxidation/ Biological aerated filter	> 90% removal of COD, TSS, total nitrogen, ammonia, oil-grease, S^{2-}, total Cr and Cr^{6+}	Liu et al. (2012)

also been integrated with coagulation and adsorption technology for removing dye, BOD and COD from tannery wastewater (Mim et al., 2023).

Nanofiltration and reverse osmosis have also been integrated for treating tannery wastewater effluent. The nano-membrane showed affinity to retain and SO_4^{2-}, Mg^{2+} and Ca^{2+} while reverse osmosis showed affinity for TDS (Suthanthararajan et al., 2004). In comparison, reserve osmosis is more suitable in treating tannery wastewater because nanofiltration suffer from higher fouling and lower rejection of contaminants (Fernández-Medrano et al., 2022). Integrated biological and chemical oxidation of tannery wastewater is characterized by low sludge production and are effective in removing COD, NH_4-N and TSS (Di Iaconi et al., 2002).

6 Conclusion

Tannery and leather industry are vital for many communities. However, the wastewater from these industries is a menace to the environment. The threat posed by these industries is even more exacerbated in rural areas due to deficient infrastructure. Among the sources of wastewater in rural areas, the effluent from tannery and leather industries are severe to the environment when discharged. Effluent from tannery and leather industries contain high concentration of chrome and salts, elevated level of COD, BOD, TS, and TDS. Appropriate technologies, which include adsorption, coagulation, electrocoagulation, membrane filtration, photocatalysis and integrated technologies, for the effective management of tannery and leather wastewater effluents should address concerns such as sustainability and simple and cheaper technological development. The performance of these technologies in treating tannery and leather wastewater effluent can be improved by optimizing the reaction conditions.

References

Aber, S., Salari, D., & Parsa, M. R. (2010). Employing the Taguchi method to obtain the optimum conditions of coagulation-flocculation process in tannery wastewater treatment. *Chemical Engineering Journal, 162*, 127–134. https://doi.org/10.1016/j.cej.2010.05.012

Aboulhassan, M. A., El Ouarghi, H., Ait Benichou, S., Ait Boughrous, A., & Khalil, F. (2018). Influence of experimental parameters in the treatment of tannery wastewater by electrocoagulation. *Separation Science and Technology, 53*, 2717–2726. https://doi.org/10.1080/01496395.2018.1470642

Afrad, M. S. I., Monir, M. B., Haque, M. E., Barau, A. A., & Haque, M. M. (2020). Impact of industrial effluent on water, soil and rice production in Bangladesh: A case of Turag River Bank. *Journal of Environmental Health Science and Engineering, 18*, 825–834. https://doi.org/10.1007/s40201-020-00506-8

Ahmad, M. I., Habib, M., Habib, U., Hai, A., & Khan, A. (2016). Analysis and treatment of tannery waste water by using the combined filtration and coagulation treatment process. *Proceedings of the Pakistan Academy of Sciences, 50*, 13.

Akyol, A., Yatmaz, H. C., & Bayramoglu, M. (2004). Photocatalytic decolorization of Remazol Red RR in aqueous ZnO suspensions. *Applied Catalysis B: Environmental, 54*, 19–24. https://doi.org/10.1016/j.apcatb.2004.05.021

Albalah, M. A., Alsabah, Y. A., & Mustafa, D. E. (2020). Characteristics of co-precipitation synthesized cobalt nanoferrites and their potential in industrial wastewater treatment. *SN Applied Sciences, 2*, 804. https://doi.org/10.1007/s42452-020-2586-6

Alharbi, F.F., Aman, S., Ahmad, N., Abdullah, M., Abid, A. G., Manzoor, S., Trukhanov, S., Sayyed, M. I., Tishkevich, D., & Trukhanov, A. (2023). Investigation of photoreduction of Cr (VI) and electrocatalytic properties of hydrothermally produced novel $CoFe_2O_4$/ZnO nanostructure. *Solid State Sciences, 143*, 107278. https://doi.org/10.1016/j.solidstatesciences.2023.107278

Amin, M. (2015). Textile waste water treatment by chemical coagulation technology. *Global Journal of Advanced Engineering Technologies and Sciences, 2*, 33–41.

Amirhandeh, S. Z. H., Salem, A., & Salem, S. (2023). Treatment of tannery wastewater by silica nanoparticles produced from rice husk ash via a green route. *Environmental Science and Pollution Research, 30*, 13039–13047. https://doi.org/10.1007/s11356-022-23052-2

Andrade de Souza, E., Cerqueira, U. M., & Silva, L. (2020). Comparative study of various advanced oxidation processes for the treatment of tannery wastewater. *Desalination and Water Treatment, 181*, 88–97. https://doi.org/10.5004/dwt.2020.25183

Apaydin, Ö., Kurt, U., & Gonüllü, M. (2009). An Investigation on the treatment of tannery wastewater by electrocoagulation. *Global NEST Journal, 11*, 546–555.

Appiah-Brempong, M., Michelle Korkor Essandoh, H., Yaw Asiedu, N., Kwame Dadzie, S., & Yao Momade, F. (2023). Performance optimization of chemical and green coagulants in tannery wastewater treatment: A response surface methodology approach. *Journal of Optimization, 2023*, 9939499. https://doi.org/10.1155/2023/9939499

Appiah-Brempong, M., Essandoh, H. M. K., Asiedu, N. Y., Dadzie, S. K., & Momade, F. W. Y. (2022). Artisanal tannery wastewater: Quantity and characteristics. *Heliyon, 8*, e08680. https://doi.org/10.1016/j.heliyon.2021.e08680

Arhin, S. G., Banadda, N., Komakech, A. J., Pronk, W., & Marks, S. J. (2019). Application of hybrid coagulation–ultrafiltration for decentralized drinking water treatment: Impact on flux, water quality and costs. *Water Supply, 19*, 2163–2171. https://doi.org/10.2166/ws.2019.097

Ay, U., & Ahmed, A. (2016). A concise review on the effect of tannery waste water on aquatic fauna. *Merit Research Journals of Medicine and Medical Sciences, 4*, 476–479. https://doi.org/10.6084/m9.figshare.4246583

Ayoub, G. M., Hamzeh, A., & Semerjian, L. (2011). Post treatment of tannery wastewater using lime/bittern coagulation and activated carbon adsorption. *Desalination, 273*, 359–365. https://doi.org/10.1016/j.desal.2011.01.045

Balaji, S., Kalaivani, T., Shalini, M., Gopalakrishnan, M., Rashith Muhammad, M. A., & Rajasekaran, C. (2016). Sorption sites of microalgae possess metal binding ability towards Cr(VI) from tannery effluents—A kinetic and characterization study. *Desalination and Water Treatment, 57*, 14518–14529. https://doi.org/10.1080/19443994.2015.1064032

Balasubramanian, S., & Pugalenthi, V. (1999). Determination of total chromium in tannery waste water by inductively coupled plasma-atomic emission spectrometry, flame atomic absorption spectrometry and UV–visible spectrophotometric methods. *Talanta, 50*, 457–467. https://doi.org/10.1016/S0039-9140(99)00135-6

Barkley, D. L., & Henry, M. S. (1997). Rural industrial development: To cluster or not to cluster? *Applied Economic Perspectives and Policy, 19*, 308–325. https://doi.org/10.2307/1349744

Barman, B., Hashem, M. A., & Juel, M. A. (2017). Tannery wastewater treatment using low cost coagulants with simple coagulation-filtration process. *Engineering International, 5*. https://doi.org/10.18034/ei.v5i2.1084

Barrera-Díaz, C. E., Balderas-Hernández, P., Bilyeu, B. (2018). Chapter 3—Electrocoagulation: Fundamentals and prospectives. In C. A. Martínez-Huitle, M. A. Rodrigo, Scialdone OBT-EW, & WT (Eds.) (pp. 61–76). Butterworth-Heinemann.

Benhadji, A., Taleb Ahmed, M., & Maachi, R. (2011). Electrocoagulation and effect of cathode materials on the removal of pollutants from tannery wastewater of Rouïba. *Desalination, 277*, 128–134. https://doi.org/10.1016/j.desal.2011.04.014

Biswas, S. K., Lisa, L. A., Banu, N. A., Islam, A., & Roy, A. K. (2015). Review paper microbial treatment of tannery effluents: A review. *Plant Environment Development, 4*, 13–20.

Borchate, S., Kulkarni, D., Kore, S., & Kore, V. (2012). Application of coagulation-flocculation for vegetable tannery wastewater. *The International Journal on Engineering, Science and Technology (IJEST), 4*, 1944–1948. 0975-5462.

Brandt, M. J., Johnson, K. M., Elphinston, A. J., & Ratnayaka, D. D. (2017). Chapter 8—Storage, clarification and chemical treatment. In M. J. Brandt, K. M. Johnson, A. J. Elphinston, Ratnayaka, DDBT-TWS, & E. Seventh (Eds.) (pp. 323–366). Butterworth-Heinemann.

Brites Alves, A. M., & Norberta de Pinho, M. (2000). Ultrafiltration for colour removal of tannery dyeing wastewaters. *Desalination, 130*, 147–154. https://doi.org/10.1016/S0011-9164(00)00082-5

Caliari, P. C., Pacheco, M. J., Ciríaco, L., & Lopes, A. (2021). Treatment of tannery effluent by chemical coagulation combined with batch-recirculated electro-oxidation at different anode materials. *Environmental Science and Pollution Research, 28*, 24138–24149. https://doi.org/10.1007/s11356-021-12436-5

Cañizares, P., Martínez, F., Jiménez, C., Sáez, C., & Rodrigo, M. A. (2008). Coagulation and electrocoagulation of oil-in-water emulsions. *Journal of Hazardous Materials, 151*, 44–51. https://doi.org/10.1016/j.jhazmat.2007.05.043

Carpenter, J., Sharma, S., Sharma, A. K., & Verma, S. (2013). Adsorption of dye by using the solid waste from leather industry as an adsorbent. *Adsorption, 2*.

Chakrabarti, S., Dutta, B. K. (2004). Photocatalytic degradation of model textile dyes in wastewater using ZnO as semiconductor catalyst. *Journal of Hazardous Materials, 112*, 269–278. https://doi.org/10.1016/j.jhazmat.2004.05.013

Chaudhary, P., Chhokar, V., Kumar, A., & Beniwal, V. (2017). Bioremediation of tannery wastewater BT—Advances in environmental biotechnology. In A. K. Sharma & S. S. Ahluwalia (Eds.), *Kumar R* (pp. 125–144). Springer Singapore.

Chaudhary, M., Jain, N., Barman, L., & Bhowmick, G. D. (2022). Chapter 8—Design and principles of adsorbent-based reactors for modular wastewater treatment. In S. Kaur Brar, P. Kumar, Cuprys, ABT-MTA for DW & W (Eds.) (pp. 129–148). Elsevier.

Cheng, F., Dai, Z., Shen, S., Wang, S., & Lu, X. (2021). Characteristics of rural domestic wastewater with source separation. *Water Science and Technology, 83*, 233–246. https://doi.org/10.2166/wst.2020.557

Chidambaram, S., Karmegam, U., Prasanna, M. V., & Sasidhar, P. (2012). A study on evaluation of probable sources of heavy metal pollution in groundwater of Kalpakkam region, South India. *The Environmentalist, 32*, 371–382. https://doi.org/10.1007/s10669-012-9398-1

Chowdhury, M., Mostafa, M. G., Biswas, T. K., Mandal, A., & Saha, A. K. (2015). Characterization of the effluents from leather processing industries. *Environ Process, 2*, 173–187. https://doi.org/10.1007/s40710-015-0065-7

Chowdhury, M., Mostafa, M. G., Biswas, T. K., & Saha, A. K. (2013). Treatment of leather industrial effluents by filtration and coagulation processes. *Water Resources and Industry, 3*, 11–22. https://doi.org/10.1016/j.wri.2013.05.002

Costa, C. R., Botta, C. M. R., Espindola, E. L. G., & Olivi, P. (2008). Electrochemical treatment of tannery wastewater using DSA® electrodes. *Journal of Hazardous Materials, 153*, 616–627. https://doi.org/10.1016/j.jhazmat.2007.09.005

Cotman, M., Zagorc-Končan, J., & Žgajnar-Gotvajn, A. (2004). The relationship between composition and toxicity of tannery wastewater. *Water Science and Technology, 49*, 39–46. https://doi.org/10.2166/wst.2004.0012

Cuartas-Uribe, B., Iborra-Clar, A., Bes-Piá, A., Mendoza-Roca, J. A., Galiana-Aleixandre, M. V., Iborra-Clar, M. I. (2006). Nanofiltration of a simulated tannery wastewater: Influence of chlorides concentration. *Desalination, 191*, 132–136. https://doi.org/10.1016/j.desal.2005.05.025

Dabai, A., & Mohammed, K. (2020). Chromium removal from tannery wastewater: A review. 3, 63–73

Das, R. K., Mizan, A., Zohra, F. T., Ahmed, S., Ahmed, K. S., & Hossain, H. (2022). Extraction of a novel tanning agent from indigenous plant bark and its application in leather processing. *Journal of Leather Science and Engineering, 4*, 18. https://doi.org/10.1186/s42825-022-00092-5

Das, C., Patel, P., De, S., & DasGupta, S. (2006a). Treatment of tanning effluent using nanofiltration followed by reverse osmosis. *Separation and Purification Technology, 50*, 291–299. https://doi.org/10.1016/j.seppur.2005.11.034

Das, P., Das, B., & Ysa, K. (2006b). Environmental assessment of tannery wastes from Chittagong, Bangladesh. *Journal of Water and Environment Technology, 3*, 83–90.

Dasgupta, P., Morton, J. F., Dodman, D., Karapinar, B., Meza, F., Rivera-Ferre, M., Sarr, A. T., Vincent, K., Carr, E., Raholijao, N., & Broecker, H. (2015). Rural areas. In *Climate change 2014 impacts, adaptation and vulnerability: Part A: Global and sectoral aspects*, pp 613–658.

Degache, É., & Hans, A.-L. (2006). The leather, chemical and mechanical treatments to bring sensuality. *Actualite Chimique* 3–10.

Deghles, A., & Kurt, U. (2016). Treatment of raw tannery wastewater by electrocoagulation technique: Optimization of effective parameters using Taguchi method. *Desalination and Water Treatment, 57*, 14798–14809. https://doi.org/10.1080/19443994.2015.1074622

Deng, Y., & Wheatley, A. (2016). Wastewater treatment in Chinese rural areas. *Asian Journal of Water, Environment and Pollution, 13*, 1–11. https://doi.org/10.3233/AJW-160033

Devi, A., Verma, M., Saratale, G. D., Saratale, R. G., Ferreira, L. F. R., Mulla, S. I., & Bharagava, R. N. (2023). Microalgae: A green eco-friendly agents for bioremediation of tannery wastewater with simultaneous production of value-added products. *Chemosphere, 336*, 139192. https://doi.org/10.1016/j.chemosphere.2023.139192

Di Iaconi, C., Lopez, A., Ramadori, R., Di Pinto, A. C., & Passino, R. (2002). Combined chemical and biological degradation of tannery wastewater by a periodic submerged filter (SBBR). *Water Research, 36*, 2205–2214. https://doi.org/10.1016/S0043-1354(01)00445-6

Dotto, J, Fagundes-Klen, M. R., Veit, M. T., Palácio, S. M., Bergamasco, R. (2019). Performance of different coagulants in the coagulation/flocculation process of textile wastewater. *Journal of Cleaner Production, 208*, 656–665. https://doi.org/10.1016/j.jclepro.2018.10.112

Durai, G., & Manivasagan, R. (2011) Biological treatment of tannery wastewater—A review. *Journal of Environmental Science and Technology, 4*. https://doi.org/10.3923/jest.2011.1.17

Durairaj, S. (2016). Biosorption of hexavalent chromium in a tannery industry wastewater using fungi species. *Glob J Environ Sci Manag, 3*, 105–124. https://doi.org/10.7508/gjesm.2016.02.002

El Mouhri, G., Merzouki, M., Kachkoul, R., Belhassan, H., Miyah, Y., Amakdouf, H., Elmountassir, R., & Lahrichi, A. (2021). Fixed-bed adsorption of tannery wastewater pollutants using bottom ash: An optimized process. *Surfaces and Interfaces, 22*, 100868. https://doi.org/10.1016/j.surfin.2020.100868

El Sharkawy, H. M., Shawky, A. M., Elshypany, R., & Selim, H. (2023). Efficient photocatalytic degradation of organic pollutants over TiO_2 nanoparticles modified with nitrogen and MoS2 under visible light irradiation. *Science and Reports, 13*, 8845. https://doi.org/10.1038/s41598-023-35265-7

Elabbas, S., Mandi, L., Berrekhis, F., Pons, M. N., Leclerc, J. P., & Ouazzani, N. (2016a). Removal of Cr(III) from chrome tanning wastewater by adsorption using two natural carbonaceous materials: Eggshell and powdered marble. *Journal of Environmental Management, 166*, 589–595. https://doi.org/10.1016/j.jenvman.2015.11.012

Elabbas, S., Ouazzani, N., Mandi, L., Berrekhis, F., Perdicakis, M., Pontvianne, S., Pons, M.-N., Lapicque, F., & Leclerc, J.-P. (2016b). Treatment of highly concentrated tannery wastewater using electrocoagulation: Influence of the quality of aluminium used for the electrode. *Journal of Hazardous Materials, 319*, 69–77. https://doi.org/10.1016/j.jhazmat.2015.12.067

Elkarrach, K., Omor, A., Atia, F., Laidi, O., Benlemlih, M., & Merzouki, M. (2023). Treatment of tannery effluent by adsorption onto fly ash released from thermal power stations: Characterisation, optimization, kinetics, and isotherms. *Heliyon, 9*, e12687. https://doi.org/10.1016/j.heliyon.2022.e12687

Elmehasseb, I., Kandil, S., & Elgendy, K. (2020). Advanced visible-light applications utilizing modified Zn-doped TiO_2 nanoparticles via non-metal in situ dual doping for wastewater detoxification. *Optik (Stuttg), 213*, 164654. https://doi.org/10.1016/j.ijleo.2020.164654

Elsheikh, M.A.-S. (2009). Tannery wastewater pre-treatment. *Water Science and Technology, 60*, 433–440. https://doi.org/10.2166/wst.2009.351

Espantaleón, A. G., Nieto, J. A., Fernández, M., & Marsal, A. (2003). Use of activated clays in the removal of dyes and surfactants from tannery waste waters. *Applied Clay Science, 24*, 105–110. https://doi.org/10.1016/S0169-1317(03)00153-4

Espinoza-Quiñones, F. R., Fornari, M. M. T., Módenes, A. N., Palácio, S. M., da Silva, F. G., Szymanski, N., Kroumov, A. D., & Trigueros, D. E. G. (2009). Pollutant removal from tannery effluent by electrocoagulation. *Chemical Engineering Journal, 151*, 59–65. https://doi.org/10.1016/j.cej.2009.01.043

Fahim, N. F., Barsoum, B. N., Eid, A. E., & Khalil, M. S. (2006). Removal of chromium(III) from tannery wastewater using activated carbon from sugar industrial waste. *Journal of Hazardous Materials, 136*, 303–309. https://doi.org/10.1016/j.jhazmat.2005.12.014

Farahat M. M., Sanad, M. M. S., & Abdel-Khalek, M. A. (2021). Decoration of serpentine with iron ore as an efficient low-cost magnetic adsorbent for Cr (VI) removal from tannery wastewater. *Powder Technology, 388*, 51–62. https://doi.org/10.1016/j.powtec.2021.04.061

Fauzi, A., Nuruddin, M. F., Malkawi, A. B., & Abdullah, M. M. A. B. (2016). Study of fly ash characterization as a cementitious material. *Procedia Engineering, 148*, 487–493. https://doi.org/10.1016/j.proeng.2016.06.535

Feng, J., Sun, Y., Zheng, Z., Zhang, J., Li, S., & Tian, Y. (2007). Treatment of tannery wastewater by electrocoagulation. *Journal of Environmental Sciences, 19*, 1409–1415. https://doi.org/10.1016/S1001-0742(07)60230-7

Fernández-Medrano, V., Cuartas-Uribe, B., Bes-Piá, M.-A., & Mendoza-Roca, J.-A. (2022). Application of nanofiltration and reverse osmosis membranes for tannery wastewater reuse. *Water, 14*.

Fouda, A., Hassan, S. E. D., Abdel-Rahman, M. A., Farag, M. M. S., Shehal-deen, A., Mohamed, A. A., Alsharif, S. M., Saied, E., Moghanim, S. A., & Azab, M. S. (2021c). Catalytic degradation of wastewater from the textile and tannery industries by green synthesized hematite (α-Fe_2O_3) and magnesium oxide (MgO) nanoparticles. *Current Research in Biotechnology, 3*, 29–41. https://doi.org/10.1016/j.crbiot.2021.01.004

Fouda, A., Hassan, S. E.-D., Saied, E., & Azab, M. S. (2021a). An eco-friendly approach to textile and tannery wastewater treatment using maghemite nanoparticles (γ-Fe_2O_3-NPs) fabricated by Penicillium expansum strain (K-w). *Journal of Environmental Chemical Engineering, 9*, 104693. https://doi.org/10.1016/j.jece.2020.104693

Fouda, A., Hassan, S. E.-D., Saied, E., & Hamza, M. F. (2021b). Photocatalytic degradation of real textile and tannery effluent using biosynthesized magnesium oxide nanoparticles (MgO-NPs), heavy metal adsorption, phytotoxicity, and antimicrobial activity. *Journal of Environmental Chemical Engineering, 9*, 105346. https://doi.org/10.1016/j.jece.2021.105346

Gaber, D., Abu Haija, M., Eskhan, A., & Banat, F. (2017). Graphene as an efficient and reusable adsorbent compared to activated carbons for the removal of phenol from aqueous solutions. *Water, Air, & Soil Pollution, 228*. https://doi.org/10.1007/s11270-017-3499-x

Galiana-Aleixandre, M. V., Iborra-Clar, A., Bes-Piá, B., Mendoza-Roca, J. A., Cuartas-Uribe, B., & Iborra-Clar, M. I. (2005) Nanofiltration for sulfate removal and water reuse of the pickling and tanning processes in a tannery. *Desalination, 179*, 307–313. https://doi.org/10.1016/j.desal.2004.11.076

Garcia, M., & Bécares, E. (1997). Bacterial removal in three pilot-scale wastewater treatment systems for rural areas. *Water Science and Technology, 35*, 197–200. https://doi.org/10.1016/S0273-1223(97)00258-8

Garrote, J. I., Bao, M., Castro, P., & Bao, M. J. (1995). Treatment of tannery effluents by a two step coagulation/flocculation process. *Water Research, 29*, 2605–2608. https://doi.org/10.1016/0043-1354(94)00312-U

Gervas, C., Mubofu, E. B., Mdoe, J. E. G., & Revaprasadu, N. (2016). Functionalized mesoporous organo-silica nanosorbents for removal of chromium (III) ions from tanneries wastewater. *Journal of Porous Materials, 23*, 83–93. https://doi.org/10.1007/s10934-015-0058-y

Ghorab, R. E. A., Pugazhendi, A., Jamal, M. T., Jeyakumar, R. B., Godon, J. J., & Mathew, D. K. (2022). Tannery wastewater treatment coupled with bioenergy production in upflow microbial fuel cell under saline condition. *Environmental Research, 212*, 113304. https://doi.org/10.1016/j.envres.2022.113304

GilPavas, E., Dobrosz-Gómez, I., & Gómez-García, M. Á. (2011). The removal of the trivalent chromium from the leather tannery wastewater: The optimisation of the electro-coagulation process parameters. *Water Science and Technology, 63*, 385–394. https://doi.org/10.2166/wst.2011.232

Goswami, S., & Mazumder, D. (2014). Scope of biological treatment for composite tannery wastewater. *International Journal of Environmental Sciences, 5*, 607–622.

Goutam, S. P., Saxena, G., Singh, V., Yadav, A. K., Bharagava, R. N., & Thapa, K. B. (2018). Green synthesis of TiO_2 nanoparticles using leaf extract of Jatropha curcas L. for photocatalytic degradation of tannery wastewater. *Chemical Engineering Journal, 336*, 386–396. https://doi.org/10.1016/j.cej.2017.12.029

Guo, Z.-R., Zhang, G., Fang, J., & Dou, X. (2006). Enhanced chromium recovery from tanning wastewater. *Journal of Cleaner Production, 14*, 75–79. https://doi.org/10.1016/j.jclepro.2005.01.005

Guo, S., Li, X., Zhu, J., Tong, T., & Wei, B. (2016). Au NPs@MoS_2 sub-micrometer sphere-ZnO nanorod hybrid structures for efficient photocatalytic hydrogen evolution with excellent stability. *Small, 12*, 5692–5701. https://doi.org/10.1002/smll.201602122

Hamza, M. F., Hamad, D. M., Hamad, N. A., Abdel-Rahman, A. A.-H., Fouda. A., Wei, Y., Guibal, E., & El-Etrawy, A.-A. S. (2022). Functionalization of magnetic chitosan microparticles for high-performance removal of chromate from aqueous solutions and tannery effluent. *Chemical Engineering Journal, 428*, 131775. https://doi.org/10.1016/j.cej.2021.131775

Hansen, É., Monteiro de Aquim, P., Hansen, A. W., Cardoso, J. K., Ziulkoski, A. L., & Gutterres, M. (2020). Impact of post-tanning chemicals on the pollution load of tannery wastewater. *Journal of Environmental Management, 269*. https://doi.org/10.1016/j.jenvman.2020.110787

Hansen, É., Monteiro de Aquim, P., & Gutterres, M. (2021). Current technologies for post-tanning wastewater treatment: A review. *Journal of Environmental Management, 294*, 113003. https://doi.org/10.1016/j.jenvman.2021.113003

Hao, X., Tao, E., Yang, S., & Li, Y. (2022). A new montmorillonite-based porous composites: Effectively removal of Cr(III)-organic complexes in tannery wastewater. *Journal of Polymers and the Environment, 30*, 308–318. https://doi.org/10.1007/s10924-021-02193-4

Hasan, M. J., Haque, P., & Rahman, M. M. (2022). Protease enzyme based cleaner leather processing: A review. *Journal of Cleaner Production, 365*, 132826. https://doi.org/10.1016/j.jclepro.2022.132826

Hasegawa, M. C., de Daniel, J. F. S., Takashima, K., Batista, G. A., & da Silva S. M. C. P. (2014). COD removal and toxicity decrease from tannery wastewater by zinc oxide-assisted photocatalysis: A case study. *Environmental Technology, 35*, 1589–1595. https://doi.org/10.1080/09593330.2013.874499

Hashem, M. A., Hasan, M., Momen, M. A., Payel, S., & Nur-A-Tomal, M. S. (2020). Water hyacinth biochar for trivalent chromium adsorption from tannery wastewater. *Environmental and Sustainability Indicators, 5*, 100022. https://doi.org/10.1016/j.indic.2020.100022

Hashem, M. A., Payel, S., Mim, S., Hasan, M. A., Nur-A-Tomal, M. S., Rahman, M. A., Sarker, M. I. (2022). Chromium adsorption on surface activated biochar made from tannery liming sludge: A waste-to-wealth approach. *Water Science and Engineering, 15*, 328–336. https://doi.org/10.1016/j.wse.2022.09.001

Hashem, M. A., Mim, M. W., Noshin, N., & Maoya, M. (2024a). Chromium adsorption capacity from tannery wastewater on thermally activated adsorbent derived from kitchen waste biomass. *Clean Water, 1*, 100001. https://doi.org/10.1016/j.clwat.2023.100001

Hashem, M. A., Mim, S., Payel, S., Husne-Al-Jabeen, Sakib, M. S. R., Mizan, A., Basaran, B., Yapici, A. N., & Sarker, M. I. (2024b). Thermally activated adsorbent over chemically modified Carica papaya tree adsorbents for removal of chromium from tannery wastewater. *Bioresource Technology Reports, 25*, 101692. https://doi.org/10.1016/j.biteb.2023.101692

Hendy, I., Zelenakova, M., Pietrucha-Urbanik, K., Salama, Y., & Abu-Hashim, M. (2023). Decentralized constructed wetlands for wastewater treatment in rural and remote areas of semi-arid regions. *Water, 15*.

Hevira, L., Ighalo, J. O., & Sondari, D. (2023). Chitosan-based polysaccharides for effective synthetic dye adsorption. *Journal of Molecular Liquids* 123604. https://doi.org/10.1016/j.molliq.2023.123604

Huang, G., Wang, W., & Liu, G. (2015). Simultaneous chromate reduction and azo dye decolourization by Lactobacillus paracase CL1107 isolated from deep sea sediment. *Journal of Environmental Management, 157*, 297–302. https://doi.org/10.1016/j.jenvman.2015.04.031

Iqbal, M., Muneer, M., Hussain, S., Parveen, B., Javed, M., Rehman, H., Waqas, M., & Abid, M. A. (2019). Using combined UV and H_2O_2 treatments to reduce tannery wastewater pollution load. *Polish Journal of Environmental Studies, 28*, 3207–3213. https://doi.org/10.15244/pjoes/92706

Jain, M., Garg, V. K., & Kadirvelu, K. (2013). Chromium removal from aqueous system and industrial wastewater by agricultural wastes. *Bioremediation Journal, 17*, 30–39. https://doi.org/10.1080/10889868.2012.731450

Jamwal, P., Mittal, A. K., & Mouchel, J.-M. (2011). Point and non-point microbial source pollution: A case study of Delhi. *Physics and Chemistry of the Earth, Parts A/B/C, 36*, 490–499. https://doi.org/10.1016/j.pce.2008.09.005

Jimenez-Paz, J., Lozada-Castro, J. J., Lester, E., Williams, O., Stevens, L., & Barraza-Burgos, J. (2023). Solutions to hazardous wastes issues in the leather industry: Adsorption of Chromium iii and vi from leather industry wastewaters using activated carbons produced from leather industry solid wastes. *Journal of Environmental Chemical Engineering, 11*, 109715. https://doi.org/10.1016/j.jece.2023.109715

Kamaraj, M., Srinivasan, N. R., Assefa, G., Adugna, A. T., Kebede, M. (2020). Facile development of sunlit ZnO nanoparticles-activated carbon hybrid from pernicious weed as an operative nano-adsorbent for removal of methylene blue and chromium from aqueous solution: Extended application in tannery industrial wastewater. *Environmental Technology & Innovation, 17*, 100540. https://doi.org/10.1016/j.eti.2019.100540

Karri, R. R., Ravindran, G., & Dehghani, M. H. (2021). Chapter 1—Wastewater—Sources, toxicity, and their consequences to human health. In R. R. Karri, G. Ravindran, Dehghani MHBT-SCT in SW and WM (Eds.), *Soft computing techniques in solid waste and wastewater manangement* (pp 3–33). Elsevier.

Kaseva, M. E., & Mbuligwe, S. E. (2010). Potential of constructed wetland systems for treating tannery industrial wastewater. *Water Science and Technology, 61*, 1043–1052. https://doi.org/10.2166/wst.2010.474

Katenta, J., Nakiguli, K. C., Mukasa, P., & Ntambi, E. (2020). Removal of Chromium (VI) from Tannery Effluent Using Bio-Char of Phoenix reclinata Seeds. *Green Sustain Chem, 10*, 91–107. https://doi.org/10.4236/gsc.2020.103007

Kazi, T., & Virupakshi, A. (2013). Treatment of tannery wastewater using natural coagulants. *Int J Innov Res Sci Eng Technol, 2*, 4061–4068.

Keerthi, S. V., Mahalakshmi, M., & Balasubramanian, N. (2013). Development of hybrid membrane bioreactor for tannery effluent treatment. *Desalination, 309*, 231–236. https://doi.org/10.1016/j.desal.2012.10.014

Khayet, M., Zahrim, A. Y., & Hilal, N. (2011). Modelling and optimization of coagulation of highly concentrated industrial grade leather dye by response surface methodology. *Chemical Engineering Journal, 167*, 77–83. https://doi.org/10.1016/j.cej.2010.11.108

Khodabandeloo, F., Shahsavarifar, S., Nayebi, B., Niavol, K. P., Nayebi, B., Varma, R. S., Cha, J. H., Jang, H. W., Kim, D., & Shokouhimehr, M. (2023). Applications of nanostructured semiconductor photocatalysts for the decontamination of assorted pollutants from wastewater. *Inorganic Chemistry Communications, 157*, 111357. https://doi.org/10.1016/j.inoche.2023.111357

Kongjao, S., Damronglerd, S., & Hunsom, M. (2008). Simultaneous removal of organic and inorganic pollutants in tannery wastewater using electrocoagulation technique. *Korean Journal of Chemical Engineering, 25*, 703–709. https://doi.org/10.1007/s11814-008-0115-1

Kubra, K., Pavithra, V., Gopal, B., & Janaswamy, S. (2024). An integrated chemical approach for the removal of hazardous Cr(VI) and other contaminants of tannery wastewater: Biopolymer-assisted adsorption separation and conversion into valuable materials. *Separation and Purification Technology, 330*, 125220. https://doi.org/10.1016/j.seppur.2023.125220

Kumar, M. M., Karthikeyan, R., Anbalagan, K., & Bhanushali, M. N. (2016). Coagulation process for tannery industry effluent treatment using Moringa oleifera seeds protein: Kinetic study, pH effect on floc characteristics and design of a thickener unit. *Separation Science and Technology, 51*, 2028–2037. https://doi.org/10.1080/01496395.2016.1190378

Kumari, H., Sonia, Suman, Ranga, R., Chahal, S., Devi, S., Sharma, S., Kumar, S., Kumar, P., Kumar, S., Kumar, A., & Parmar, R. (2023). *A review on photocatalysis used for wastewater treatment: Dye degradation*. Springer International Publishing.

Kuokkanen, V., Lassi, U., Kuokkanen, T., & Rämö, J. (2013). Recent applications of electrocoagulation in treatment of water and wastewater—A review. *Green and Sustainable Chemistry, 3*, 89–121. https://doi.org/10.4236/gsc.2013.32013

Kurt, U., Apaydin, O., & Gonullu, M. T. (2007). Reduction of COD in wastewater from an organized tannery industrial region by Electro-Fenton process. *Journal of Hazardous Materials, 143*, 33–40. https://doi.org/10.1016/j.jhazmat.2006.08.065

Kuspanov, Z., Bakbolat, B., Baimenov, A., Issadykov, A., Yeleuov, M., & Daulbayev, C. (2023). Photocatalysts for a sustainable future: Innovations in large-scale environmental and energy applications. *Science of the Total Environment, 885*, 163914. https://doi.org/10.1016/j.scitotenv.2023.163914

Leta, S., Assefa, F., Gumaelius, L., & Dalhammar, G. (2004). Biological nitrogen and organic matter removal from tannery wastewater in pilot plant operations in Ethiopia. *Applied Microbiology and Biotechnology, 66*, 333–339. https://doi.org/10.1007/s00253-004-1715-2

Li, T., Yang, T., Yu, Z., Xu, G., Han, Q., Luo, G., Du, J., Guan, Y., & Guo, C. (2021). Trivalent chromium removal from tannery wastewater with low cost bare magnetic Fe_3O_4 nanoparticles. *Chemical Engineering and Processing—Process Intensification, 169*, 108611. https://doi.org/10.1016/j.cep.2021.108611

Li, X., Chen, Y., Tao, Y., Shen, L., Xu, Z., Bian, Z., & Li, H. (2022). Challenges of photocatalysis and their coping strategies. *Chem Catalysis, 2*, 1315–1345. https://doi.org/10.1016/j.checat.2022.04.007

Li, Z., Yu, D., Wang, X., Liu, X., Xu, Z., & Wang, Y. (2024). A novel strategy of tannery sludge disposal—Converting into biochar and reusing for Cr(VI) removal from tannery wastewater. *Journal of Environmental Sciences, 138*, 637–649. https://doi.org/10.1016/j.jes.2023.04.014

Liu, W., Zhang, C., Gao, P., & Liu, H. (2012). Combined advanced oxidation and biological treatment of tannery effluent. *Clean Technologies and Environmental Policy, 14*(2), 251–256.

Liu, J., Zhao, Y., Ma, J., Dai, Y., Li, J., & Zhang, J. (2016). Flower-like ZnO hollow microspheres on ceramic mesh substrate for photocatalytic reduction of Cr(VI) in tannery wastewater. *Ceramics International, 42*, 15968–15974. https://doi.org/10.1016/j.ceramint.2016.07.098

Liu, Z., Lei, M., Chen, G., & Yuan, J. (2022). Treatment of chromium removal wastewater from tanning by a new coupling technology. *Processes, 10*.

Lofrano, G., Belgiorno, V., Gallo, M., Raimo, A., & Meric, S. (2006). Toxicity reduction in leather tanning wastewater by improved coagulation flocculation process. *Glob NEST Journal, 8*, 151–158.

Lofrano, G., Meriç, S., Zengin, G. E., & Orhon, D. (2013). Chemical and biological treatment technologies for leather tannery chemicals and wastewaters: A review. *Science of the Total Environment, 461–462*, 265–281. https://doi.org/10.1016/j.scitotenv.2013.05.004

Lofrano, G., Belgiorno, V., Gallo, M., Raimo, A., & Meric, S. (2018). Toxicity reduction in leather tanning wastewater by improved coagulation flocculation process. *Glob NEST Journal, 8*, 151–158. https://doi.org/10.30955/gnj.000386

Ma, J., Yang, N., Shi, J., & Yang, X. (2022). Tailoring the structure of layered double oxide for enhancing adsorption of anionic chemicals on collagen fibers in chrome-free leather processing. *Clean Technologies and Environmental Policy, 24*, 2947–2955. https://doi.org/10.1007/s10098-022-02366-2

Mahmood, S., Khalid, A., Mahmood, T., Arshad, M., & Ahmad, R. (2013). Potential of newly isolated bacterial strains for simultaneous removal of hexavalent chromium and reactive black-5 azo dye from tannery effluent. *Journal of Chemical Technology and Biotechnology, 88*, 1506–1513. https://doi.org/10.1002/jctb.3994

Makos, J. D., & Hrncir, D. C. (1995). Chemistry of Cr(VI) in a constructed wetland. *Environmental Science and Technology, 29*, 2414–2419. https://doi.org/10.1021/es00009a038

Marichamy, M. K., Kumaraguru, A., & Jonna, N. (2021). Particle size distribution modeling and kinetic study for coagulation treatment of tannery industry wastewater at response surface optimized condition. *Journal of Cleaner Production, 297*, 126657. https://doi.org/10.1016/j.jclepro.2021.126657

De Martino, A., Iorio, M., & Capasso, R. (2013). Sustainable sorption strategies for removing Cr^{3+} from tannery process wastewater. *Chemosphere, 92*, 1436–1441. https://doi.org/10.1016/j.chemosphere.2013.03.051

Masindi, V., Tekere, M., & Foteinis, S. (2023). Treatment of real tannery wastewater using facile synthesized magnesium oxide nanoparticles: Experimental results and geochemical modeling. *Water Resources and Industry, 29*, 100205. https://doi.org/10.1016/j.wri.2023.100205

Mekonnen, E., Yitbarek, M., & Soreta, T. R. (2015). Kinetic and thermodynamic studies of the adsorption of Cr(VI) onto some selected local adsorbents. *South African Journal of Chemistry, 68*, 45–52.

Mella, B., Benvenuti, J., Oliveira, R. F., & Gutterres, M. (2019). Preparation and characterization of activated carbon produced from tannery solid waste applied for tannery wastewater treatment. *Environmental Science and Pollution Research, 26*, 6811–6817. https://doi.org/10.1007/s11356-019-04161-x

Mella, B., Puchana-Rosero, M. J., Costa, D. E. S., & Gutterres, M. (2017). Utilization of tannery solid waste as an alternative biosorbent for acid dyes in wastewater treatment. *Journal of Molecular Liquids, 242*, 137–145. https://doi.org/10.1016/j.molliq.2017.06.131

Mella, B., de Barcellos, B. S. C., da Silva Costa, D. E., & Gutterres, M. (2018). Treatment of leather dyeing wastewater with associated process of coagulation-flocculation/adsorption/ozonation. *Ozone: Science & Engineering, 40*, 133–140. https://doi.org/10.1080/01919512.2017.1346464

Mendoza-Roca, J. A., Galiana-Aleixandre, M. V., Lora-García, J., & Bes-Piá, A. (2010). Purification of tannery effluents by ultrafiltration in view of permeate reuse. *Separation and Purification Technology, 70*, 296–301. https://doi.org/10.1016/j.seppur.2009.10.010

Mim, S., Hashem, M. A., & Payel, S. (2023). Coagulation-adsorption-oxidation for removing dyes from tannery wastewater. *Environmental Monitoring and Assessment, 195*, 695. https://doi.org/10.1007/s10661-023-11309-3

Moges, A., Nkambule, T. T. I., & Fito, J. (2022). The application of GO-Fe_3O_4 nanocomposite for chromium adsorption from tannery industry wastewater. *Journal of Environmental Management, 305*, 114369. https://doi.org/10.1016/j.jenvman.2021.114369

Mohammed, K., & Sahu, O. (2019). Recovery of chromium from tannery industry waste water by membrane separation technology: Health and engineering aspects. *Scientific African, 4*, e00096. https://doi.org/10.1016/j.sciaf.2019.e00096

Monteiro, F. C., Guimaraes, I. D. L., de Almeida Rodrigues, P., da Anunciação de Pinho, J. V., & Conte-Junior, C. A. (2023). Degradation of PAHs using TiO_2 as a semiconductor in the heterogeneous photocatalysis process: A systematic review. *Journal of Photochemistry and Photobiology A-Chemistry, 437*, 114497. https://doi.org/10.1016/j.jphotochem.2022.114497

Moradi, M., & Moussavi, G. (2019). Enhanced treatment of tannery wastewater using the electrocoagulation process combined with UVC/VUV photoreactor: Parametric and mechanistic evaluation. *Chemical Engineering Journal, 358*, 1038–1046. https://doi.org/10.1016/j.cej.2018.10.069

Morris, B. (2003). The components of the wired spanning forest are recurrent. *Probability Theory and Related Fields, 125*, 259–265. https://doi.org/10.1007/s00440-002-0236-0

Munz, G., Gori, R., Cammilli, L., & Lubello, C. (2008). Characterization of tannery wastewater and biomass in a membrane bioreactor using respirometric analysis. *Bioresource Technology, 99*, 8612–8618. https://doi.org/10.1016/j.biortech.2008.04.004

Mustapha, S., Tijani, J. O., Ndamitso, M. M., Abdulkareem, S. A., Shuaib, D. T., Mohammed, A. K., & Sumaila, A. (2020). The role of kaolin and kaolin/ZnO nanoadsorbents in adsorption studies for tannery wastewater treatment. *Science and Reports, 10*, 13068. https://doi.org/10.1038/s41598-020-69808-z

Mustapha, S., Tijani, J. O., Egbosiuba, T. C., Sumaila, A., Amigun, T. A., Salihu, A. B., Ibrahim, Y. O., Ndamitso, M. M., & Abdulkareem, S. A. (2022). Sol–gel synthesis of Kaolin/TiO_2 nanocomposites for photocatalytic degradation of tannery wastewater BT—Industrial wastewater treatment: Emerging technologies for sustainability. In R. R. Karri & M. H. Dehghani (Eds.), *Karchiyappan T* (pp. 323–343). Springer International Publishing.

Mustapha, S., Ndamitso, M. M., Abdulkareem, A. S., Tijani, J. O., Mohammed, A. K., & Shuaib, D. T. (2019). Potential of using kaolin as a natural adsorbent for the removal of pollutants from tannery wastewater. *Heliyon, 5*, e02923. https://doi.org/10.1016/j.heliyon.2019.e02923

Nayeri, D., & Mousavi, S. A. (2022). A comprehensive review on the coagulant recovery and reuse from drinking water treatment sludge. *Journal of Environmental Management, 319*, 115649. https://doi.org/10.1016/j.jenvman.2022.115649

Nazer, D. W., Al-Sa'ed, R. M., & Siebel, M. A. (2006). Reducing the environmental impact of the unhairing–liming process in the leather tanning industry. *Journal of Cleaner Production, 14*, 65–74. https://doi.org/10.1016/j.jclepro.2005.04.002

Nigam, M., Rajoriya, S., Rani Singh, S., & Kumar, P. (2019). Adsorption of Cr (VI) ion from tannery wastewater on tea waste: Kinetics, equilibrium and thermodynamics studies. *Journal of Environmental Chemical Engineering* 7, 103188. https://doi.org/10.1016/j.jece.2019.103188

Nure, J. F., Mengistu, A., Abewaa, M., Angassa, K., Moyo, W., Phiri, Z., Mafa, P. J., Kuvarega, A. T., & Nkambule, T. T. I. (2023). Adsorption of Black MNN reactive dye from tannery wastewater using activated carbon of Rumex Abysinicus. *Journal of the Taiwan Institute of Chemical Engineers, 151*, 105138. https://doi.org/10.1016/j.jtice.2023.105138

Nur-E-Alam, M., Abu Sayid Mia, M., Ahmad, F., & Mafizur Rahman, M. (2018). Adsorption of chromium (Cr) from tannery wastewater using low-cost spent tea leaves adsorbent. *Applied Water Science, 8*, 129. https://doi.org/10.1007/s13201-018-0774-y

Nur-E-Alam, M., Mia, M. A. S., Ahmad, F., & Rahman, M. M. (2020). An overview of chromium removal techniques from tannery effluent. *Applied Water Science, 10*, 205. https://doi.org/10.1007/s13201-020-01286-0

Oladipo, A. A., Adeleye, O. J., Oladipo, A. S., & Aleshinloye, A. O. (2017). Bio-derived MgO nanopowders for BOD and COD reduction from tannery wastewater. *Journal of Water Process Engineering, 16*, 142–148. https://doi.org/10.1016/j.jwpe.2017.01.003

Oladoja, N. A. (2017). Appropriate technology for domestic wastewater management in under-resourced regions of the world. *Applied Water Science, 7*, 3391–3406. https://doi.org/10.1007/s13201-016-0495-z

Oladoja, N. A., Saliu, T. D., Ololade, I. A., Anthony, E. T., & Bello, G. A. (2017). A new indigenous green option for turbidity removal from aqueous system. *Separation and Purification Technology, 186*, 166–174. https://doi.org/10.1016/j.seppur.2017.05.054

Oller, I., Malato, S., & Sánchez-Pérez, J. A. (2011). Combination of advanced oxidation processes and biological treatments for wastewater decontamination—A review. *Science of the Total Environment, 409*, 4141–4166. https://doi.org/10.1016/j.scitotenv.2010.08.061

Ortega, L. M., Lebrun, R., Noël, I. M., & Hausler, R. (2005). Application of nanofiltration in the recovery of chromium (III) from tannery effluents. *Separation and Purification Technology, 44*, 45–52. https://doi.org/10.1016/j.seppur.2004.12.002

Ortiz-Monsalve, S., Valente, P., Poll, E., Jaramillo-García, V., Pegas Henriques, J. A., & Gutterres, M. (2019). Biodecolourization and biodetoxification of dye-containing wastewaters from leather dyeing by the native fungal strain Trametes villosa SCS-10. *Biochemical Engineering Journal, 141*, 19–28. https://doi.org/10.1016/j.bej.2018.10.002

Paisio, C. E., Quevedo, M. R., Talano, M. A., González, P. S., & Agostini, E. (2014). Application of two bacterial strains for wastewater bioremediation and assessment of phenolics biodegradation. *Environmental Technology (United Kingdom), 35*, 1802–1810. https://doi.org/10.1080/09593330.2014.882994

Pal, M., Malhotra, M., Mandal, M. K., Paine, T. K., & Pal, P. (2020). Recycling of wastewater from tannery industry through membrane-integrated hybrid treatment using a novel graphene oxide nanocomposite. *Journal of Water Process Engineering, 36*, 101324. https://doi.org/10.1016/j.jwpe.2020.101324

Palani, Y., Rao Jonnalagadda, R., & Fathima Nishter, N. (2017). Adsorption on activated carbon derived from tannery fleshing waste: Adsorption isotherms, thermodynamics, and kinetics. *Environmental Progress & Sustainable Energy, 36*, 1725–1733.

Pan, T., Zhu, Z. L., Xu, G. M., & Chen, X. J. (2007). Vertical bio-filter technique for decentralized wastewater system in a case of Sino-German technical cooperation pilot project in Changzhou city. In *6th Minist Forum Infrastruct Dev Asia-Pacific Reg 2nd International Conference on China Urban Water Development*, pp. 700–706.

Parthasarathy, P., & Narayanan, S. K. (2014). Effect of hydrothermal carbonization reaction parameters on. *Environmental Progress & sustainable Energy, 33*, 676–680.

Pasciucco, F., Pecorini, I., & Iannelli, R. (2022). Planning the centralization level in wastewater collection and treatment: A review of assessment methods. *Journal of Cleaner Production, 375*, 134092. https://doi.org/10.1016/j.jclepro.2022.134092

Paul, S., Dey, T. K., & Raha, R. (2018). Tannery waste water treatment through aggregation of bio-sorption and chemical coagulation aided by indigenous coagulant. *International Journal of Current Engineering and Technology, 8*. https://doi.org/10.14741/ijcet/v.8.4.14

Pearson, R. G. (1963). Hard and soft acids and bases. *Journal of the American Chemical Society, 85*, 3533–3539. https://doi.org/10.1021/ja00905a001

Precious Sibiya, N., Rathilal, S., & Kweinor Tetteh, E. (2021). Coagulation treatment of wastewater: Kinetics and natural coagulant evaluation. *Molecules, 26*.

Rahaman, A., Hosen, M. R., Hena, M., Naher, U., & Moniruzzaman, M. (2016). A study on removal of chromium from tannery effluent treatment of chrome tanning waste water using tannery solid waste, 237–242. https://doi.org/10.22034/ijhcum.2016.04.001

Rambabu, K., & Velu, S. (2016). Modified polyethersulfone ultrafiltration membrane for the treatment of tannery wastewater. *International Journal of Environmental Studies, 73*, 819–826. https://doi.org/10.1080/00207233.2016.1153900

Ranjan, R., Kumar, L., & Sabumon, P. C. (2020). Process performance and reuse potential of a decentralized wastewater treatment system. *Water Science and Technology, 80*, 2079–2090. https://doi.org/10.2166/wst.2020.046

Rees, C. A., Provis, J. L., Lukey, G. C., & van Deventer, J. S. J. (2007). In situ ATR-FTIR study of the early stages of fly ash geopolymer gel formation. *Langmuir, 23*, 9076–9082. https://doi.org/10.1021/la701185g

Religa, P., Kowalik, A., Gierycz, P. (2011). Application of nanofiltration for chromium concentration in the tannery wastewater. *Journal of Hazardous Materials, 186*, 288–292. https://doi.org/10.1016/j.jhazmat.2010.10.112

Renugadevi, R., Venkatachalam, T., Narayanasamy, R., & Dinesh Kirupha, S. (2016) Preparation of Co doped TiO_2 nano thin films by Sol Gel technique and photocatalytic studies of prepared films in tannery effluent. *Optik (Stuttg), 127*, 10127–10134. https://doi.org/10.1016/j.ijleo.2016.07.090

Reyes-Serrano, A., López-Alejo, J. E., Hernández-Cortázar, M. A., & Elizalde, I. (2020). Removing contaminants from tannery wastewater by chemical precipitation using CaO and $Ca(OH)_2$. *Chinese Journal of Chemical Engineering, 28*, 1107–1111. https://doi.org/10.1016/j.cjche.2019.12.023

Rezgui, S., Ghazouani, M., Bousselmi, L., & Akrout, H. (2022). Efficient treatment for tannery wastewater through sequential electro-Fenton and electrocoagulation processes. *Journal of Environmental Chemical Engineering, 10*, 107424. https://doi.org/10.1016/j.jece.2022.107424

Riyat, M. R. I., Salam, A., Molla, M. T. H., Islam, M. S., Bashar, M. A., Chandra, D., Ahsan, S., Roy, D., & Ahsan, M. S. (2022). Magnetically recyclable core–shell structured $Co_{0.5}Zn_{0.5}Fe_2O_4$@polyaniline nanocomposite: High stability and rapid photocatalytic degradation of commercial azo dyes and industrial effluents. *Reaction Kinetics, Mechanisms and Catalysis, 135*, 1077–1098. https://doi.org/10.1007/s11144-022-02182-1

Romero-Dondiz, E. M., Almazán, J. E., Rajal, V. B., & Castro-Vidaurre, E. F. (2015). Removal of vegetable tannins to recover water in the leather industry by ultrafiltration polymeric membranes. *Chemical Engineering Research and Design, 93*, 727–735. https://doi.org/10.1016/j.cherd.2014.06.022

Romero-Dondiz, E. M., & Almazán, J. E., Rajal, V. B., & Castro-Vidaurre, E. F. (2016) Comparison of the performance of ultrafiltration and nanofiltration membranes for recovery and recycle of tannins in the leather industry. *Journal of Cleaner Production, 135*, 71–79. https://doi.org/10.1016/j.jclepro.2016.06.096

Roy, K., Dey, T. K., Zuha, S. T., Jamal, M., Srivastava, M., & Uddin, M. E. (2023b). Removal of turbidity from tannery wastewater using graphene oxide-ferric oxide nanocomposites as an adsorbent. *International Journal of Environmental Science and Technology, 20*, 5597–5608. https://doi.org/10.1007/s13762-022-04301-w

Roy, K., Dey, T. K., Jamal, M., Rathanasamy, R., Chinnasamy, M., & Uddin, M. E. (2023a). Fabrication of graphene oxide–keratin–chitosan nanocomposite as an adsorbent to remove turbidity from tannery wastewater. *Water Science and Engineering, 16*, 184–191. https://doi.org/10.1016/j.wse.2022.12.003

Sa'ad, A. M., Abubakar, U. A., Usman, Y. M., Suleiman, FUI, & H S L (2022). Adsorption potentials of rice husk for treatment of tannery effluent in Zaria, Nigeria. *International Journal of Information Technology, II*, 99–115.

Sabur, M. A. (2012). Treatment of textile wastewater by coagulation precipitation method. *Journal of Scientific Research, 4*, 623–633. https://doi.org/10.3329/jsr.v4i3.10777

Saira, G. C., & Shanthakumar, S. (2023). Zero waste discharge in tannery industries—An achievable reality? A recent review. *Journal of Environmental Management, 335*, 117508. https://doi.org/10.1016/j.jenvman.2023.117508

Sankaranarayannan, S., & Gowdam, A. (2020). An experimental study on treatment of industry waste water by natural coagulants 7, 81–84.

Saranya, D., & Shanthakumar, S. (2020). An integrated approach for tannery effluent treatment with ozonation and phycoremediation: A feasibility study. *Environmental Research, 183*. https://doi.org/10.1016/j.envres.2020.109163

Sawalha, H., Alsharabaty, R., Sarsour, S., & Al-Jabari, M. (2019). Wastewater from leather tanning and processing in Palestine: Characterization and management aspects. *Journal of Environmental Management, 251*, 109596. https://doi.org/10.1016/j.jenvman.2019.109596

Sawalha, H., Al-Jabari, M., Elhamouz, A., Abusafa, A., & Rene, E. R. (2020). Chapter 25—Tannery wastewater treatment and resource recovery options. In: T. Bhaskar, A. Pandey, E. R. Rene, & Tsang, DCWBT-WB (Eds.) (pp. 679–705). Elsevier.

Schrank, S. G., José, H. J., Moreira, R. F. P. M., & Schröder, H. F. (2004b). Comparison of different advanced oxidation process to reduce toxicity and mineralisation of tannery wastewater. *Water Science and Technology, 50*, 329–334. https://doi.org/10.2166/wst.2004.0345

Schrank, S. G., José, H. J., Moreira, R. F. P. M., & Schröder, H. F. (2004a). Elucidation of the behavior of tannery wastewater under advanced oxidation conditions. *Chemosphere 56*, 411–423. https://doi.org/10.1016/j.chemosphere.2004.04.012

Schrank, S. G., José, H. J., Moreira, R. F. P. M., & Schröder, H. F. (2005). Applicability of Fenton and H_2O_2/UV reactions in the treatment of tannery wastewaters. *Chemosphere, 60*, 644–655. https://doi.org/10.1016/j.chemosphere.2005.01.033

Shahbazi, R., & Pedram, M. Z. (2021). Continuous photocatalytic set-up assisted with nano TiO_2 plate for tannery wastewater treatment. *Water Science and Technology, 83*, 2732–2743. https://doi.org/10.2166/wst.2021.164

Sivagami, K., Sakthivel, K. P., & Nambi, I. M. (2018) Advanced oxidation processes for the treatment of tannery wastewater. *Journal of Environmental Chemical Engineering, 6*, 3656–3663. https://doi.org/10.1016/j.jece.2017.06.004

Sivaprakasam, S., Mahadevan, S., Sekar, S., & Rajakumar, S. (2008). Biological treatment of tannery wastewater by using salt-tolerant bacterial strains. *Microbial Cell Factories, 7*, 1–7. https://doi.org/10.1186/1475-2859-7-15

Sivaram, N. M., & Barik, D. (2019). Chapter 5—Toxic waste from leather industries. In: Barik, DBT-E from TOW for H and PG (Eds.) Woodhead Publishing Series in Energy (pp. 55–67). Woodhead Publishing.

Solanki, M., Sundaramurthy, S., Das, S. N., & Shukla, K. (2013). Treatment of real textile wastewater using coagulation technology. *International Journal of ChemTech Research, 5*, 610–615.

Song, Z., Williams, C. J., & Edyvean, R. G. J. (2004). Treatment of tannery wastewater by chemical coagulation. *Desalination, 164*, 249–259. https://doi.org/10.1016/S0011-9164(04)00193-6

Song, Z., Williams, C. J., & Edyvean, R. G. J. (2000) Sedimentation of tannery wastewater. *Water Research, 34*, 2171–2176. https://doi.org/10.1016/S0043-1354(99)00358-9

Srinivasan, S. V., Prea Samita Mary, G., Kalyanaraman, C., Sureshkumar, P. S., Sri Balakameswari, K., Suthanthararajan, R., & Ravindranath, E. (2012). Combined advanced oxidation and biological treatment of tannery effluent. *Clean Technologies and Environmental Policy, 14*, 251–256. https://doi.org/10.1007/s10098-011-0393-x

Streit, K. F., Ferreira, J. Z., Bernardes, A. M., & Norberta De Pinho, M. (2009). Ultrafiltration/nanofiltration for the tertiary treatment of leather industry effluents. *Environmental Science and Technology, 43*, 9130–9135. https://doi.org/10.1021/es902105q

Suman, H., Sangal, V. K., & Vashishtha, M. (2021). Treatment of tannery industry effluent by electrochemical methods: A review. *Materials Today: Proceedings, 47*, 1438–1444. https://doi.org/10.1016/j.matpr.2021.03.300

Sungur, Ş., & Özkan, A. (2017). Characterization of wastewaters obtained from Hatay tanneries. *International Journal of Natural and Engineering Sciences, 2*, 111–118. https://doi.org/10.28978/nesciences.330599

Suthanthararajan, R., Ravindranath, E., Chits, K., Umamaheswari, B., Ramesh, T., & Rajamam, S. (2004). Membrane application for recovery and reuse of water from treated tannery wastewater. *Desalination, 164*, 151–156. https://doi.org/10.1016/S0011-9164(04)00174-2

Sutherland, R. (1947). Tanning industry. *Industrial & Engineering Chemistry, 39*, 628–631. https://doi.org/10.1021/ie50449a017

Tahir, S.S., & Naseem, R. (2007). Removal of Cr(III) from tannery wastewater by adsorption onto bentonite clay. *Separation and Purification Technology, 53*, 312–321. https://doi.org/10.1016/j.seppur.2006.08.008

Tao, E., Ma, D., Yang, S., Sun, Y., Xu, J., & Kim, E. J. (2019). Zirconium dioxide loaded montmorillonite composites as high-efficient adsorbents for the removal of Cr^{3+} ions from tanning wastewater. *Journal of Solid State Chemistry, 277*, 502–509. https://doi.org/10.1016/j.jssc.2019.07.002

Tianzhi, W., Weijie, W., Hongying, H., & Khu, S.-T. (2021). Effect of coagulation on bio-treatment of textile wastewater: Quantitative evaluation and application. *Journal of Cleaner Production, 312*, 127798. https://doi.org/10.1016/j.jclepro.2021.127798

Tripathi, A., & Narayanan, S. (2018). Impact of TiO_2 and TiO_2/g-C_3N_4 nanocomposite to treat industrial wastewater. *Environmental Nanotechnology, Monitoring & Management, 10*, 280–291. https://doi.org/10.1016/j.enmm.2018.07.010

Tsamo, C., Zangue, A. H., Herbaud, E. E., & Nchang, S. T. D. (2023). Investigating the potential of using tannery process solid hair waste for spent engine and vegetable oils removal from water and toxicity assessment using maize. *Case Studies in Chemical and Environmental Engineering, 8*, 100412. https://doi.org/10.1016/j.cscee.2023.100412

Tünay, O., Kabdasli, I., Orhon, D., & Ates, E. (1995). Characterization and pollution profile of leather tanning industry in Turkey. *Water Science and Technology, 32*, 1–9. https://doi.org/10.1016/0273-1223(96)00132-1

Turunen, J., Karppinen, A., & Ihme, R. (2019). Effectiveness of biopolymer coagulants in agricultural wastewater treatment at two contrasting levels of pollution. *SN Appl Sci, 1*, 210. https://doi.org/10.1007/s42452-019-0225-x

Urbina-Suarez, N. A., Machuca-Martínez, F., & Barajas-Solano, A. F. (2021). Advanced oxidation processes and biotechnological alternatives for the treatment of tannery wastewater. *Molecules, 26*.

Vadasarukkai, Y. S., & Gagnon, G. A. (2017). Influence of the mixing energy consumption affecting coagulation and floc aggregation. *Environmental Science and Technology, 51*, 3480–3489. https://doi.org/10.1021/acs.est.6b06281

Vaiano, V., & Iervolino, G. (2018). Facile method to immobilize ZnO particles on glass spheres for the photocatalytic treatment of tannery wastewater. *Journal of Colloid and Interface Science, 518*, 192–199. https://doi.org/10.1016/j.jcis.2018.02.033

Vaiano, V., Sacco, O., Stoller, M., Chianese, A., Ciambelli, P., & Sannino, D. (2014). Influence of the photoreactor configuration and of different light sources in the photocatalytic treatment of highly polluted wastewater 12, 63–75. https://doi.org/10.1515/ijcre-2013-0090

Vankar, P. S., & Bajpai, D. (2008). Phyto-remediation of chrome-VI of tannery effluent by Trichoderma species. *Desalination, 222*, 255–262. https://doi.org/10.1016/j.desal.2007.01.168

Varank, G., Erkan, H., Yazýcý, S., Demir, A., & Engin, G. (2014). Electrocoagulation of tannery wastewater using monopolar electrodes: Process optimization by response surface methodology. *International Journal of Environmental Research, 8*, 165–180.

Vijayalakshmi, P., Raju, G. B., & Gnanamani, A. (2011). Advanced oxidation and electrooxidation as tertiary treatment techniques to improve the purity of tannery wastewater. *Industrial and Engineering Chemistry Research, 50*, 10194–10200. https://doi.org/10.1021/ie201039z

Wahyuni, E. T., Wahyuni, S., Lestari, N. D., & Suherman, S. (2023). Utilization of tannery wastewater as a source of Cr doped into TiO_2 for improving its activity under visible light in the Congo red degradation. *Reaction Kinetics, Mechanisms and Catalysis, 136*, 1067–1084. https://doi.org/10.1007/s11144-023-02367-2

Wang, M., Webber, M., Finlayson, B., & Barnett, J. (2008). Rural industries and water pollution in China. *Journal of Environmental Management, 86*, 648–659. https://doi.org/10.1016/j.jenvman.2006.12.019

Wang, L., Huang, X., Han, M., Lyu, L., Li, T., Gao, Y., Zeng Q, Hu C (2019a) Efficient inhibition of photogenerated electron-hole recombination through persulfate activation and dual-pathway degradation of micropollutants over iron molybdate. Appl Catal B Environ 257:117904. https://doi.org/10.1016/j.apcatb.2019.117904

Wang, S., Li, L., Zhu, Z., Zhao, M., Zhang, L., Zhang, N., Wu, Q., Wang, X., & Li, G. (2019b). Remarkable improvement in photocatalytic performance for tannery wastewater processing via SnS2 modified with N-doped carbon quantum dots: Synthesis, characterization, and 4-nitrophenol-aided Cr(VI) photoreduction. *Small, 15*, 1804515. https://doi.org/10.1002/smll.201804515

Wang, X., Faungnawakij, K., & Chareonpanich, M. (2019c). Editorial: photocatalysis—From solar power to sustainable chemical production. *ChemCatChem, 11*, 5838–5841. https://doi.org/10.1002/cctc.201902248

Wang, Y., & Hu, A. (2014). Carbon quantum dots: Synthesis, properties and applications. *Journal of Materials Chemistry C, 2*, 6921–6939. https://doi.org/10.1039/C4TC00988F

Wei, H., Gao, B., Ren, J., Li, A., & Yang, H. (2018). Coagulation/flocculation in dewatering of sludge: A review. *Water Research, 143*, 608–631. https://doi.org/10.1016/j.watres.2018.07.029

Wiggins, S., & Proctor, S. (2001). How special are rural areas? The economic implications of location for rural development. *Development Policy Review, 19*, 427–436. https://doi.org/10.1111/1467-7679.00142

Yadav, A., Singh, K., Panda, A. B., Labhasetwar, P. K., & Shahi, V. K. (2021). Membrane distillation crystallization for simultaneous recovery of water and salt from tannery industry wastewater using TiO_2 modified poly(vinylidene fluoride-co-hexafluoropropylene) nanocomposite membranes. *Journal of Water Process Engineering, 44*, 102393. https://doi.org/10.1016/j.jwpe.2021.102393

Yadav, A., & Pandey, J. (2017). Contribution of point sources and non-point sources to nutrient and carbon loads and their influence on the trophic status of the Ganga River at Varanasi, India. *Environmental Monitoring and Assessment, 189*, 475. https://doi.org/10.1007/s10661-017-6188-8

Yahya, M. D., Obayomi, K. S., Abdulkadir, M. B., Iyaka, Y. A., & Olugbenga, A. G. (2020b). Characterization of cobalt ferrite-supported activated carbon for removal of chromium and lead ions from tannery wastewater via adsorption equilibrium. *Water Science and Engineering, 13*, 202–213. https://doi.org/10.1016/j.wse.2020.09.007

Yahya, M. D., Abubakar, H., Obayomi, K. S., Iyaka, Y. A., & Suleiman, B. (2020a). Simultaneous and continuous biosorption of Cr and Cu (II) ions from industrial tannery effluent using almond shell in a fixed bed column. *Results in Engineering, 6*, 100113. https://doi.org/10.1016/j.rineng.2020.100113

Yang, F., Wang, X.-B., Shan, Y., Wu, C., Zhou, R., Hengl, N., Pignon, F., & Jin, Y. (2023). Research recap of membrane technology for tannery wastewater treatment: A review. *Collagen and Leather, 5*, 24. https://doi.org/10.1186/s42825-023-00132-8

Yang, F., Huang, Z., Huang, J., Wu, C., Zhou, R., & Jin, Y. (2021). Tanning wastewater treatment by ultrafiltration: Process efficiency and fouling behavior. *Membranes (Basel), 11*.

Yargeau, V. (2012). 17—Water and wastewater treatment: Chemical processes. In F. B. T.-M. S. Zeman (Ed.), Woodhead Publishing Series in Energy (pp. 390–405). Woodhead Publishing.

Younas, F., Niazi, N. K, Bibi, I., Afzal, M., Hussain, K., Shahid, M., Aslam, Z., Bashir, S., Hussain, M. M., & Bundschuh, J. (2022). Constructed wetlands as a sustainable technology for wastewater treatment with emphasis on chromium-rich tannery wastewater. *Journal of Hazardous Materials, 422*, 126926. https://doi.org/10.1016/j.jhazmat.2021.126926

Zapana, J. S. P., Arán, D. S., Bocardo, E. F., & Harguinteguy, C. A. (2020). Treatment of tannery wastewater in a pilot scale hybrid constructed wetland system in Arequipa, Peru. *International Journal of Environmental Science and Technology, 17*, 4419–4430. https://doi.org/10.1007/s13762-020-02797-8

Zhang, H., He, X., Zhao, W., Peng, Y., Sun, D., Li, H., & Wang, X. (2017). Preparation of Fe_3O_4/TiO_2 magnetic mesoporous composites for photocatalytic degradation of organic pollutants. *Water Science and Technology, 75*, 1523–1528. https://doi.org/10.2166/wst.2017.002

Zhao, C., & Chen, W. (2019). A review for tannery wastewater treatment: Some thoughts under stricter discharge requirements. *Environmental Science and Pollution Research, 26*, 26102–26111. https://doi.org/10.1007/s11356-019-05699-6

Zheng, D., Qin, L., Wang, T., Ren, X., Zhang, Z., & Li, J. (2014). Coagulation pretreatment of highly concentrated acrylonitrile wastewater from petrochemical plants. *Water Science and Technology, 70*, 345–351. https://doi.org/10.2166/wst.2014.235

Zhou, H., Tan, Z., & Li, X. (2012). Assessment of wastewater pollution in pig leather industry in China. *Water and Environment Journal, 26*, 521–529. https://doi.org/10.1111/j.1747-6593.2012.00312.x

Zouboulis, A., Traskas, G., & Samaras, P. (2008). Comparison of efficiency between poly-aluminium chloride and aluminium sulphate coagulants during full-scale experiments in a drinking water treatment plant. *Separation Science and Technology, 43*, 1507–1519. https://doi.org/10.1080/01496390801940903

Strategic Management of Wastewater from Natural Rubber Processing Industries

Sunday Chukwuemeka Agbo and Chimaine Feudjio Tsague

Abstract With rapid industrialization, the agro-based rubber business provides raw materials for numerous industries. Wastewater pollution from rubber processing industries has been on the increase in the countries of Africa and Asia because of the growing number of mature rubber trees. During the various production processes, a lot of chemicals and water are used, which leads to the generation of large amounts of highly polluted wastewater. This wastewater is characterized by high toxicity, strong odour and colour, low pH, high suspended solids, high temperature, and high chemical and biological oxygen demand (i.e. COD and BOD). These undesirable characteristics make it imperative to treat the highly polluted raw rubber processing wastewater before it is discharged into water bodies. This chapter is a discourse on the importance of efficient management of wastewater from natural rubber processing industries. It highlights the physicochemical characteristics of the wastewater and its impact on the environment. Special emphasis is laid on the conventional and non-conventional technologies that are being used in the management of wastewater from the natural rubber industry.

Keywords Waste water · Natural rubber · Industries · Management · Environmental pollutions

S. C. Agbo (✉)
Ceramics Research and Production Department (CRPD), Projects Development Institute (PRODA), Emene Industrial Layout, Emene, Enugu State, Nigeria
e-mail: sc.agbo@stu.unizik.edu.ng

C. F. Tsague
Department of Inorganic Chemistry, Faculty of Sciences, University of Yaoundé 1, Yaoundé, Cameroon
e-mail: feutsachi@gmail.com

N. A. Oladoja and E. I. Unuabonah (eds.), *Strategic Management of Wastewater from Intensive Rural Industries*, Springer Water,
https://doi.org/10.1007/978-3-031-90314-4_7

1 Introduction

The rubber industry (Fig. 1) is one of the steadily growing industries in the world because it serves as a raw material for many industries, such as automobiles, footwear, construction industries, and agro- and agricultural industries (Pillai & Girish, 2014). The rubber processing industry is one of the highest wastewater generators that pollutes water among other industries (Agtas et al., 2021). The rubber waste is composed of rubber molecules, which is an isoprene polymer (Cis-l, 4-polyisoprene [C5H8] n) (Fig. 2), protein, sugar, and organic salts that originate from natural rubber and processing operations (Viet., 1999). The wastewater from the rubber industry generates odour and contamination due to the degradation of proteins, lipids, Lutoid, Frey-Wyssling particles, and other inorganic and organic salts (Fig. 3) (Rodríguez Urbina et al., 2023; Ruangdit, 2022). Rubber industries use a lot of chemicals for the processing of rubber; among them are caustic soda, ammonia, and sulfuric acid, which are generally used in latex for coagulation, preservation, and creaming processes. The use of ammonia in natural rubber liquid latex is for the preservation and non-decomposition of the latex. It acts as a bactericide and stops the activities of the microorganisms (International Rubber Study Group, 2021). Other chemicals from the rubber processing industries that exert biochemical oxygen demand are carbonaceous materials, sulfide, and sulfate salts (Agtas et al., 2021; Ordóñez et al., 2014; Hamid et al., 2016). Consequently, wastewater discharged from natural rubber processing usually contains high levels of biological oxygen demand (BOD), chemical oxygen demand (COD), and suspended solids (SS). These characteristics may vary from country to country due to the differences in raw latex and the techniques applied in the processing. A serious threat from rubber wastewater is a high concentration of ammonia, which contributes to the undesirable phenomenon in surface water known as "eutrophication" (Mohammadi et al., 2010).

Water pollution from natural rubber processing and rubber-based industries is huge and deleterious to the environment including streams, rivers, and other water bodies (Chaudhury & Sarkar, 2023). The wastes generated from these industries are inevitable by-products of continuous production processes. Waste generated is generally classified as solid, liquid, or gaseous. Liquid waste is commonly discharged into sewers or rivers, which in many countries are subjected to treatment before

Fig. 1 Natural rubber trees and natural rubber sheet (Watari Takahiro, 2018)

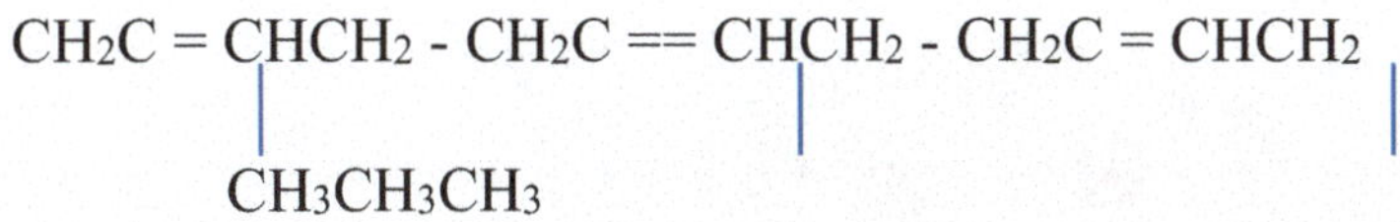

Fig. 2 Chemical structure of natural rubber (Ho et al., 2023a, 2023b)

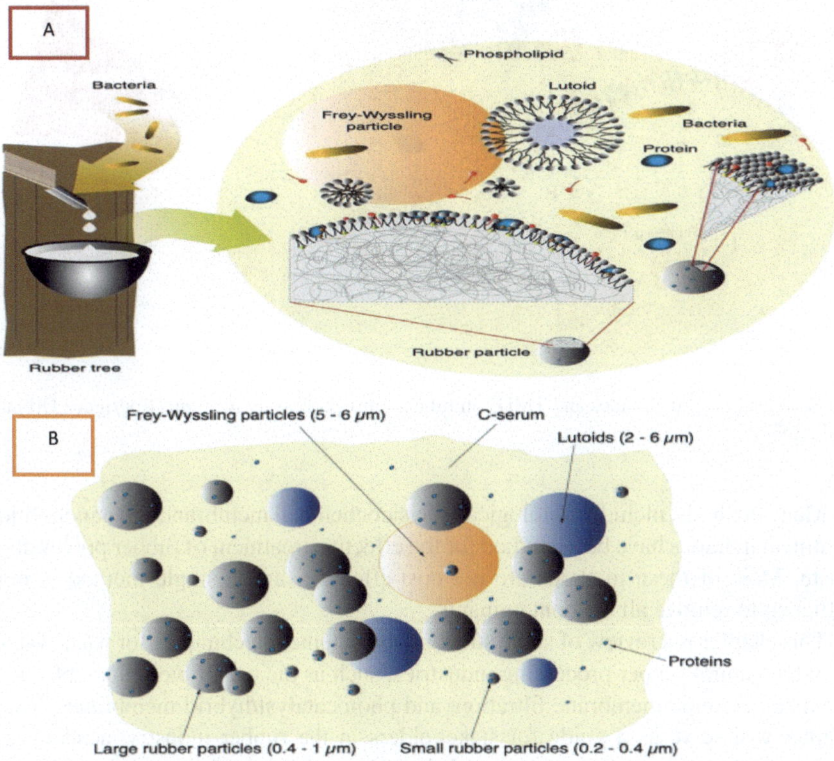

Fig. 3 **a**, **b** Liquid natural rubber latex and its composition (Rodríguez Urbina et al., 2023)

discharge. Gaseous waste is normally vented into the atmosphere, either with or without treatment, depending on the composition and the specific regulations of the country involved. Solid waste can be an asset when properly managed, but it poses the greatest threat to life because of its potential to pollute the terrestrial, aquatic, and aerial environments (Bishop, 2000). The strategic management of wastewater from natural rubber processing industries remains one of the major challenges in developing countries. The growing global concern for a healthy environment requires that wastewater from natural rubber processing industries is appropriately managed to mitigate and eradicate any potential harm to public health and the environment (Pillai & Girish, 2014).

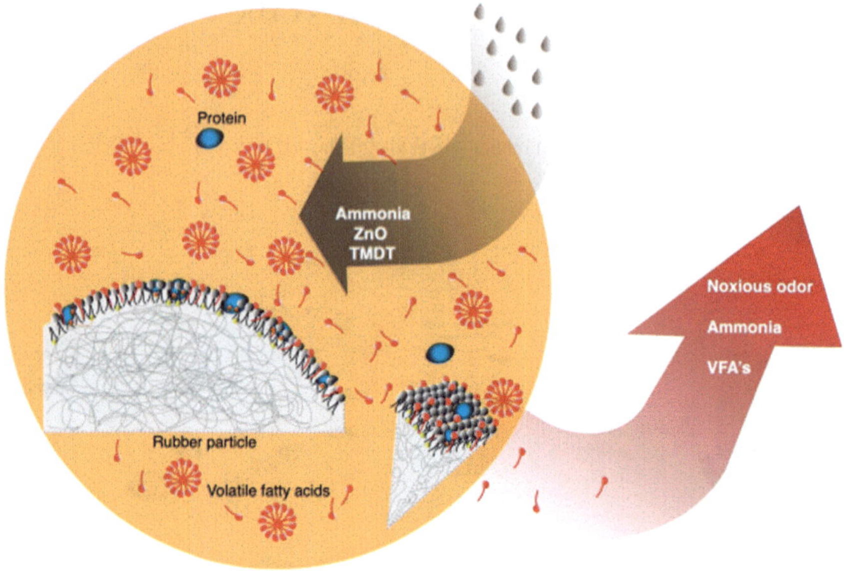

Fig. 4 Ammonia, zinc oxides and TMDT in rubber latex processing medium (Rodríguez Urbina et al., 2023)

Many methods, including biological, physicochemical membrane and nanohybrid treatment methods have been studied for the effective treatment of rubber processing waste. Most of these methods are not cost-effective, and a single method is not sufficient to remove all the contaminants.

This chapter is a review of some strategic management techniques for wastewater from the natural rubber processing industries, such as the use of biological, physiochemical, ceramic membrane filtration, and photocatalyst/hybrid membranes. This chapter will serve as a guide for stakeholders in the rubber industry, academics, professional engineers, operators, and decision-makers in the strategic management of wastewater from natural rubber processing industries. This is to enhance sustainable water quality and environmental protection.

2 Characteristics of Wastewater from Rubber Processing Industries

The characteristics of wastewater from rubber processing industries and discharge standard limits of their effluent are summarized in Table 1 shown.

The characteristics of the wastewater from natural rubber processes are based on the compositions of the raw materials and additives used in the various units of the operations. They are composed of suspended solids and organic materials such

Table 1 Summary of principal characteristics of the wastewater from rubber processing industries (Molina et al. (2021), Chaudhury and Sarkar (2021), Kantachote et al. (2004), Hadiyanto et al. (2020), Sulaiman et al. (2010), Rotimi et al. (2021), Mohamadi et al. (2020))

Parameter	Values	Discharge standard limits
TDS (mg/l)	80–200	
pH	5.0–3.8	6–9
TSS (mg/l)	258–500	≤ 30
BOD (mg/l)	1398–7000	≤ 50
Turbidity (NTU)	45–90	< 50
COD (mg/l)	3488–13,000	≤ 100
Ammonia (mg/l)	250–326	≤ 40
Sulphide(mg/l)	16.5	
Sulphate(mg/l)	375–1500	
Total nitrogen (mg/l)	468–1500	300
Chloride	149.40	350
Colour	Yellowish	
Odour	Offensive	

as sugar, protein, and lipids from the remains of the latex. Also contained in the liquid waste are mineral salts such as sulfides, sulphates, Chloride, ammoniums, heavy metals, and other nitrogen-containing compounds (Nguyen & Luong, 2012; Kantachote et al., 2004; Pillai & Girish, 2014).

3 Environmental Impact of Rubber Processing Wastewater

The additives used in natural rubber operations, such as tetramethyltriuram disulfide (TMTD), sodium sulphide, sulphuric acid, formic acid, ammonia and zinc oxides (ZnO), (as shown in Fig. 3) are usually the major constituents of rubber processing wastewater (Rodríguez Urbina et al., 2023; Kantachote et al., 2004). The discharge of this wastewater into the water system causes eutrophication because of the bacterial multiplication during the biodegradation of ammonium compounds from the remains of ammonia in the natural rubber processing plants (Nguyen & Luong, 2012). The ammonia present in the wastewater leads to high concentrations of nitrogen in effluents which has been linked to methemoglobinemia in children (Mohammadi et al., 2010). The wastewater from natural rubber operations such as washing, cooling water bleed, washing latex storage, sanitary coagulation, and tank wash results in offensive odour in the vicinity of industries, while runoff issues during the wet season jeopardize the health of residents downstream (Motina et al., 2021; Pillai & Girish, 2014).

The discharge of wastewater from rubber processing industries into water bodies in Malaysia is usually eighty million litres per day and the associated environmental pollution such as bacteria, viruses and disease-causing pathogens, which have polluted the surface waters and contaminated shellfish populations, also led to restrictions on human recreation, drinking water consumption have been reported (Mohamadi et al., 2020; Massoudinejad et al., 2015; Napi & Tamo, 2007). The rubber industry wastewater contains toxic metals that affect both terrestrial and aquatic organisms and has prolonged consequences on the environment (Motina et al., 2021; Chaudhury & Sarkar, 2021). Zinc oxide (ZnO) is the major metal oxide used in the rubber processing industry to stabilize and preserve the liquid natural rubber latex (Rodríguez Urbina et al., 2023). The wastewater contains zinc oxide contaminants, which are absorbed and stored in the roots, fruits, leaves, and every other part of the plants, and can cause the death of plants when accumulated in excess (Das et al., 2016; Massoudinejad et al., 2015). According to Rojruthai et al., (2021), zinc oxide is said to be among the category of environmentally damaging substance that creates an acute hazard risk in aquatic environment (Rodríguez Urbina et al., 2023).

4 Rubber Processing Wastewater Treatment Technologies

A typical rubber process effluent has a pH value that ranges between 3.7 and 5.5, BOD between 1500 and 7000 mg/L, COD between 3500 and 14,000 mg/L, suspended solid between 200 and 700 mg/L, total nitrogen between 200 and 1800 mg/L and sulfate between 500 and 2000 mg/L (Mohammadi et al., 2010). Water is used extensively by the rubber industry to clean equipment and raw materials used in the creation of high-quality rubber (Chittella et al., 2021). Untreated rubber effluent is malodorous and rich in organic contaminants such as carotene, inorganic minerals, lipids, organic acids, proteins, nitrogen, suspended solids and other ancillary chemicals (Alazaiza et al, 2022; MohdUdaiyappan et al., 2017). The removal of these compounds would permit the unrestricted discharge of treated wastewater into water bodies. The conventional methods for the treatment of such wastewater components released by factories as well as the non-conventional methods are presented below:

4.1 Conventional Technology

4.1.1 Biological Treatment

Pond Technology

Biological ponds are commonly used for the treatment of wastewater from the rubber industry as they are cheap and efficient for removing organic loads. However, the process requires a large land mass (Das et al., 2023). Even though the pond system is cheap, the process also involves a large assembly space and hence, produces a strong odour. One of the shortcomings of the pond technology is that it is capable of contaminating groundwater if the pond layering and waterproofing are not properly done. The biological pond treatment process is appropriate for less polluted wastewater and it should not be installed near domestic regions (Augustina et al., 2017). The pond systems usually include high-loaded anaerobic ponds and low-loaded aerobic ponds. Before the pond treatment system, there is a latex trap and a neutralization stage as a pretreatment step. If the ponds are effectively designed and optimally operated, they can remove over 95% of the BOD from the wastewater.

Watari, (2018) examined the wastewater treatment pond systems used in the production of natural rubber. He conducted a study to assess the process performance of the pond system or algal tank, used as an ABT post-treatment. The overall COD removal efficiency and total BOD removal efficiency for the duration of the experiment were 18.1 ± 21.4% and 49.4 ± 28.1%, respectively, according to the data. Because low-settlement ability algae were included in the effluent, the TSS of the algal tank effluent frequently increased. Algae absorb phosphorus and ammonia; hence, the algal tank provides benefits for nutrient removal. During the entire testing period, the algal tank demonstrated 43.3 ± 15.2% total nitrogen (TN) removal and 51.5 ± 11.8% ammonia removal efficiency.

In general, settling ponds have few drawbacks, as it takes sixty days for the treatment process to be fully operational. That means a great amount of water has to be kept over time for treatment. This method needs a large area of land, huge operating costs, and time. Conventional treatment options have been favoured in the past for the management of rubber effluent but biological treatment (especially aerobic, anaerobic, and facultative ponds) (Fig. 5) is becoming increasingly popular since they are inexpensive and have a high performance for organic load reduction (Mohammadi et al., 2010).

(i) The aerobic pond contains bacteria and algae in suspension, and the aerobic conditions (the presence of dissolved oxygen) prevail throughout its depth. There are two types of aerobic ponds: shallow ponds and aerated ponds. The shallow oxidation ponds obtain their dissolved oxygen (DO) via two phenomena; oxygen transfer between air and water surface; and that produced by photosynthetic algae (AFM, 1988). An aerated pond is similar to an oxidation pond except that it is deeper and mechanical aeration devices are used to transfer oxygen into the wastewater, which also mixes the wastewater and bacteria. The aerated pond system is also designed to reduce wastewater solids and BOD through settling and decomposition by the system bacteria. In an aerated pond, the DO required by the bacteria is supplied by mechanical aerators or diffused aeration (Williard et al., 2013). The disadvantage is that mechanical aeration devices require maintenance and energy for operation (Shilton, 2001).
(ii) Anaerobic ponds, the smallest units in the series, are sized according to their "volumetric organic loading" (i.e., the quantity of organic matter, expressed in grams of BOD_5 per day) applied to each cubic metre of pond volume. Ponds may receive volumetric organic loadings in the range of 100–350 g

Fig. 5 Picture showing a series of anaerobic, facultative and aerobic ponds (Al-Hashimi et al. 2013)

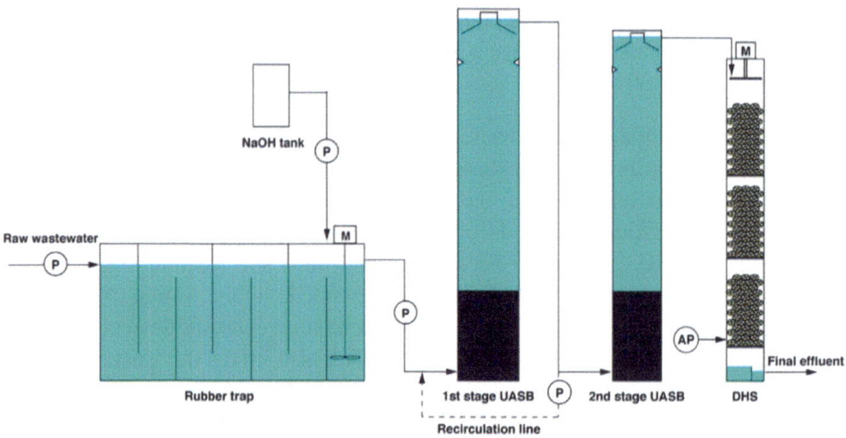

Fig. 6 Schematic diagram of the two-staged system. P—pump; M—motor; AP—air pump (Tanikawa et al., 2016)

BOD_5/m^3 per day, depending on the design temperature. An anaerobic pond is designed to treat incoming waste under anaerobic (no dissolved oxygen) conditions throughout its entire depth. The process can be very odorous, but it is highly efficient in destroying organic waste (Williard et al., 2013).

(iii) Facultative ponds are of two types: primary facultative ponds that receive raw wastewater (after screening and grit removal) and secondary facultative ponds that receive settled wastewater from the primary stage (usually the anaerobic ponds effluent). Facultative ponds are designed for BOD_5 removal based on their "surface organic" loading. Facultative ponds are characterized by having an upper aerobic and lower anaerobic zone, with active purification occurring in both (Hassan, 2011).

Anaerobic and Aerobic Microbial Treatment

The fermentation process in which organic material is degraded and biogas (composed of mainly methane and carbon dioxide) is produced is referred to as anaerobic digestion. Anaerobic digestion processes occur in many places where organic material is available and redox potential is low (zero oxygen) (Van et al., 2008). As an energy-generating process, the anaerobic treatment technology has been receiving growing interest since its first application, primarily due to the simplicity of the technology, low space requirement, and low excess sludge production (Van et al., 2015). This biodegradation occurs in many places where organic material is available and in anaerobic or anoxic conditions. Anaerobic treatment is a more attractive treatment process compared with aerobic treatment. Anaerobic treatment can effectively remove biodegradable organic compounds leaving mineralized compounds such as NH_4^+, $PO_4{}^{3-}$, and S^{2-} in the solution. The bioreactor of anaerobic treatment is a very simple system and can be applied at any scale and almost any place. The greatest benefit of anaerobic treatment is useful energy in the form of methane can be recovered by anaerobic digestion. In general, 40–45 m^3 of biogas can recover from 100 kg-COD of influent. In addition, anaerobic wastewater treatment can reduce excess sludge production (Watari, 2018).

The biological process of anaerobic digestion (AD) requires a reactor, which excludes air, and enables the flow of an organic feed through this reactor. The AD consists of a series of actions by different consortia of microbes that convert organic compounds into methane, carbon dioxide, and microbial biomass (John et al., 2019). The anaerobic degradation pathway of organic matter occurs in four successive stages, namely: (i) hydrolysis, (ii) acidogenesis, (iii) acetogenesis and (iv) methanogenesis (Van Lier et al., 2008).

In hydrolysis: carbohydrates are classified into monosaccharides and disaccharides, while proteins and fats are classified as amino acids, and fatty acids, respectively.

In acidogenesis: the acidifying bacteria convert hydrolysis products to short-chain organic acids.

In acetogenesis: the acetogenic bacteria produce acetic acid, H_2, and CO_2 from fermentation products ("dark" fermentation is the fermentative conversion of organic substrate to hydrogen). The acetogens fall into two main groups:

(i) Hydrogen-producing acetogens breakdown volatile fatty acids to CO_2 and H_2 [butyrate (butanoic acid)] $CH_3CH_2CH_2COOH + 4H_2O \rightarrow CH_3COOH + 2CO_2 + 6H_2$.
(ii) Homoacetogens: $4H_2 + 2CO \rightarrow CH_3COOH + 2H_2O$.

In methanogenesis, which is the end of the degradation chain, two groups of methanogenic archaea produce methane from acetate or hydrogen and carbon dioxide.

(i) Acetoclastic methanogenesis: $CH_3COOH \rightarrow CH_4 + CO_2$
(ii) Autotrophic or hydrogenotrophic methanogenesis $4H_2 + CO_2 \rightarrow CH_4 + 2H_2O$.

(John et al., 2019 and Van Lier et al., 2008).

In general, an anaerobic process is preferred as biogas such as biomethane and biohydrogen are produced via a rapid breakdown process of organic substances from rubber industry wastewater (Kah Chun Ho et al., 2023a, 2023b). However, the anaerobic process is usually affected by ammonia levels in rubber processing wastewater; therefore, spillway is applied to reduce the amount of acid used for the coagulation stage as well as the concentration of ammonia in the wastewater (Nguyen et al., 2012). In order to enhance the production of biogas (hydrogen and methane), Kongjan et al. (2014) treated skim latex serum (SLS) using two-stage up-flow anaerobic sludge blanket (UASB) reactors (Fig. 2) operated under thermophilic conditions. It was reported that the two-stage UASB generated an optimum hydrogen production yield of 2.25 ± 0.09 L H_2/L SLS at 36 h hydraulic residence time (HRT). Meanwhile, the methane production yield of 6.41 ± 0.52 L CH_4/L SLS was attained at 9 d of HRT. Both hydrogen and methane yield obtained by the two-stage UASB was greater than the yield achieved by batch fermentation (Kongjanet al., 2014).

Watari et al. (2017) used a pilot-scale up-flow anaerobic sludge blanket to reduce the levels of COD and BOD in rubber wastewater by up to 55.6% and 77.8%, respectively. Besides, Rosman et al. (2013), treated rubber wastewater using anaerobic granular sludge and removed the COD up to 96.5%. Moreover, the addition of electron donors into the anaerobic system enhanced the methanogenesis stage during the anaerobic process of rubber industry wastewater. Paepatung et al. (2020), investigated the effectiveness of zero-valent iron (ZVI) in improving anaerobic digestion and biogas production from synthetic wastewater containing COD/SO_4^{2-} ratio (2 to 3), which is similar to skimmed rubber latex wastewater. It was found that the ZVI enhanced the electron transfer activities of microbes in the reactor, leading to higher activities of methane-forming microbes and increased methane production of 0.25 L CH_4/g COD (Paepatung et al., 2020).

Boonchua (1998), investigated the performance of an anaerobic filter and anaerobic digester in the treatment of wastewater from the rubber processing industry and found that the anaerobic filter is more efficient for several values of hydraulic retention time (HRT) in BOD removal.

For the aerobic treatment of rubber industry wastewater, aeration plays a crucial role in the rate of aerobic breakdown of organic substances by bacteria. Chaiprapat et al. investigated the effects of retention time (RT) and air mix ratio on the performance and kinetics of the aerobic biofilm reactor (1.8 L) in H_2S removal from wastewater rubber latex plants. At the highest air mix ratio (biogas-to-air ratio = 1:4), H_2S was reduced to below 500 mg/L with the highest H_2S removal of 94.7% at RT of 160 s. Besides, high removal of COD, ammonia and total nitrogen (TN) was attained at 98.4%, 92.7%, and 89.5%, respectively (Chaiprapat et al., 2011). Yang et al., (2022) studied the effect of aeration strategies in aquaculture wastewater treatment by sequencing batch biofilm reactors. It was found that post-aeration was beneficial to the consistency of the nitrifier abundance and activity of the key nitrifying enzyme, hence enhancing the removal efficiency of ammonium (NH_4^+) (Yang et al., 2022; Kah Chun Ho et al., 2023a, 2023b). It should be noted that aerobic treatment consumes high energy for aeration and mixing, affecting the overall sustainability of the treatment system (Chang et al., 2023).

Watari (2018), noted that the natural rubber processing factories in Southeast Asian countries commonly utilize a combination of anaerobic–aerobic lagoon systems for treating natural rubber processing wastewater. The treatment systems have shown a high chemical oxygen demand (COD) removal efficiency of 65–90% (Madhu et al., 1994). It has also been reported that an up-flow anaerobic sludge blanket (UASB) reactor is one of the most promising systems, given its high organic loading rate (OLR) capacity, low operational costs, and energy recovery in the form of methane for the treatment of different types of industrial wastewater (Lettinga, 1991). Other authors have reported the application of the UASB system to treat natural rubber processing wastewater (Mohammidi et al., 2010; Nguyen & Luong, 2012, Jawjit and Liengchaeranist, 2010).

In the research of Watari (2018), a laboratory-scale experiment was conducted to develop an appropriate wastewater treatment system for natural rubber processing wastewater consisting of a baffled reactor (BR), a UASB reactor, and a DHS reactor (a BR–UASB–DHS system). In addition, 16S rRNA gene sequencing of microorganisms in the retained sludge in the UASB reactor was undertaken to identify the microbial communities present. The study demonstrated that the proposed BR– UASB–DHS system can efficiently treat natural rubber processing wastewater (Watari, 2018).

Oxidation Ditch

The oxidation ditch is an advanced process based on biologically activated sludge with long sludge retention time (SRT). This method is used to treat wastewater with high levels of contaminants, particularly wastewater with high levels of ammonia. The long SRT creates favourable conditions for the occurrence of nitrification and the elimination of biological organic matter. An air supply system is installed in the oxidation ditch to ensure that wastewater and activated sludge are thoroughly mixed and are given the oxygen they need for microbial activity. For the removal of organic matter and nitrogen-containing contaminants from rubber production effluent, the oxidation ditch technique is reliable and very effective. For instance, Ibrahim (1980), reported that a laboratory-scale oxidation ditch process showed > 90% nitrogen removal efficiency with natural rubber processing wastewater. The oxidation ditch has been applied for the wastewater treatment derived from the processing of concentrated rubber latex and Ribbed Smoked Sheet RSS (Nguyen et al., 2012).

However, there are some requirements for using and running this process, including the need for highly qualified labour, high operating costs, and recurring maintenance.

4.2 Non-conventional Technology

4.2.1 Advanced Technologies

By the fact that the conventional processes used in wastewater treatment are becoming less and less effective due to the changing nature of the wastewater strength with the increased use of chemicals in the rubber production processes. Many advanced technologies are being researched worldwide for wastewater treatment from the natural rubber latex processing industry. Hence, these techniques are being suggested to eliminate issues faced with conventional systems, therefore the following paragraphs will shortly introduce the most common ones (Tanikawa et al., 2019).

Electrochemical Method

Nowadays, the electrochemical method receives a lot of consideration for wastewater treatment because of its cost, easy control, short HRT and high efficiency (Chopra et al., 2011; Vijayaraghavan et al., 2008). The electrochemical treatment requires only 30–45 min compared to the conventional biological treatment (such as the activated sludge process) and the other aerobic oxidation methods (Vimalamma et al., 2009). Another advantage of the electrochemical process is emulsion breaking and bleaching by oxygen ions produced in the reaction chamber oxidising bacteria, viruses, sulphides and other biohazards. One of the main attractions of electrochemical treatment is the use of solar energy for electrolysis, which can be of great help to rural areas to reduce the running costs of the treatment processes (Nguyen et al., 2012). Electrolytic treatment of rubber wastewater has been studied and identified as an effective technique that could efficiently remove organic matter with the use of hypochlorous acid generation inside the treatment unit. Optimum conditions for the unit were found to be an initial pH of 4.5, sodium chloride content of 3%, and a current density of 74.5 mA/cm^2 (Wijerathna et al., 2023).

One of the chemical wastewater treatments without coagulants is electrocoagulation. Electrocoagulation is an electrochemical method for waste treatment, where the release of active coagulant in the form of metal ions (usually aluminium or iron) into a solution occurs in the anode, whereas in the cathode an electrolysis reaction occurs in the form of hydrogen gas release (Hassani et al., 2016; Jumaah and Othman., 2015). According to Rusdianasari et al. (2021), the electrocoagulation process using aluminium-stainless steel electrodes effectively reduced the value of TSS, BOD_5, COD, and NH_3, and significantly increased the pH value of rubber processing wastewater. After the electrocoagulation process, the pH value increased to near neutral pH of 7.04 and the values of TSS levels (10.8 mg/L), BOD (5 17 mg/L), COD (50 mg/L), and NH_3 (0.53 mg/L) decreased. The results obtained showed that the electrocoagulation method was able to reduce the levels of pollutants in the wastewater below the environmental quality standard, which means the liquid waste cannot pollute the surrounding environment (Bow et al., 2021).

4.2.2 Advanced Treatment Technologies

Membrane Technology

Ceramics Membrane

Due to the economic and technical limitations of conventional wastewater treatment methodologies, many industries turn to membrane technologies to perform more reliable wastewater treatment operations. The membrane is a selective barrier that allows some selective substances to pass through but prevents others from doing so. The membrane is made from organic or inorganic material. Although inorganic material-made membranes have better separation capabilities than organic material-made membranes, inorganic material-made (Ceramics) membranes are more expensive (Ruangdit, 2022). The membrane technology is one of the advanced methods with many advantages for wastewater treatment (Sulaiman et al., 2010; Kusworo et al., 2021; Kusworo & Utomo, 2022), and have been studied for the treatment of rubber latex processing wastewater. Sulaiman et al., (2010) showed that the optimum flux of the process was obtained at 0.009 m^3 /m^2h and this flux rate; the process could be continuously operated for more than one month without the chemical cleaning of membranes. The removal efficiency of COD was 96.99% at the initial concentration of 3,500 mg/L COD, BOD was 96.78%, total-N 65.17% and N-NH3 61.35%.

Among the various membrane materials, ceramic membranes have gained attention for industrial wastewater treatment because of their high durability, superior chemical, mechanical, and thermal stability, bacteria resistance, ability to back flush, ease of cleaning, sterilization, robustness, lower operational costs, and high porosity (Luque et al., 2008). Polymeric membranes are more generally used in the membrane process industry because of their lower capital cost, scalability, and good separation characteristics (Robeson, 2010). However, they have low fouling resistance, low lifespan, low temperature ranges and low corrosion resistance (Kaniganti et al., 2015). Despite the advances achieved in this technology, the potential of ceramic membranes for industrial wastewater treatment has not yet been fully realized (Samaei et al., 2018). The major disadvantage of ceramic membranes is the high cost of capital in fabrication (Azaman et al., 2021). Ceramic membranes are typically made of inorganic non-metallic and metallic oxides such as aluminium oxide, zirconium oxide, titanium oxide, and silicon carbide (Asi & Zhang, 2021).

Ceramic membranes are produced with earthy mineral powders such as alumina, zirconia, titania, silica, kaolin, apatite, graphite, phosphate, dolomite, diatomites, organic burnt materials such as sawdust, rice husk, fly ash (Han et al., 2013a, 2013b; Qin et al., 2014; Rekik et al., 2016; Schillo et al., 2010; Zhu et al., 2010).

A study by Witttayakul et al. *as* cited in Yusof et al., (2017) has said that a ceramics filtering system can deliver great effectiveness in lowering the organic load from latex plant effluent and can lower the concentration of hydrogen disulfide.

In a study, a ceramics membrane was used in the first stage of filtration to remove large particles of pollutants (e.g., rubber particles, dirt, and sludge) in the natural rubber skimmed latex (Yusof et al., 2017). The permeate from the first stage (latex serum) was channelled to the second stage filtration, as the feed, where a polyamide membrane nanofiltration (NF) was used to filter out proteins that are naturally present in the latex serum. The result of the first filtration showed that ceramic membrane ultrafiltration (UF) was able to achieve 75% and 37% removal of suspended solid (SS) and dissolved solid (DS) respectively, while the second stage which uses NF membrane gave 80% SS and 75% DS rejection. The highest SS and DS removal was recorded at 97% and 96% respectively using a first-stage NF membrane. Membrane filtration is also found very efficient in reducing BOD and COD of the natural rubber skimmed latex. From the experimental results, the BOD and COD of the effluent discharged after the final stage membrane filtration were reduced by 99% and 84% respectively.

Nanohybrid Membrane

The major disadvantage of membrane processes is fouling, which results from the deposit of particles known as foulants on membrane surfaces, such as hydrophilic dissolved organic matter, salts, and colloids. To improve the membrane technology processes, the nanohybrid membrane has been introduced. The membrane surface modulation with nanomaterials such as clay nanoparticles, carbon nanotubules, and metal and non-metal oxide nanoparticles is called nanohybrid membrane (Mohamadi et al., 2020; Sun et al., 2018) The development of nanohybrid membrane is the current attractive method for improving the performance of microfiltration (MF), ultrafiltration (UF), nanofiltration (NF), and reverse osmosis (RO) which are usually applied in water/wastewater treatment and desalination. The nanohybrid membrane has great potential to resolve the drawbacks of the conventional pressure-driven membrane in its application for natural rubber wastewater treatment (Ezugbe et al., 2022; Ezugbe & Rathilal, 2020).

The application of nanohybrid membranes in high-organic-content wastewater treatment also showed better antifouling behaviour (Kusworo et al., 2018). A study of nanohybrid membrane development and experimental application in natural rubber wastewater has been reported by Kusworo & Utomo (2022).

The nanomaterials/photocatalytic (TiO_2, ZnO, SnO_2, and other semiconductor metal oxides) with self-cleaning ability have been reported as membrane nanofillers in the treatment of natural rubber wastewater.

The use of zinc oxide, titanium dioxide, Silicon dioxide and graphene oxide nanoparticles, as nanofillers, has been reported for the treatment of rubber processing wastewater [(Kusworo et al. (2020a); Kusworo, et al. (2020); Kusworo*et al* (2017)]. The nanofillers were used because of their strong hydrophilicity, high mechanical and chemical stability, antibacterial activity, and low-cost material. They improved the structural properties, and antifouling, prevented the growth of biofouling and enhanced both the permeability and selectivity. The presence of ZnO nanoparticles contributed to reducing the surface negative charge, thus enhancing the removal of NH_4^+ in wastewater. The TiO_2-blended Polyethersulfone/Sulfonated (PES) membrane also exhibited lower fouling resistance in rubber wastewater treatment (Ayyaru & Ahn, 2018; Yang *et al.*, 2019). Kusworo et al., (2017) showed that the SiO_2 nanohybrid membrane improved performance in terms of permeability and selectivity in natural rubber wastewater treatment. The combination of two or more nanomaterials has been reported to be effective in the removal of organic and ammonia in natural rubber wastewater. According to the study of Kusworo et al., (2021), the mixture of TiO_2/GO in polysulfone (PSf) membrane has remarkable enhancement of permeate flux and removal of organic and ammonia from natural rubber wastewater. The combination of nanomaterials can help to reduce the fabrication cost. For instance, the combination of TiO_2 and ZnO can be selected to overcome the high cost of TiO_2 without a decrease in the functionality. Natural rubber wastewater treatment using membrane is a viable means to generate water quality that can be reused as process water in the industry.

Kusworo et al., (2023) fabricated a nanohybrid membrane with a non-solvent-induced phase separation (NIPS) method. The addition of TiO_2 and reduced graphene oxide (rGO) increased the hydrophilic properties of the membrane. The filtration of natural rubber wastewater revealed that the permeate flux was achieved up to 22.7 L/m^2/h and the rejection efficiency of COD, and ammonia were up to 90% and 81%, respectively. The fouling resistance evaluation showed that the dual filler nanohybrid membrane has a total resistance that is four times lower than that of the pristine PSf membrane.

Ozonation Method

Ozone radiation treatment has several benefits when used in the treatment sequence, including the ability to operate in ambient global conditions, the decomposition of volatile and semi-volatile organic compounds in the aqueous phase, the absence of additional chemical requirements, the absence of the need to remove an excessive amount of toxic compounds before discharge, and the transformation of refractory organic pollutants into highly degradable products (Ye et al., 2010; Wijerathna et al., 2023). Furthermore, the ozonation method can convert large molecular-weight materials (non-biological organic materials) into small molecular-weight materials. Ozone is one of the most active and widely available oxidizing agents, capable of transferring ammonia into nitrate. Furthermore, the ozonation process can eliminate odours, reduce oxygen-demanding matter, turbidity, and surfactants, increase dissolved oxygen, and reduce suspended solids. Ozonation has a high capital cost, a high electricity consumption, and is highly corrosive. The relatively high cost of ozone generation, combined with ozone's extremely short half-life period, is a major limitation of the ozonation process (Nguyen et al., 2012). Some of the non-biodegradable materials and the ammonia nitrogen fraction of natural rubber wastewater cannot be completely removed using biological processes. Rungruang and Babel (2008) studied the efficiency of batch activated sludge process (BAS) with or without ozonation for rubber wastewater treatment at Chonburi, Thailand. The effects of different contact times (0–90 min) and ozone dosages (37.20, 56.90 and 66.44 mg O_3/ L O_2) at various pH (7.4, 9.0 and 11.0) on wastewater treatment were studied. It was found that 66.44 mg O_3/ L O_2 of ozone dosage; pH of 9.0 and 30 min of contact time were the optimum conditions for the reduction of pollutants. The combination of BAS and ozonation ensured higher removal efficiency for all parameters when compared to other systems (Rungruang & Babel, 2008).

Advanced Oxidation Processes (AOPs)

Advanced oxidation processes (AOPs) have been used for the removal of wastewater pollution by turning the pollutants into less-toxic and readily biodegradable products. It has been investigated to couple the ozonation method with Fenton reagents to overcome the limitations of ozonation methods. When used as oxidants, hydroxyl radicals created by the Fenton reagent react more quickly than ozone and H_2O_2. Studies by Ghezzar et al. indicate that pulsed high voltage is used to further improve advanced oxidation processes to produce highly active oxidants. This process is known as Non-Thermal Plasma (NTP) and it comprises a gliding arc discharge system device (Ghezzar et al., 2008). Nevertheless, because of the hazardous nature of the sludge produced by the Fenton process, according to Gamaralalage et al. studies, the possibility of reusing the sludge in the Fenton process as a Fe source could be an environmentally viable and economically promising method (Gamaralalage et al., 2019).AOPs can break down a variety of refractory chemicals, but they are expensive, require chemicals and complicated processes, and occasionally produce secondary sludge that may be difficult to dispose of (Wijerathna et al., 2023). Mohamadreza et al., (2015), combined the ozonation process with the physicochemical process as a pretreatment process, to improve the efficiency of COD removal by up to 90.6%. This combined process was very effective in removing the pollutants present in rubber industry wastewater, as it produced a final effluent with low suspended solids, clear, and odourless (Massoudinejad et al., 2015).

5 Resources Recovery from Rubber Processing Wastewater

The applications of rubber processing effluents for crop irrigation have been known and applied as a waste management method with the additional benefit of acting as an organic fertilizer (Swaminathan and Valdheeswaran, 1991). Tekasakul and Tekasakul, (2006) reported an increase in soil nitrogen, phosphorus, potassium, magnesium, calcium and organic carbon with increasing rubber processing effluent. According to Maliki et al (2020), the physicochemical properties of effluents from natural rubber processing industries revealed the presence of plant nutrients, such as nitrate, sulphate, phosphorus, potassium, magnesium, calcium and sodium in appreciable amounts to enhance crop yield. The number of leaves, leaf length and total biomass yield were better with rubber processing effluents amended soils than the unamended soil.

Sdoodee et *al.*, (2005), utilized wastewater from the last pond in rubber effluent treatment pond in watering several kinds of plants. The plants treated with mixtures of wastewater and tap water had higher yields than those treated with tap water.

Eifediyi et al., (2013), collected rubber wastewater and applied it to enhance the physical properties of the soil by raising the soil's organic matter content and nutrient potential. Applying 150,000 L per hectare of rubber wastewater resulted in the maximum production and vegetative characteristics of cucumbers. However, the growth and yield metrics of cucumbers decreased when rubber effluent was applied over 150,000L/ha.

6 Conclusion

A significant proportion of wastewater generated by the natural rubber industry has been either underutilized or untreated before being properly disposed of, which has put the water resources under stress. High organic content, a high acidic pH, a high ammonia concentration, and fouling that reduces selectivity, productivity, and membrane longevity are features of natural rubber effluent. Industrial wastewater reclamation and treatment are vital strategies for providing and maintaining safe water to meet global water needs.

From this chapter, it is observed that some novel strategies for the management of wastewater from natural rubber processing have been developed to overcome the limitations of the conventional methods.

Furthermore, this chapter reviews, recent progress in the strategic management of wastewater from the natural rubber industry and shows that there are many useful aspects to the application of membrane technology in the management of wastewater from natural rubber processes to increase treatment efficiency, environmental awareness, sustainability of clean water sources, and economics.

References

Agustina, T. E., Sirait, E. J., & Silalahi, H. (2017). Treatment of rubber industry wastewater by using Fenton reagent and activated carbon 79(7), 31–37.

Air Force Manual. (1988). *Domestic wastewater treatment*. Department of the Army and Air Force, TM 5-814-3/AFM 88-11 (Vol. III).

Alazaiza, M. Y. D., Albahnasawi, A., Ahmad, Z., Bashir, M. J. K., Al-Wahaibi, T., Abujazar, M. S. S., Abu Amr, S. S., & Nassani, D. E. (2022). Potential use of algae for the bioremediation of different types of wastewater and contaminants: Production of bioproducts and biofuel for green circular economy. *Journal of Environmental Management, 324*, Article 116415. https://doi.org/10.1016/j.jenvman.2022.116415

Al-Hashimi, M. A. I., & Hussain, H. T. (2013). Stabilization pond for wastewater treatment. *European Scientific Journal, 9*(14), 278–294.

AshiqurRahman, ShahriarHabib, MustafijurRahman, SymonJahanSajib& Abu Yousuf (2019). A novel multi-phase treatment scheme for odorous rubber effluent. Environmental Technology, https://doi.org/10.1080/09593330.2019.1668965

Asif, M. B., & Zhang, Z. (2021). Ceramic membrane technology for water and wastewater treatment: A critical review of performance, full-scale applications, membrane fouling and prospects. *Chemical Engineering Journal, 418*, Article 129481.

Ayyaru, S., & Ahn, Y.-H. (2018). Fabrication and Separation Performance of Polyethersulfone/Sulfonated TiO2 (PES–STiO2) Ultrafiltration Membranes for Fouling Mitigation. *Journal of Industrial and Engineering Chemistry, 67*(November), 199–209. https://doi.org/10.1016/j.jiec.2018.06.030

Bousbih Saida, Emna Errais, FadilaDarragi, JoelleDuplay, Malika Trabelsi-Ayadi, et al..Treatment of Textile Wastewater using monolayeredUltrafiltation Ceramic Membrane Fabricated from Natural Kaolin Clay. Environmental Technology, In press, ff10.1080/09593330.2020.1729242ff. ffhal-02618860

Bishop, P. L. (2000). *Pollution prevention: fundamentals and practice*. Singapore: McGraw-Hill; ISBN 0-07-366147-3.

Bow, Y., Syakdani, A.,&Mayasari, D. I. (2021). The Effectiveness of Electrocoagulation Process in Rubber Wastewater Treatment using Combination Electrodes. In *IOP Conference Series: Earth and Environmental Science* (Vol. 709, No. 1, p. 012009). IOP Publishing.

Chaiprapat, S., Mardthing, R., Kantachote, D., & Karnchanawong, S. (2011). Removal of hydrogen sulfide by complete aerobic oxidation in acidic biofiltration. *Process Biochemistry, 46*, 344–352. https://doi.org/10.1016/j.procbio.2010.09.007

Chang, Z. H., Ting, L. L. H., & Sum, J. Y. (2023). A review on membrane separation in natural rubber processing: Concentration, Recovery and Treatment. *Chemical Engineering and Processing-Process Intensification*, 109541.

Chaudhury, S. K., & Sarkar, S. (2021). *A Study On Pollution Associated With Rubber Processing Centers In Mohanpur Block*. Tripura.

Chittella, H., Yoon, L. W., Ramarad, S., & Lai, Z. W. (2021). Rubber waste management: A review on methods, mechanism, and prospects. *Polymer Degradation and Stability, 194*, Article 109761. https://doi.org/10.1016/j.polymdegradstab.2021.109761

Chopra, A. K., Sharma, A. K., & Kumar, V. (2011). Overview of electrolytic treatment: An alternative technology for purification of wastewater. *Archives of Applied Science Research, 3*(5), 191–206.

Joydeep, D., & Mondal, A. (2022). SrijitBiswas and Soma Nag, (2022), The eco-friendly treatment of rubber industry effluent by using adsorbent derived from Moringaoleifera bark and Pseudomonas sp, cultured from effluent. *Water Science & Technology, 86*(11), 2808. https://doi.org/10.2166/wst.2022.387)

Edirisinghe, DilharaGithanjali. “Chemical modification of rubber waste and characterization of their blends with natural rubber.“PhD diss., 2014.

Eifediyi, E. K., Ihenyen, J. O., &Ojiekpon, I. F. (2012).Evaluation of the effects of rubber factory effluent on soil nutrients, growth and yield of cucumber (Cucumissativus L.). *Niger Ann Nat Sci, 12*(1), 21–28.

Ezugbe, E. O., Tetteh, E. K., Rathilal, S., Armah, E. K., Amo-Duodu, G., & Asante-Sackey, D. (2022), Direct Membrane Filtration for Wastewater Treatment. *Advanced Materials for Wastewater Treatment and Desalination*, 167.

Ezugbe, ElormObotey, and SudeshRathilal. (2020), "Membrane Technologies in Wastewater Treatment: A Review." Membranes 10 (5): 89. https://doi.org/10.3390/membranes10050089.

Gamaralalage, D., Sawai, O., & Nunoura, T. (2019). Reusing the generated sludge as Fe source in Fenton process for treating crepe rubber wastewater.*J. Mater. Cycles Waste Manag., 21*, 248–257. https://doi.org/10.1007/s10163-018-0784-8

Ghezzar, M. R., Belhadj, M., Abdelmalek, F., Rais, A., & Addou, A. (2008). Non-Thermal plasma degradation of wastewater in presence of titanium oxide by gliding arc discharge. *Int. J. Environ. Waste Manag., 2*, 458–470. https://doi.org/10.1504/IJEWM.2008.021786

Hamid, M. F., Idroas, M. Y., Ishak, M. Z., ZainalAlauddin, Z. A., Miskam, M. A., & Abdullah, M. K. (2016), An experimental study of briquetting process of torrefied rubber seed kernel and palm oil shell. *BioMed research international, 2016.*

Han JH, Oh E, Bae B, Song IH (2013a) The effect of kaolin addition on the characteristics of a sintered diatomite composite support layer for potential microfiltration applications. Ceram Int 39: 8955-8962

Han JH, Oh E, Bae B, Song IH (2013b) The fabrication and characterization of sintered diatomite for potential microfiltration for applications. Ceram Int 39: 7641-7648.

HadiyantoHadiyanto, MarcelinusChristwardana, Detian Indah Pratiwi, SilvianaSilviana, Muhammad Syarifudin&AdianKhoironi | (2020) Rubber wastewater treatment using UV, ozone, and UV/ozone and its effluent potency for microalgae Spirulinaplatensis cultivation medium, Cogent Engineering, 7:1, 1797980, https://doi.org/10.1080/23311916.2020.1797980

Hassani, G., Alinejad, A., Sadat, A., Esmaeili, A., Ziaei, M., Bazrafshan, A. A., & Sadat, T. (2016). Optimization of landfill leachate treatment process by electrocoagulation, electroflotation and sedimentation sequential method. *International Journal of Electrochemical Science, 11*(8), 6705–6718.

Hassan, A. F. (2011). Waste Stabilization Pond. *An-Najah National University.*

Ho, K. C., Chan, M. K., Chen, Y. M., &Subhramaniyun, P. (2023). Treatment of rubber industry wastewater review: Recent advances and future prospects. *Journal of Water Process Engineering, 52*, 103559.https://doi.org/10.1016/j.jwpe.2023.103559

Ibrahim, A. B. 1980 Laboratory evaluation of removal of nitrogen from rubber processing effluent using the oxidation ditch process. Journal of the Rubber Research Institute of Malaysia, 28(1), 26–31)

Ismail, A. F., Goh, P. S., Hasbullah, H., & Aziz, F. (Eds.). (2022). *Advanced Materials for Wastewater Treatment and Desalination: Fundamentals to Applications.*

International Rubber Study Group (2021).Rubber Statistical Bulletin.Vol 76, No. 7–9.

John J. Milledge, Elinor P. Thompson, Andrés Sauvêtre, (2019), Peter Schroeder, Patricia J. Harvey, Novel developments in biological technologies for wastewater processing, Sustainable Water and Wastewater Processing, Elsevier, https://doi.org/10.1016/B978-0-12-816170-8.00008-9)

Jumaah, M. A., & Othman, M. R. (2015).COD removal from landfill leachate by electrochemical method using charcoal-PVC electrode. *International Journal of ChemTech Research, 8*(12), 604–609.

Kaniganti, C. M., Emani, S., Thorat, P., Uppaluri, R. (20150, Microfiltration of synthetic bacteria solution using low-cost ceramic membranes. Sep. Sci. Technol. 50(1), 121–135.

Kantachote, D., &Innuwat, W. (2004).Isolation of Thiobacillus sp. for use in the treatment of rubber sheet wastewater. *Songklanakarin Journal of Science and Technology, 26*(5), 649–657.

Kah Chun Ho, MieowKeeChan, Yui Moon Chen, PretibaaSubhramaniyun (2023), Treatment of rubber industry wastewater review: Recent advances and future prospects, *Journal of Water Process Engineering* 52, 103559.

Kongjan, R. Jariyaboon, S. O-Thong, (2014), Anaerobic digestion of skim latex serum (SLS) for hydrogen and methane production using a two-stage process in a series of up-flow anaerobic sludge blanket (UASB) reactor, Int J Hydrogen Energy 39 (2014) 19343–19348, https://doi.org/10.1016/j.ijhydene.2014.06.057

Kusworo, T. D., Kumoro, A. C., Utomo, D. P., &Wulandari, L. M. (2023, February).Dual filler nanohybrid membrane PSf-TiO2/rGO for enhancing nanofiltration performance in rubber wastewater treatment.In *AIP Conference Proceedings* (Vol. 2667, No. 1). AIP Publishing.

Kusworo, AndriCahyoKumoro, Nita Aryanti, and DaniPujiUtomo (2021), "Removal of Organic Pollutants from Rubber Wastewater Using Hydrophilic NanocompositeRGOZnO/PES Hybrid Membranes." Journal of Environmental Chemical Engineering 9 (6): 106421. https://doi.org/10.1016/j.jece.2021.106421

Kusworo, T. D., &Utomo, D. P. (2022).Nanohybrid Membrane for Natural Rubber Wastewater Treatment. *Advanced Materials for Wastewater Treatment and Desalination*, 243.

Kusworo, T.D., and Lutfi Mia Wulandari. 2021. "Fabrication of High-Performance PSf-RGO/TiO2 UF Membrane for Ruberry Wastewater Treatment." IOP Conference Series: Materials Science and Engineering 1053 (1): 012025. https://doi.org/10.1088/1757-899X/1053/1/012025.

Kusworo, Tutuk, Nita Aryanti, DaniUtomo, and EnnyNurmala. 2020. "Performance Evaluation of PES-ZnONanohybrid Using a Combination of UV Irradiation and CrossLinking for Wastewater Treatment of the Rubber Industry to Clean Water." Journal of Membrane Science and Research 7 (1): 4–13. https://doi.org/10.22079/jmsr.2020.120490.1334.

Kusworo, Q., DaniPujiUtomo, I., & Ramadhan, I. R. (2018). Enhancement of Separation Performance of Nano Hybrid PES –TiO2 Membrane Using Three Combination Effects of Ultraviolet Irradiation, Ethanol-Acetone Immersion, and Thermal Annealing Process for CO2 Removal. *Journal of Environmental Chemical Engineering, 6*(2), 2865–2873. https://doi.org/10.1016/j.jece.2018.04.023

Luque, S., Gómez, D., & Álvarez, J. R. (2008). Industrial applications of porous ceramic membranes (pressure-driven processes). *Membrane Science and Technology, 13*, 177–216.

Massoudinejad, M., Mehdipour-Rabori, M., & Dehghani, M. H. (2015). Treatment of Natural Rubber Industry Wastewater through a Combination of Physicochemical and Ozonation Processes. *J. Adv. Environ. Health Res., 3*(4), 242–249.

Motina, L., Rex, A., & Yvonne, E. (2021). Designing a Latex Effluent Treatment System Using Acid Coagulation Rubber Recovery Tank in Series with a Constructed Wetland. *International Journal of Sciences: Basic and Applied Research (IJSBAR), 58*(1), 64–81.

Mohammadi, M., Man, H. C., Hassan, M. A., & Yee, P. L. (2010).Treatment of wastewater from the rubber industry in Malaysia. *African Journal of Biotechnology*, *9*(38), 6233-6243.

Mohamadi, S., H. Hazrati, and J. Shayegan (2020) Influence of a new method of applying adsorbents on membrane fouling in MBR systems and Environment Journal. 34: p. 355–366.

Mohd Udaiyappan, A.F., Abu Hasan H., Takriff, M.S., Sheikh Abdullah, S.R. (2017), A review of the potentials, challenges and current status of microalgae biomass applications in industrial wastewater treatment, *J. Water Process Eng.* 20,8–21, https://doi.org/10.1016/j.jwpe.2017.09.006.

NapiWouapi, Tamo V. Maimo (2007)."Towards a comprehensive analysis of cleaner technology potentials to address industrial pollution arising from natural rubber processing industry: A case study of Cameroon development corporation -rubber factories "Msc. Thesis Project. Roskilde University, Denmark.

Nguyen, H. N., & Luong, T. T. (2012). Situation of wastewater treatment of natural rubber latex processing in the Southeastern region, Vietnam. *Journal of Vietnamese Environment, 2*(2), 58–64.

Paepatung, W. Songkasiri, H. Yasui, C. Phalakornkule, Enhancing methanogenesis in fed-batch anaerobic digestion of high-strength sulfate-rich wastewater using zero-valent scrap iron, J. Environ. Chem. Eng. 8 (2020), https:// doi.org/https://doi.org/10.1016/j.jece.2020.104508.

Pillai, H. P. J. S., &Girish, K. (2014).Rubber processing industry effluent treatment using a bacterial consortium. *International Journal of Current Microbiology and Applied Sciences, 3*(10), 775–782.

Qin, W., Peng, C., Lv, M., & Wu, J. (2014). Preparation and properties of high-purity porous alumina support at low sintering temperature. *Ceramics International, 40*, 13741–13746.

Rekik, S. B., Bouaziz, J., Deratani, A., & Baklouti, S. (2016). Development of an Asymmetric Ultrafiltration Membrane from Naturally Occurring Kaolin Clays: Application for the Cuttlefish Effluents Treatment. *J MembraSciTechnol, 6*, 159. https://doi.org/10.4172/2155-9589.1000159

Robeson, L. M. (2010). Polymer blends in membrane transport processes. *Industrial & Engineering Chemistry Research, 49*(23), 11859–11865.

Rotimi, I.M., Adewumi, J.R. and Ajibade, F.O (2021) Treatment of Rubber Latex Processing Wastewater using Activated Carbon Derived from Palm Kernel Shell, *Nigerian Research Journal of Engineering and Environmental Sciences* 6(2) 2021 pp. 741–752

Rodríguez Urbina, J. C., Osswald, T. A., Estela Garcia, J. E., &Román, A. J. (2023).Environmentally safe preservation and stabilization of natural rubber latex in an acidic environment. *SPE Polymers*.

Rojruthai P, Rodgerd P, Ho C-C, Sakdapipanich J.(2021). The use of 1,2-benzisothiazolin-3-one (BIT) in preparation of low ammonia and zinc-free natural rubber latex concentrate. J Rubber Res. 2021;24(5):783–795.

Rosman, N. H., Nor Anuar, A., Othman, I., Harun, H., Sulong Abdul Razak, M. Z., Ujang, Z., Mat Hassan, M. A. H., Chelliapan, S., &Ujang, Z. (2013). Cultivation of aerobic granular sludge for rubber wastewater treatment.Bioresource Technology, 129(1), 620–623. https://doi.org/10.1016/j.biortech.2012.12.113).

Ruangdit, S. (2022). *Modification of Polysulfone Membrane for Separation of Compound Substance in Natural Rubber Serum* (Doctoral dissertation, Prince of Songkla University).

Rungruang N, Babel S (2008).Treatment of natural rubber processing wastewater by the combination of ozonation and activated sludge process. In: International Conference on Environmental Research and Technology (ICERT 2008), Parkroyal Penang, Malaysia, pp. 259- 263).

Mohsen, S. S., Gato-Trinidad, S., & Altaee, A. (2018). The Application of Pressure-Driven Ceramic Membrane Technology for the Treatment of Industrial Wastewaters -A Review. *Separation and Purification Technology*. https://doi.org/10.1016/j.seppur.2018.02.041

Schillo, M. C., Park, L. S., Chiu, W. V., & Verweij, H. (2010). Rapid thermal processing of inorganic membranes. *Journal of Membrane Science, 362*, 127–133.

Sdoodee, S., Chaiprapat, S. and Ovatlarnporn, C. (2005). Assessment of the Impact of Using Partially Treated Rubber Sheet Wastewater for Cultivation, The Thailand Research Fund(TRF)Research Report.

Shilton, A. (2001). *Studies into the hydraulics of waste stabilization ponds*. Massey University.

Sulaiman, N. M. N., Ibrahim, S., & Abdullah, S. L. (2010). Membrane bioreactor for the treatment of natural rubber wastewater. *International Journal of Environmental Engineering, 2*(1–3), 92–109.

Sun, H., B. Tang, and P. Wu, (2018), Hydrophilic hollow zeolitic-imidazolate framework-8 modified ultrafiltration membranes with significantly enhanced water separation properties.*Journal of Membrane Science*, 551: p. 283–293.

Tanikawa, D., Syutsubo, K., Hatamoto, M., Fukuda, M., Takahashi, M., Choeisai, P. K., & Yamaguchi, T. (2016). Treatment of natural rubber processing wastewater using a combination system of a two-stage up-flow anaerobic sludge blanket and down-flow hanging sponge system. *Water Science and Technology, 73*(8), 1777–1784.

Tanikawa, D., Watari, T., Mai, T.C., Fukuda, M., Syutsubo, K., Nguyen, N.B., Yamaguchi, T., (2019a), Characteristics of greenhouse gas emissions from an anaerobic wastewater treatment system in a natural rubber processing factory, *Environ.Technol.* 40, 2954–2961. https://doi.org/10.1080/09593330.2018.1459872

Tekasakul, P., & Tekasakul, S. (2006). Environmental problems related to natural rubber production in Thailand. *Journal of Aerosol Research, 21*(2), 122–129.

Van Lier, J. B., Mahmoud, N., & Zeeman, G. (2008).Anaerobic wastewater treatment.biological wastewater treatment, principles, modelling and design, 415–456.)

Van Lier, J. B., Van der Zee, F. P., Frijters, C. T. M. J., &Ersahin, M. E. (2015). Celebrating 40 years of anaerobic sludge bed reactors for industrial wastewater treatment. Reviews in Environmental Science and Bio/Technology, 14(4), 681–702).

Vijayaraghavan, K., Ahmad, D., Yazid, A.Y.A., (2008), Electrolytic treatment of latex wastewater.*Desalination* 219(1–3):214–2.

Viet, N. T. (1999). *Sustainable treatment of rubber latex processing wastewater: the UASB-system combined with aerobic post-treatment*. Wageningen University and Research.

Vimalamma, T. A., Nair, R. N., & Madhu, G. (2009). Electrochemical treatment of skim serum effluent from natural rubber latex centrifuging units. *Journal of Hard Materials, 167*(1–3), 494–499.

Watanabe, K. (2001) "Microorganisms relevant to bioremediation", Current Opinion in Biotechnology, vol 12, pp 237–241).

Watari Takahiro (2018), Development of a Novel Treatment System for Natural Rubber Processing Wastewater in Vietnam, doctoral thesis.

Watari, T., Mai, T. C., Tanikawa, D., Hirakata, Y., Hatamoto, M., Syutsubo, K., Fukuda, M., Nguyen, N. B., & Yamaguchi, T. (2017). Performance evaluation of the pilot scale up-flow anaerobic sludge blanket–downflow hanging sponge system for natural rubber processing wastewater treatment in South Vietnam. Bioresource Technology, 237(2), 204–212. https://doi.org/10.1016/j.biortech.2017.02.058)

Williard N. Sexauer, Roger V. Karn (2013), Introduction to Stabilization Ponds, Stabilization Pond Systems Operation, Maintenance, Management. Nancy Ellefson Minnesota

Wijerathna, W. S. M. S. K., Wimalaweera, T. I. P., Samarajeewa, D. R., Lindamulla, L. M. L. K. B., Rathnayake, R. M. L. D., Nanayakkara, K. G. N., Jegatheesan, V., Wei, Y., & Jinadasa, K. B. S. N. (2023). Imperative assessment on the current status of rubber wastewater treatment: Research development and future perspectives. *Chemosphere, 338*, Article 139512.

Yang, X., Yuan, J., Guo, W., Tang, X., Zhang, S. (2022), The enzymes-based intermediary model explains the influence mechanism of different aeration strategies on nitrogen removal in a sequencing batch biofilm reactor treating simulated aquaculture wastewater, *J. Clean. Prod.* 356, https://doi.org/10.1016/j.jclepro.2022.131835.

Yusofa, K. M. M., Sabria, M. P. A., Veellua, J. K., Asisa, A. J., Arouac, M. K., &Sulaimanc, N. M. N. Improvement On Environmental Pollution In Latex Processing Factory Using Membrane Filtration.

Zhu, J., Fan, Y., & Xu, N. (2010). Preparation and characterization of alumina membranes on capillary supports: Effect of film-coating on crack-free membrane preparation Chin. *J ChemEng, 3*, 377–383.

Sustainable Management of Wastewater from Oil Palm Processing Industry

Olukemi Aromolaran, Oyeshina Gideon Oyeku, Olutayo Israel Falodun, and Emmanuel I. Unuabonah

Abstract Palm oil tree (*Elaeis guineensis*) is an important oil crop commonly grown for its edible oil and other valued products. The exponential population growth has placed a higher demand on palm oil consumption, which has resulted in the release of increased palm oil mill effluent (POME) into the environment. While researchers have focused on the economic importance of the tree, the increased generation of POME is contributing to environmental degradation, loss of biodiversity and global release of greenhouse gases. This chapter presents the environmental impact resulting from the release of POME and sustainable management practices (vermicomposting, cultivation of micro-algal, bioremediation, phytoremediation) that are utilized to mitigate these impact as well as recover resources that are useful to man from POME.

Keywords Palm oil mill effluent · Environmental pollution · Sustainable management · Biogas generation · Value-added products

---

O. Aromolaran (✉)
Microbiology Programme, Bowen University, Iwo, Osun State, Nigeria
e-mail: olukemi.aromolaran@bowen.edu.ng

O. Aromolaran · E. I. Unuabonah
African Centre of Excellence for Water and Environmental Research (ACEWATER), Redeemer's University, Ede, Osun State, Nigeria
e-mail: unuabonahe@run.edu.ng

O. G. Oyeku
Pure and Applied Biology Programme, Bowen University, Iwo, Osun State, Nigeria

Department of Marine Sciences, University of Georgia, Athens, GA, USA

O. G. Oyeku
e-mail: gideon.oyeku@bowen.edu.ng

O. I. Falodun
Department of Microbiology, University of Ibadan, Ibadan, Oyo State, Nigeria
e-mail: falod2013@gmail.com

E. I. Unuabonah
Department of Chemical Sciences, Redeemer's University, Ede, Osun State, Nigeria

N. A. Oladoja and E. I. Unuabonah (eds.), *Strategic Management of Wastewater from Intensive Rural Industries*, Springer Water,
https://doi.org/10.1007/978-3-031-90314-4_8

1 Introduction

The African oil palm tree (*Elaeis guineensis*) belongs to the family Arecaceae. It is a perennial plant that is usually straight, non-branching stem and long leaves growing at the top (Fig. 1). The leaf and stem are useful for making broom, basket and constructions. The sap gives the highly nutritious alcoholic palm wine, while the red palm oil is used for cooking and other industrial purposes (for producing margarine, ice cream, confectionaries, and pharmaceuticals). The second-grade oil from the palm fruit is used for making soap, cosmetics, and greases. Although, oil palm tree originated from West Africa, Indonesia and Malaysia have become the major producers of palm oil, globally (Cheah et al., 2023). Palm oil is rich in oleic acid, beta-carotene and vitamin E, which gives its many nutritional and health benefits, such as prevention of heart disease, anti-cancer, improve brain cells and aids in good vision (Oguntibeju et al., 2009). Extraction of palm oil from the tree usually begins with the harvesting of the palm fruits, sterilization, then squeezing out of the oil, purifying and finally drying to reduce the oil's moisture. Different kinds of wastes, such as the empty fruit bunch (EFB), fiber and shell, palm kernel meal (PKM), and palm oil mill effluent (POME) are generated in the production flow.

Among all the wastes from the palm oil industry, POME constitutes a major environmental challenge as it is the most voluminous waste generated in the production flow (Dominic & Baidurah, 2022). It is an acidic, viscous, brownish liquid, with high fraction of total solids, biological oxygen demand, chemical oxygen demand, oil and grease (Lam & Lee, 2011). For every tone of palm oil produced, up to 5.0–7.5 tonnes

Fig. 1 Oil palm tree (*Elaeis guineensis*)

of POME are generated (Ahmad et al., 2003), since about half of the quantity of water required in the processing turn out to become POME. In most rural areas, the POME is usually released into the nearby water bodies without prior treatment, and this, undeniably cause serious environmental degradation, loss of biodiversity and deaths of aquatic organisms due to oxygen depletion.

In countries like Malaysia, Indonesia, and Taiwan, remarkable efforts towards the development of related research, policy development and compliance have yielded notable success in the pollution abatement of POME, but adequate handling and management still remains a serious challenge in most west African countries (Iskandar et al., 2018; Okereke & Ginikanwa, 2020). In the west African countries, informal oil mills account for the majority of the palm oil refinery, and many are situated in rural or remote areas with poor accessibility (Izah & Ohimain, 2014; Izah & Ohimain, 2016; Awere et al., 2022). Treatment facilities are basically absent, and raw POME is either stored in open evaporation ponds, discharged into nearby drainage channels, agricultural lands and/or water ways (Awotoye et al., 2011). It is pertinent to note that proper treatment is required before POME is discharged into the environment, but this is not without an additional cost, as highly sophisticated machine is required for proper management. Therefore, sustainable management of POME is advocated. This chapter focuses on the generation of POME from the milling process, its environmental impacts, and the sustainable management practices.

2 Palm Oil Processing

The production yield from oil palm in many rural industries is limited due to poor financial capability (lack of access to credit and loans), low literacy level, lack of required training and skills, unavailability of processing facilities and technological expertise (Tambi et al., 2021; Tening et al., 2023), and poor or sometimes no electricity supply to power the automated engines. This has resulted in low quality and low production yield of palm oil and excessive manual labour compared with the automated industries.

Palm fruits are embedded in spikelet, usually known as the bunch (Fig. 2) with matured fruits about half the weight of the whole mesocarp and the bunch is approximately 25 kg.

Usually, in many rural areas, after harvesting, the bunches are transported, using baskets or trucks, to the processing site, the fruit is first separated from the bunch using an axe or a cutlass and can then collected from the spikelet. In a bigger scale industry, rotary beater bars are employed to separate the fruits from the bunch. The empty bunches are used as cooking fuel in some small-scale industry, while bigger industry burn the empty bunches into ash (FAO, 2023). Since the ashes are rich in potassium, they are often used as fertilizer for the oil palm plantation. The palm fruits are then cooked/sterilized using hot wet heat at a very high temperature(140 °C) or pressurized steam (3×105 Pa) (Wu et al., 2010). This stage will detach the nut from the shell (mesocarp) (Lam & Lee, 2011), breakdown the enzyme responsible

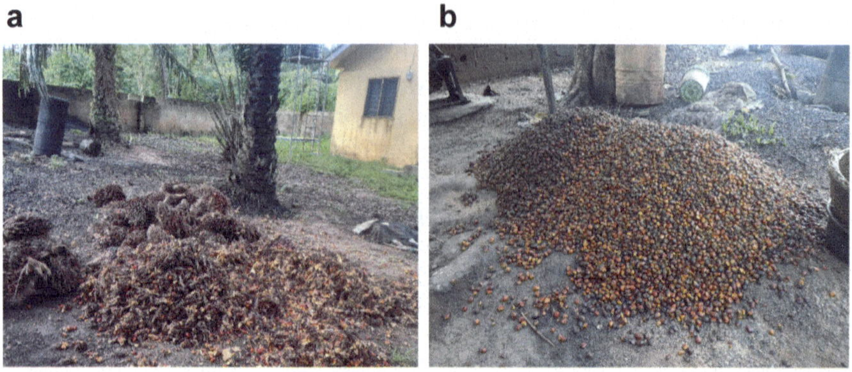

Fig. 2 Oil palm bunches (a) Oil palm fruits (b)

for the build-up of free fatty acids in the crude oil extract (Hassan et al., 2006), coagulate the protein and soften the pulp, while breaking down the gums and resins (to prevent foaming of the oil when it is fried). The condensate from this stage is the first source of POME. The sterilized fruits are then passed into a digester, where the fruits are pounded, to reduce the viscosity of the oil and breakdown the outer layer of the fruit. In many rural settings, this is usually done manually, while in bigger industries, cylindrical vessel fitted with a central rotating shaft, carrying a number of mechanical arms at high temperature (80–90 °C) are used (Wu et al., 2010). The next step is pressing, which can be the dry or the wet method. In the dry method, the fruits are mechanically pressed to squeeze out the oil, while the hot method involves the use of hot water to bring out the oil. At this point, the oil is viscoid (sludge), and composed of oil, water, debris, fiber and other non-oily components. The separation of the oil from the viscoid medium is carried out through a process known as clarification. This involves the addition of hot water into the crude oil, which causes the insoluble solids to settle at the bottom and the oil droplet to move to the surface. The coarse fibre is removed by screening, the lighter mixture is boiled for about 1–2 h, and the crude oil is separated from every form of impurities, such as water, debris, fibrous material and other solids. The insoluble solids at the base, which represent the sludge is drained, and can be further processed before discharge. Figure 3 shows a woman scooping out the oil from the clarifier basin in an oil processing outlet in southwestern Nigeria (Fig. 4).

The clear oil is decanted into a receptacle container, reheated to reduce the moisture content, and the oil is finally scooped into clean containers, from where it is filled into plastic kegs and ready for sale.

The press remaining after the sterilization stage is made up of fibre and nuts. The two are separated and the fibre is used as boiler fuel, while the nuts are cracked and sent to the hydrocyclone (to separate the kernels from empty shells). The wastewater from this process is also released as POME.

The three major streams of wastes generated during the production of palm oil are: solid, liquid waste (effluents) and gaseous waste (emissions).

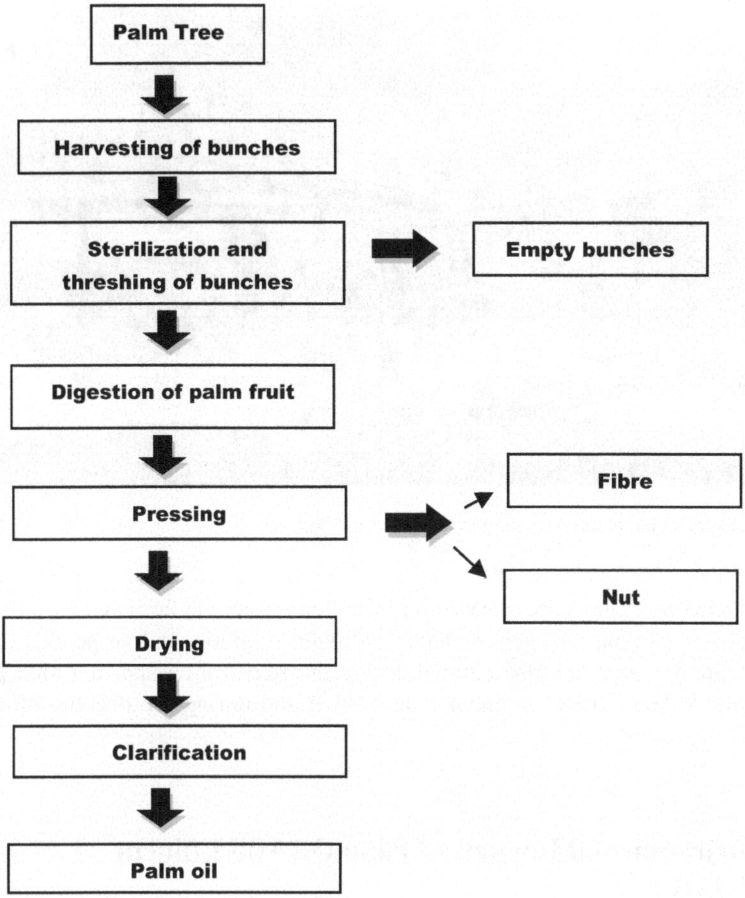

Fig. 3 Palm oil processing flow diagram

Solid wastes—These are generated during threshing, pressing and kernel cracking. It includes the empty fruit bunch (EFB), palm press fiber (PPF), palm kernel shell (PKS), palm kernel meal (PKM) and chaff.

Liquid waste—This is mainly the POME, which is gotten during the sterilization of fresh palm fruits (36%), oil clarification (60%) and the kernel processing (4%) (Ahmed et al., 2015; Suksong et al., 2020). It appears as slurry, thick brownish viscous liquid and usually contains high colloidal suspension with unpleasant odour. The raw POME usually contains high proportion of water (95–96%), constituting 0.6–0.7% oil and 4–5% total solids, and the pH ranges between 4 and 5 (Yap & Chan, 2021). Also present are biological oxygen demand (18.23–23.9 g/L), and chemical oxygen demand (45.81–54.86 g/L) (Yap & Chan, 2021).

Fig. 4 A typical palm oil processing industry in rural Nigeria

Gaseous waste—Gases are released into the atmosphere via farm vehicles, in form of oxides of carbon, nitrogen, sulphur, hydrogen sulphide, and suspended particulate matters. Gases are also emitted during the sterilization and digestion phase (Ohimain & Izah, 2013; Ohimain et al., 2013) and during the decomposition of POME.

3 Environmental Impacts of Palm Oil Mill Effluent (POME)

The environmental impacts of POME are devastating, and the direct discharge can be destructive to the air, land and water (Ogunyemi et al., 2010; Ibrahim et al., 2012).

3.1 Impact on Air Quality

Odorous gases, including ammonia and hydrogen sulphide, are produced during the degradation of POME by microbes, to produce foul odour that pollute the environment. Chronic exposure to ammonia may result in bronchiectasis, asphyxiation, and optic neuropathy (Maries & Maries, 2018). Long-term exposure to hydrogen sulphide can cause visual complications, respiratory irritation, and headaches (Chou et al., 2016). Similarly, greenhouse gases (GHGs), such as methane (CH_4) and carbon dioxide (CO_2) are released during the decomposition of POME (Enstrom et al., 2019). These gases are associated with global warming, which is one of the major challenges confronting global environmental sustainability (Ohimain & Izah, 2014; Anyaoha & Zhang, 2021, 2022).

It is estimated that more than 90% of the GHGs produced during the palm oil milling processes are derived from POME degradation processes by microbes (Rahayu et al., 2015). Palm oil effluent discharged into open ponds by the small-scale mills (commonly situated in rural areas) may serve as an important source of GHGs, due to minimal/no efforts to prevent the escape of these GHGs (Edward et al., 2015). Up to 90% of these released GHGs are captured in bigger industries, where there are facilities for such (Hosseini & Wahid, 2015).

In the West Africa, the small scale and artisanal mills occupy a larger share of the palm processing industry; nearly 80% in Nigeria (Biodun et al., 2021) and between 60 and 80% in Ghana (Osei-Amponsah et al., 2012; Khatun, 2020). Although, there is paucity of information on the GHG emissions from POME by rural-based palm milling industries, it is considered to be a remarkable source of GHG emissions into the atmosphere (Ohimain & Izah, 2014). The continuous release of GHGs into the environment has adverse impact on public health, economic growth, energy prices, the behavioral pattern of biological diversity, and seasonal temperature fluctuations and ecosystem deterioration (Weiskopf et al., 2020) (Fig. 5).

3.2 Impact on the Terrestrial Environment

Palm oil effluent has diverse impacts on the terrestrial system viz:

a. *Soil physicochemical characteristics*

The discharge of raw POME onto surrounding soils has been reported to cause changes in the physicochemical properties of the receiving soil, especially at the areas close to the point of release. For instance, in Southern Nigeria, significantly higher temperature (27–32 °C) was observed around POME soils, compared to unexposed soils (25 °C) (Chinyere et al., 2018). The higher soil temperature observed in the POME-exposed soil is attributed to the degradation of organic matter in the effluent by microbes. Soil temperature influences plant root growth, nutrient uptake and water viscosity (Rai & Chatrath, 2019; Ali & Omofunmi, 2021). Water viscosity in turn affects metabolic activities in the plant roots, through changes in membrane lipids and enzymes associated with nutrient uptake (Dong et al., 2001). Chinyere et al. (2013) noted a remarkable change in soil pH (from alkaline pH 7–10 to acidic pH 5–6), following continuous discharge of POME unto the soil. Similarly, a reduction in the pH of contaminated agricultural soil was reported by Nnaji et al. (2016) and Chikwendu and Ogbonna (2018) from the same region. However, Okwute and Isu (2007a), Eno et al. (2017), Chinyere et al. (2018); Nwachukwu et al. (2018) and Nta et al. (2020) found a significant increase in the pH of contaminated soils, in comparison with the non-contaminated soils.

Naturally, crude POME is an acidic colloidal suspension (Opasina et al., 2018), however, when it is discharged into the soil, the pH gradually increases to alkaline, as

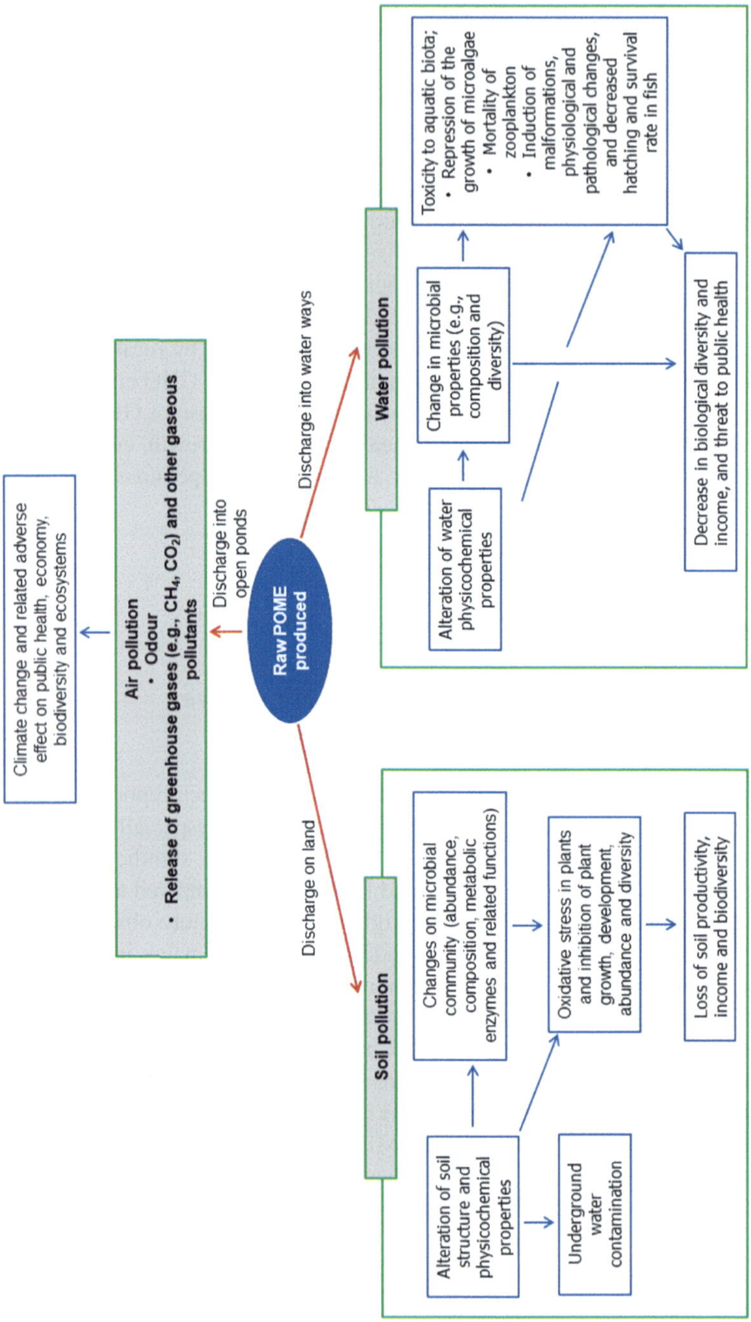

Fig. 5 Environmental impacts of palm oil mill effluent

a result of biodegradation that takes place over time (Ako & Anegbe, 2022). Hence, sites freshly contaminated by the discharged effluent and/or nearer to the point of effluent discharge may exhibit acidic pH than those irregularly impacted and/or farther away from the point of discharge (Ohimain et al., 2012). The availability of plant nutrients in the soil is affected by the pH of the soil. Typically, most plants grow better within soil pH range of 6.5–7.5, but their growth can be adversely affected when there is alteration in the soil pH (Nwachukwu et al., 2018). At lower pH, excess hydrogen ions can combine with the hydroxide ion of insoluble aluminum minerals to produce water and aluminum ion (Al^{3+}) in solution (Brix, 2008). In line with this fact, significantly higher Al^{3+} level was recorded in POME-polluted soil samples in comparison with the unperturbed soils (Chinyere et al., 2018). Aluminum ions commonly interfere with enzymes in several plant, inhibit DNA replication, modify nitrate uptake by plants and the structure of the cell membranes (Bojorquez-Quintal et al., 2017). High levels of metals, such as sodium (Na^{+}), potassium (K^{+}), calcium (Ca^{2+}), magnesium (Mg^{2+}), cadmium (Cd), copper (Cu), chromium (Cr), zinc (Zn), iron (Fe), nickel (Ni) and arsenic (As) have been documented in POME-impacted soils (Nnaji et al., 2016; Ubani et al., 2017; Nwachukwu et al., 2018; Maliki et al., 2020; Ikechukwu & Uzoma, 2021; Ezeomedo & Ogbogu, 2022). In addition, accumulation of toxic metals (Cd, Cr, Pb, Ni and As), beyond permissible limits, in exposed soils and cultivated crops have also been documented (Ikechukwu & Uzoma, 2021). Accumulation of toxic metals in plants can lead to repression of photosynthesis, respiration, transpiration, intake of water and essential minerals for growth, root and shoot growth; inhibition of ATP production; as well as induction of necrosis, lipid peroxidation, and DNA damage by overproduction of ROS (Hermans et al., 2011; Pourrut et al., 2011; Mishra & Bharagava, 2016; Abbas et al., 2018; Hassan et al., 2019). Overconsumption of such food crops by humans could cause gastrointestinal distress, hypoglycemia, damage of vital organs, erratic heartbeat, and in some cases, cancers (Wani et al., 2015; Genchi et al., 2020; Sonone et al., 2020; Rahaman et al., 2021).

Indiscriminate discharge of POME on soils has also been found to increase their electrical conductivity, due to high amount of soluble salts in POME (USDA, 2011). High amount of nitrate, phosphate, oil and grease in soils have also been reported to be as a result of POME (Okwute & Isu, 2007a; Chinyere et al., 2013; Nnaji et al., 2016; Chikwendu & Ogbonna, 2018; Nwachukwu et al., 2018; Maliki et al., 2020; Nta et al., 2020; Ako & Anegbe, 2022). Increase in the phosphorus amount could be attributed to high adsorption in the soil, or precipitation in the soil, or from gradual degradation of POME (Mustapha, 2018). Additionally, increase in soil moisture due to the POME discharge, have been reported (Chinyere et al., 2013; Chikwendu & Ogbonna, 2018; Chinyere et al., 2018; Nwachukwu et al., 2018) and this increases the solubility of cations (Fe^{2+} and Mn^{2+}), which subsequently combine with phosphate to form insoluble complexes, thereby, reducing the available phosphate in the soil solution. This condition could lead to phosphate stress and the stimulation of phosphatases (Chinyere et al., 2018). Several literature reports have suggested that the microbial degradation of sugars (glucose, xylose, arabinose, manose and galactose), fatty acids and amino acids present in untreated POME increases the organic matter level in the soil (Maliki et al., 2020; Nta et al., 2020; Ako & Anegbe, 2022; Ezeomedo & Ogbogu, 2022).

b. *Soil biological characteristics*

Soil serves as habitat for diverse microorganisms, such as bacteria, fungi, and protozoans, which influence its productivity for agricultural and other uses (Ramesh et al., 2019). Many researchers have revealed that raw POME could alter the microbiological characteristics of a soil. Okwute et al. (2007) reported that POME pollution significantly increased the soil aerobic bacterial population, but resulted in the exclusion of nitrifying bacteria (ammonium oxidizers) from the soil, indicating that the effluent could have adverse impact on nitrification. They attributed this impact to increased moisture and organic content, as well as decreased aeration in the soil, making it difficult for ammonium oxidizers to grow. Ibe et al. (2014) also observed that POME negatively impacted the microbial community in the soil, as well as various enzymes, which enable them to carry out metabolic activities that ensure the continuity of elemental cycles in nature. POME-polluted soil samples had significant lower numbers of nitrifying, phosphate solubilizing, total heterotrophic, cellulolytic and lipolytic bacteria, but higher palm oil degrading/utilizing bacteria, in comparison to unexposed soil samples. POME contamination provided favorable environment and utilizable components for palm oil-utilizing bacteria to flourish. The remarkably lower number of various bacterial groups at the POME-contaminated soil may be attributed to high concentration of POME, excess water, depletion of oxygen as well as antibacterial activity of oil and grease, certain metals, phenols and other organic acids in the POME (Pascual et al., 2007; Nwuago et al., 2008; Nwoko & Ogunyemi, 2010; Arisht et al., 2020). The observed reduction in the activity of most enzymes could be attributed to their subjection to stress in the effort to oxidize organic matter contained in the POME-contaminated soil (Nwaogu et al., 2012).

Orji et al. (2006) and Opasina et al. (2018) reported that POME-contaminated soils may exhibit lower diversity microorganisms, but have abundance of bacteria and fungi in comparison to older and/or abandoned soils. The findings of Okwute and Isu (2007), Eno et al. (2018), and Williams (2018) agree with this, and they suggest that in addition, factors including distance from the point of discharge, soil characteristics, and environmental adaptation of the native microbes also play a role in determining the response of microorganisms to the effluent. In general, adverse changes in the soil bacterial community and enzymes activity can make the soil to lose its productive potentials (Maila & Cloete, 2005), and as such, plants, animals, and man who depend on the soil would be adversely affected (Osuji & Nwoye, 2007). Experimental and field-based studies have shown that raw POME is capable of posing serious threat to crops and other plants growing on the affected lands. Oikeh et al. (2014) indicated that raw POME induces oxidative stress in plants. Similarly, POME has inhibitory effects (lower germination rate, growth rate, plant height, shoot fresh weight and dry weight, root, leaf and stalk length) on the development of crops (Ifediora & Ifediora, 2016; Amoke & Amadi, 2019; Okolie et al., 2020; Onuorah et al., 2022; Enujeke et al., 2023). In severe conditions of pollution, POME reduces the vegetation and/or growth of crop plants (Opasina et al., 2018; Awotoye et al.,

2019; Onuorah et al., 2022; Enujeke et al., 2023) because of the clogging of the seeds which could adversely affect the respiratory process in the seed embryo.

3.3 *Impact on the Aquatic Environment*

a. *Impact on Water Quality*

The degradation of water quality as a result of the release of untreated POME has been reported by many researchers. In a study, Edward et al. (2017) reported that River Awemu, Ijero-Ekiti, Nigeria which was subjected to POME discharge exhibited pH, TS, TSS, TDS, chloride, BOD, metals (Na, Mg and Pb) and PO_4^{3-} which had values higher than WHO limit and Nigeria Standard for Drinking Water Quality. Awotoye et al. (2019) also observed the impact of POME on Oluwa River, in Ondo State, Nigeria which showed high levels of NO_3^-, Zn^{2+}, BOD_5 and COD values above the WHO limits.

b. *Impact on aquatic biota*

A few studies have assessed the toxicity of POME to aquatic biota using different types of organisms and biological methods. Sharuddin et al. (2017) reported reduction in bacterial diversity in a surface water, as a result of POME discharge. The main bacteria families detected were the Chromatiaceae and Alcaligenaceae. Jasni et al. (2020) also reported that the growth of the microalgae—*Coelastrella* sp., *Chlorella* sp. and *Scenedesmus* sp. exposed to constituents of POME were suppressed. In addition, based on the combination of whole effluent toxicity (WET) and toxicity identification evaluation (TIE) tests using *Daphnia magma*, a a study was able to establish that POME is toxic and that its constituents including phenol, 2,6-bis (1,1-dimethylethyl)-, copper and zinc were major toxicants in the effluent (Hashiguchi et al. 2020). In another study, the majority of *Danio rerio* fish embryos exposed to raw POME were reported to exhibit malformation, coagulation/mortality within 24 to 96 h after hatching. Also, embryonic development in the species was observed to be inhibited at 0.12 mg/L of Cu, and 4.9 mg/L of Fe for 96 h (Hashiguchi et al., 2021).

Toxic effects including vacuolation, necrosis and diffused congestion of cells in liver, decrease in red blood cells, packed cell volume and haemoglobin as well as increase in white blood cells have been reported in African catfish (*Clarias gariepinus*) exposed to sub-lethal concentrations of POME between 0.01 and 0.09 ml/L for 14 days (Afolabi et al., 2023). There is also a documentation of POME-mediated oxidative stress, lipid peroxidation, haemato-biochemical, enzymatic and ionic changes in juvenile *Heteroclarias* following sub-lethal exposure to POME at levels between 4 and 12 mg/L (Owolabi et al. 2021). These effects were evident as concentration-dependent reduction in haemoglobin, red blood cell, packed cell volume, mean cell volume, mean cell haemoglobin and mean cell haemoglobin concentration, as well as increase in white blood cell, and enzymes (alanine aminotransaminase, aspartate aminotransaminase, lactate dehydrogenase, glutathione peroxidase, superoxide dismutase and catalase) activities in the gill and liver of the fish. Akinsorotan (2013) assessed the potential impact of untreated POME on the early life stage of Nile tilapia (*Oreochromis niloticus*). The results showed that the POME was highly toxic to this fish after 96 h of exposure ($LC_{50} = 9.19$ mgL^{-1}) with complete mortality at 50 mg/L after 24 h. Other sub-lethal effects were: behavioral abnormalities (air gulping, erratic movement, loss of reflex, barbell deformation, molting, excessive mucus secretion and hemorrhage), physiological effects (cell proliferation, epithelial lifting, lamellar fusion and lamellar cell hyperplasia, vacuation of hepatocytes and necrosis), degenerative effects (vacuation and necrosis) and changes in the liver and gill (Akinsorotan 2013). Prolonged (up to 45 days) exposure to 1.6–3.1 mg/L of POME has also been documented to induce pathological changes, such as erythrocyte infiltration, hyperplasia, epithelial lifting and lamellar fusion in the gill of Nile tilapia (Muliari et al., 2018). A short-term (96 h) exposure to the same concentrations of POME significantly decreased hatching and survival rate, with induced malformations, such as kyphosis, lordosis and curved tail, in juvenile stages of the fish (Muliari et al., 2020a). Muliari et al. (2019) and Muliari et al. (2020b) also reported the toxic effects of POME on the female and male reproductive organs of fish.

4 Sustainable Management of Palm Oil Mill Effluent

The ability of POME to negatively impact the soil and water property, biodiversity as well as increase greehhouse gases reveals the need for its proper treatment before it is discharged into the environment. Appropriate treatment of POME before discharge into the surroundings is highly essential, due to the risks it poses to the environment and to public health. In some instances, the POME wastes are converted into other valuable products, but these are yet to be adopted in many rural milling setups (Asamoah & Awuku, 2020). It is therefore necessary to sensitize rural millers on good and appropriate POME management and sustainable practices to minimise the impact of the effluent on the environment (Okere et al., 2021). Effective and sustainable management options for POME should be typically inexpensive, accessible, users and eco-friendly (Nwoko et al., 2010). Normally, any sustainable management approach should be able to reduce the hazards associated with the release of POME, generate useful products from it, help conserve the environment and mitigate air, land and water pollution (Rupani et al., 2019). Utilizing a sustainable approach in the oil palm milling process will not only meet the standard discharge limits for wastewater but will ensure sustainability in the milling process, creation of value-added product and reduced effluent treatment cost. Some of these useful and sustainable methods are discussed below:

(a) ***Vermicomposting and Fertilizer***: This process involves the deployment of earthworms to decompose organic matter in the POME (Sim & Wu, 2010). The earthworm conditions and fragments the waste for further biochemical degradation by microorganisms (John-Paul et al., 2011; Okereke & Ginikanwa, 2020). As the earthworm biomass declines, microbial enzyme activity also declines, hence, resulting in the stabilization of organic matter. Vermicomposting is affordable, reduces waste volume and provides the opportunity for resource recovery (Rupani et al., 2018). It offers a double advantage in that the earthworm biomass can be subsequently processed into proteins and use as animal feeds (Singh et al., 2011). The vermicompost or casting, which is non-toxic and rich in nutrients can be applied as fertilizer and soil conditioner in oil palm plantations or in any other farmland (Rupani et al., 2013; Zainal, 2014; Lim & Wu, 2015; Lee et al., 2019). In comparison to common but expensive physical, and chemical methods used for POME treatment, vermicomposting offers the advantage of a viable cost-effective technique, with the potential to generate products having lower mass, high humus content, low processing time, less phyto-toxicity, greater fertilizer value and earthworm as a significant additional product that can be used as feed for fish (Rupani et al., 2017; Parolini et al., 2020; Awere et al., 2022). Vermicompost is also rich in nitrogen, phosphorus, potassium, and others nutrients in available forms for plant use (Lim et al., 2014; Lim & Wu, 2016).

Syirat et al. (2014) used epigeic earthworms (*Eudrillus eugeniae*) to decompose POME sludge and it yielded high quality compost within 60 days. Other studies have blended POME with lignocellulosic waste materials (e.g., empty fruit bunch), which are usually produced in huge amounts during the process of extraction of palm oil. Rupani et al. (2017) carried out vermicomposting of a mixture of POME and palm pressed fibre (1:1) using *Lumbricus rubellus* within a 30 days period. The compost obtained showed remarkable improvement in phosphorus, nitrogen and potassium levels and promoted the growth and germination of mung bean (using 75% of the vermicompost extract obtained). In a similar study, different mixtures of empty fruit bunch and POME were successfully vermi-composted using earthworms for 84 days period. A remarkable increase in the total phosphorus (0.2–1.4 mg/kg), total nitrogen (0.4–1.7 mg/kg) and total potassium (0.06–0.5 mg/kg) of amended soil was observed, with maximum amounts recorded for vermicompost obtained from 1:1 empty fruit bunch-POME combination (Nahrul et al., 2012). Also, Karim et al. (2022) observed that fortified vermicompost from POME sludge improved soil physicochemical properties, such as moisture content, aggregate stability, porosity, soil shear stress, exchangeable cation capacity, as well as macro and micronutrients. Significant improvement in the growth and performance of maize (*Zea mays*) plant was evident as increased leaf area index, dry weight, root weight density, photosynthesis process rate and chlorophyll content were recorded.

Palm oil mill effluent can be improved and use as fertilizers in agricultural soils. This can be possible by making use of the ponding system of treatment, whereby the effluent is allowed to undergo natural decomposition. The organic nitrogen in the POME is converted to ammonia, and later to nitrate by the indigenous nitrifying bacteria in the POME. The POME can be applied as liquid fertilizer or allowed to dry and use as manure. Supriatna et al. (2022) reported the use of POME as compost for soil amendment. Compost from POME is rich in nutrients which are useful for plant growth and enhances the soil structure and fertility. This also improves soil ability to retain moisture and nutrients, especially in areas where the soil quality is poor (Supriatna et al., 2022; Goldan et al., 2023). The co-composting of EFB and POME (EFB-POME) has also been reported to mitigate the emission of greenhouse gases into the air by more than 70% and the indigenous microorganisms in the EFB-POME have potential ligninolytic and cellulolytic enzymes, which can further enhance the quality of the compost (Krishnan et al., 2017).

(b) ***Micro-algal cultivation***: POME contains high nutrients (nitrogen, phosphorus) and minerals (calcium, potassium, phosphorus, magnesium, zinc, and iron) which can be utilized by microalgae for their growth (Nur et al., 2022). It is an efficient and cost-effective treatment technique (Hariz & Takrif, 2017). Microalgae generally require nutrient rich medium to grow adequately and its cultivation has been demonstrated for the treatment of pulp mill effluent (Usha et al., 2016), textile effluent (Ghazal et al., 2018) and agricultural wastes (La Bella et al., 2022) among others. The microalgal biomass obtained from this treatment process can also be used for desirable value-added products (Hadiyanto & Danny, 2013; Al-Jabri et al., 2021). However, because of the colloidal, turbid and viscous nature of raw POME, dilution of the POME may be necessary to allow light penetration into the culture media for growth of the algae (Bello et al., 2013). Haruna et al. (2017) reported the cultivation of *Chlorella sorokiniana* in diluted POME (20–80%) and reduction of COD, nitrate, phosphate and ammonium at the rate of 19–91%, 33–71%, 29–64% and 70–98%, respectively was observed. A similar study also recorded up to 79% COD, 90% BOD, 100% NH_3-N and 86% total phosphorus removal by *Chlorella vulgaris* cultured in 10% POME (Emparam et al., 2020). A green microalga was also observed to reduce phosphate by 97.7%, ammonium by 95.08%, fatty oil by 99.7% and nitrate by 80.74% (Elvitriana et al., 2021). Other microalgae already reported with the potential to remediate POME include: *Spirulina plantesis* (Maulana et al., 2012), *Chlamydomonas incerta* (Kamyab et al., 2015a), *Chlamydomonas pyrenoidosa* (Kamyab et al., 2015b), *Chlorella* sp., *Chlamydomonas* sp., (Hadiyanto et al., 2012; Hazman et al., 2018), *Nanochloropsis oculata* (Shah et al., 2016), *Chlorella sorokiniana* (Cheah et al., 2018), *Nannochloropsis* sp. (Emparan et al., 2019), *Tetraselmis suecica* (Nur et al., 2020), *Spirulina* sp., *Dunaliella* sp., and *Chaetoceros calciltrans* (Nur et al., 2022). The period of treatment was between 7 and 28 days. The microalgal biomass obtained from the POME treatment is useful in the production of lipids and biodiesel, as well as for CO_2 sequestration (Hazman et al., 2018; Hariz et al., 2019).

(c) ***Bioremediation***: Bioremediation is a waste degradation process that makes use of living organisms, such as bacteria and fungi, and/or their enzymes to restore a contaminated environment back to its original condition. The efficiency of this method for the reclamation of wastewater has drawn considerable attention in recent times. This technique could work as a simple, feasible and cost-effective biological treatment for mitigating environmental pollution associated with POME. The oil effluent is able to provide carbon as well as other required nutrients, such as nitrogen and phosphorus, which are required for the growth of microbes with little or no supplementation (Okwute & Ijah, 2014). Diverse microorganisms with the capability to breakdown organic compounds in POME, thus reducing its BOD, COD and colour have been reported. These include some indigenous microorganisms, such as bacteria (*Geobacillus zalihae, Bacillus* sp., *Pseudomonas* sp., *Staphylococcus aureus*), and fungi (*Aspergillus, Trichoderma, Penicillium, Mucor, Candida, Saccharomyces, Rhizopus, Geotrichum, Candida* and *Fusarium*) (Ebana et al., 2017; Wakil & Osesusi, 2017; Imo & Ihejirika, 2021; Athoillah & Ahmad, 2022; Ibiyemi & Julius, 2022). The degradation of POME by these microorganisms is possible because of their ability to produce lipolytic enzymes (lipase and cellulose), which are able to break down the lipids in the effluent (Iwuagwu & Ugwuanyi, 2014). A study carried out by Bala et al. (2014) identified *Bacillus cereus* isolated from POME as a viable candidate for the degradation of POME. The TSS was reduced by 71.63% and oil and grease down by 85.14%. Lanka and Pydipalli (2018) also established that *Emericella nidulans* and *Trichoderma reesei* isolated from POME were capable of removing BOD, COD, oil and grease in the POME. Oleaginous fungal (*A. niger, A. oryzae* and *Rhizopus* spp.) were isolated from POME as a low-cost carbon resource for lipid and biodiesel production. One of these isolates exhibited high lipid yield for biodiesel production as well as 70% COD removal efficiency (Athoillah & Ahmad, 2022).

Co-culturing of bacteria and fungi to remediate POME has also been considered as a promising alternative for achieving maximum biomass and/or bio-products. For instance, the co-culture of a bacterium (*Bacillus cereus*) and fungus (*Lipomyces starkeyi*) has been reported to achieve a COD removal of 83.66% in POME. The bioremediation efficiency of this procedure was confirmed by the promotion of *Vigna radiata* germination (Karim et al., 2021). Similarly, mixed cultures of bacteria (*B. cereus* and *B. subtilis*) were observed to produce high degradative activity as evident by remarkable reduction of BOD by 93.11% and COD by 90.64% (Bala et al., 2015). Bala et al. (2018) proposed the use of mixed consortium of indigenous microbes (*Micrococcus luteus, Stenotrophomonas maltophila, Bacillus cereus, B. subtilis, Providencia vermicola, Klebsiella pneumonia, Aspergillus niger, A. fumigatus, A. nomius* and *Meyerozyma guilliermondii*) for the reduction of organic load in POME as well as its biodegradation. They recorded a reduction of BOD (90.23%), COD (91.06%), and TSS (92.23%). Similarly, Thegarathah and Simarani (2020) have psotulated that the immobilization of remediating microbe on cheap and sustainable substrate like the coconut husk could provide larger surface area for cell growth and also facilitate faster growth when compared with the use of cell free remediation.

(d) ***Phytoremediation***: Some plant species can bio-accumulate pollutants in their system, thus purifying the environment. This is referred to as phytoremediation.

A suitable plant is expected to have high uptake ability for organic and inorganic contaminants and/or degrade or volatize them, high tolerance to polluted water, rapid growth rate, be easy to harvest and easily controlled in dispersion, as well as have high adaptability to the climate of the concerned area (Ali et al., 2013). Generally, young plants grow faster and have higher nutrient uptake rates than old ones. Aquatic macrophytes are good candidates for use due to their fast growth rate, large biomass and ability to extract and accumulate pollutants (Guittony-Philippe et al., 2014). Some plant species, such as water hyacinth (*Eichhornia crassipes*), water spinach (*Ipomoea aquatica*), water fern (*Azolla filiculoides*), water lettuce (*Pistia stratiotes*), vetiver grass (*Chrysopogon zizanioides*) have been successfully used in phytoremediation. Phytoremediation of POME is an environmentally friendly technique, as reported by Wei et al. (2019) who applied water hyacinth (*Eichhornia crassipes*) to reclaim POME. In a similar study, Zulfahmi et al. (2021) used water spinach (*Ipomoea* aquatic) in the remediation of POME and the result showed reduction in COD by 86.3%, nitrate by 21.5%and phosphate by 91.0%. These plant species can be adopted for the treatment of POME before it is being discharged to the environment.

(e) ***Animal feed meal***: Scientific investigations have also shown that the remediation of POME using bacteria and fungi could also serve as a means of cultivating useful biomass without the addition of nutrient supplementation. The biomass produced can be used as enriched feed directly to animals, be made part of feed composition or animals, or for producing value-added bio-products, such as biofuels and fatty acids. This can provide extra income and improve the sustainability of the palm mill industries (Athoillah & Ahmad, 2022). *Lipomyces starkeyi* (a fungus) was to attain waste-to-wealth by producing food-grade yeast (Iwuagwu & Ugwuanyi, 2014). Adebiyi et al. (2020) have reported the use of palm oil mill sludge to feed pigs by some small-scale palm oil millers. Palm oil mill effluent is usually rich in crude proteins, crude fibre, fat, ash, potassium, phosphorus, calcium, copper and zinc in varying proportions. Nugroho et al. (2023) included 10% POME in the diet of post-weaning goats and the result showed improvement in the growth performance of the kids. High body weight gain and growth performance was observed in goats fed with the combination of palm acid oil, Napier grass, coconut water and water lettuce in a Malaysian study (Amalina et al., 2021). Similarly, increased production was found in broiler, pekin ducks and rabbit fed with up to 10–15% incorporated POME feed (Tartrakoon et al., 2016; Adebiyi et al., 2020).

(f) ***Bioenergy production***: Studies have shown that POME can be useful in the production of biogas because of its high organic content (Sankaran et al., 2010; Lam & Lee, 2011; Ahmed et al., 2015; Rajani et al., 2019). Biogas is a renewable energy, which can be produced by anaerobic digestion of organic matter by microorganisms. It is a mixture of methane, carbon-dioxide and hydrogen gas. The biogas generated in the anaerobic pond of POME can be captured and utilize for power (heat and electricity) generation. Microbes present in POME are able to breakdown the organic constituents of the POME to produce biogas. The process involves in biogas production include: hydrolysis, acidiogenesis, acetogenesis and methanogenesis (Ohimain & Izah, 2017). Hydrolysis is the conversion of complex organic molecules in POME to simpler products, such as fatty acids, amino acids and sugars. Acidogenesis which is also refers to fermentation is the further breakdown of the products formed in hydrolysis phase to acetate, carbon-dioxide and hydrogen gas. Other products, such as ethanol, lactate, priopionate and butyrate are formed during the acetogenesis phase. The two bacteria groups involved in the last phase, methanogenesis are the acetotrophic methanogens and the hydrogenotrophic methanogens. While the former uses formate as the electron donor, the later make use of hydrogen as the electron acceptor to produce methane gas (Ohimain & Izah, 2017). Without prior optimization, it is possible to generate 28 m^3 of biogas from 1 m^3 of POME in a treatment plant operated at 30–40 °C as reported by Ahmed et al. (2015). In a similar research, Rashid et al. (2019) were able to generate 3.8 L/d of biogas by loading POME in an anerobic bioreactor with optimized varying organic loading rate, C-N ratio and hydraulic retention time. Sidik et al. (2013) found that the co-digestion of POME (70%) and cow dung (30%) was able to generate approximately 2 L of biogas. Since POME contains high level of organic matter, adoption of anaerobic digestion in the first stage of the treatment process is a necessity to convert the bulk of the wastes to biogas. The treated effluent is further subjected to an aerobic treatment in order to meet the required discharge standards. Some microorganisms, such as *Clostridium butyricum* EB6 (Chong et al., 2009), *Stenotrophomonas rhizopila* strain E-P10, *Bacillus toyonensis* strain BCT-7112 (Utami et al., 2019) have been successfully adopted to enhance production of biogas from POME.

(g) ***Value-added products***: Several studies have shown that microorganisms can convert some of the constituents of POME to more useful products, such as ethanol, citric acid, enzymes (lignin peroxidase, protease, lipase and cellulase) (Okoli, 2023). Kalil et al. (2003) produced acetone-butanol-ethanol (ABE) from 90% sedimented POME after fermenting with *Clostridium acetobutylicum* NCIMB13357 and *C. saccharoperbutylacetonicum* N1-4 in batch culture system and it was found that up to 4 g L^{-1} ABE was produced at initial pH of 5.8 with *C. acetobutylicum* NCIMB13357. Production of citric acid from POME has been proven effective. In a study by Alam et al. (2008), up to 5.2 g/L of citric acid was produced from POME within 7 days. Research has shown that POME can be converted to organic acids (acetic and propionic acids) anaerobically. The organic acids can then be converted to biodegradable plastic, such as polyhydroxyalkanoates (PHA) by the bacterium, *Rhodobacter sphaeroides* (Hassan et al., 1996; Salihu & Alam, 2012). Hydrolytic enzymes have been captured from particular species of fungi with diverse industrial uses and the possibility of extracting hydrolytic enzymes from POME has been reported (Razak et al., 1997). Fermentation of POME by *T. reesei* resulted in the production of cellulose (0.917 U/ml) and carboxymethylcellulose (2.51 U/ml) after 5 days in a laboratory study (Rashid et al., 2009). The use of *Aspergillus niger* and augmentation of the medium with 0.6 g/l NH_4NO_3 increased the produced enzyme (Prasertsan et al., 1997). Razak et al. (1997) isolated *Rhizopus oryzae* and *Rhizopus rhizopodiformis* from POME with the capability to produce extracellular lipases at optimized conditions. In a similar study by Salihu et al. (20011a), *Candida cylindracea* produced up to 4.02 U/ml lipase in POME supplemented medium.

5 Conclusion

The high organic content of POME and its large volume is of significant environmental and public health concern, when released into the environment without prior treatment. Some of the environmental effects of POME include the release of greenhouse gases and some other detrimental gases into the atmosphere, adverse effect on the physicochemical properties of soil, induced oxidative stress on plant, loss of biodiversity and degenerative effects as well as the death of some aquatic organisms. Sustainable management of POME will not only reduce waste, it will also enhance resource conservation and recovery, promote soil fertility, fight climate change and convert waste to valued products. Utilization of bio-energies from POME will also promote the United Nations Sustainable Goal 7 (Affordable and Clean Energy). Provision of financial interventions for modernize processing, especially to the rural millers, as well as promulgate and effectively enforce laws on sustainable and eco-friendly POME management practices is thereby advocated.

References

Adebiyi, O. A., Oboli, U. T., Adejumo, I. O., Osinowo, O. A., Olumide, M., & Adeshola, A. T. (2020). Nutritive value of palm oil sludge and its influence on fat composition and deposition in grower pigs. *Nigerian Journal Animal Science, 22*(1), 253–326.

Ahmad, L., Ismail, A., & Bhatia, S. (2003). Water recycling from palm oil mill effluent (POME) using membrane technology. *Desalination, 157*, 87–95.

Ahmed, Y., Yaakob, Z., Akhtar, P., & Sopian, K. (2015). Production of biogas and performance evaluation of existing treatment processes in palm oil mill effluent (POME). *Renewable and Sustainable Energy Reviews, 42*, 1260–1278.

Alam, M. Z., Jamal, P., & Nadzir, M. M. (2008). Bioconversion of palm oil mill effluent for citric acid production: Statistical optimization of fermentation media and time by central composite design. *World Journal of Microbiology and Biotechnology, 24*, 1177–1185.

Amalina, I. F., Haziq, J. M., Syukor, A. R. A., Ridwan, A. F. A., & Rashid, A. H. M. (2021). Study of palm acid oil (PAO) from sludge palm oil mill effluent (pome) as goat's feed. *Materials Today: Proceedings., 41*, 96–101.

Cheah, W. Y., Siti-Dina, R. P., Leng, S. T. K., Er, A. C., & Show, P. L. (2023). Circular bioeconomy in palm oil industry: Current practices and future perspectives, *Environmental Technology & Innovation, 30*,103050. https://doi.org/10.1016/j.eti.2023.103050

Chong, M., Rahman, N. A. A., Rahim, R. A., Aziz, S. A., Shirai, Y., & Hassan, M. A. (2009). Optimization of biohydrogen production by *Clostridium butyricum* EB6 from palm oil mill effluent using response surface methodology. *International Journal of Hydrogen Energy., 34*(17), 7475–7482. https://doi.org/10.1016/j.ijhydene.2009.05.088

Dominic, D., & Baidurah, S. (2022). Recent developments in biological processing technology for palm oil mill effluent treatment—a review. *Biology, 11*(4), 525. https://doi.org/10.3390/biology11040525

Hassan, M. A., Shirai, Y., Kusubayashi, N., Abdulkarim, M. I., Nakanishi, K., & Hashimoto, K. (1996). Effect of organic acid profiles during anaerobic treatment of palm oil mill effluent on the production of Polyhydroxyalkanoates by *Rhodobacter sphaeroides. Journal of Fermention Bioengineering, 82*(2), 151–156.

Kalil, M. S., Kit, P. W., Yusoff, W. M. W., Sadazo, Y., & Rahman, R. A. (2003). Direct fermentation of palm oil mill effluent to acetone-butanol-ethanol by solvent producing clostridia. *Pakistan Journal of Biological Sciences, 6*, 1273–1275. https://doi.org/10.3923/pjbs.2003.1273.1275

Krishnan, Y., Bong, C. P. C., Azman, N. F., Zakaria, Z., Othman, N., Abdullah, N., et al. (2017). Co-composting of palm effluent bunch and palm oil mill effluent: Microbial density and potential mitigation of greenhouse gas emission. *Journal of Cleaner Production, 146*, 94–100.

Lam, M. K., & Lee, K. T. (2011). Renewable and sustainable bioenergies production from palm oil mill effluent (POME): Win–win strategies toward better environmental protection. *Biotechnology Advances, 29*, 124–141.

Nugroho, N., Alamsyah, R., Wardani, A. S., Maulana, H., Wiratni, M. L., Tanto, D., & Baliarti, E. (2023). The effect of palm oil mill effluent supplementation on post-weaning goat performance. *IOP Conference Series: Earth and Environmental Science, 1174*, Article 012018.

Oguntibeju, O. O., Esterhuyse, A. J., & Truter, E. J. (2009). Red palm oil: Nutritional, physiological and therapeutic roles in improving human wellbeing and quality of life. *British Journal of Biomedical Science 66*(4), 216–22. https://doi.org/10.1080/09674845.2009.11730279

Prasertsan, P., Kittikul, A. H., Kunghae, A., Maneesri, J., & Oi, S. (1997). Optimization for xylanase and cellulase production from *Aspergillus niger* ATTC 6275 in palm oil mill wastes and its application. *World Journal Microbiology Biotechnology, 13*, 555–559.

Rajani, A., Kusnadi, S. A., Saepudin, A., Gobikrishnan, S., & Andriani, D. (2019). Review on biogas from palm oil mill effluent (POME): Challenges and opportunities in Indonesia. *IOP Conference Series: Earth and Environmental Science, 293*, Article 012004.

Rashid, M., Shakib, N., & Rahman, T. (2019). Biogas production from POME by optimum level of

Rashid, S. S., Alam, M. Z., Abdulkarim, M. A., & Salleh, M. H. (2009). Management of palm oil mill effluent through production of cellulases by filamentous fungi. *World Journal Microbiology Biotechnology, 25*(12), 2219–2226.

Razak, C. N. A., Salleh, A. B., Musani, R., Samad, M. Y., & Basri, M. (1997). Some characteristics of lipases from thermophilic fungi isolated from palm oil mill effluent. *Journal of Molecular Catalysis b: Enzyme, 3*, 153–159.

Suksong, W., Tukanghan, W., Promnuan, K., Kongjan, P., Reungsang, A., Insam, H., & O-Thong, S. (2020). Biogas production from palm oil mill effluent and empty fruit bunches by coupled liquid and solid state anaerobic digestion. *Bioresources Technology, 296*, 122304.

Yap, C. C., & Chan, Y. J. (2021). Chapter 27-Approching zero liquid discharge concept using high-rate integrated pilot-scale bioreactor in palm oil mill effluent (POME) treatment. In A. W. Mohammad & W. L. Ang (Eds.), *Integrate and hybrid process technology for water and wastewater treatment* (pp. 641–667). Elsevier.

Zulfahmi, I., Kandi, R. N., Huslina, F., Rahmawati, L., Muliari, M., Sumon, K. A., et al. (2021). Phytoremediation of palm oil mill effluent (POME) using water spinach (*Ipomoea aquatica*). *Environmental Technology and Innovation, 21*, Article 101260.

Management of Poultry and Poultry Wastewater

Abidemi A. Sangoremi, Opeyemi K. Olayanju, Daniel Terlanga Koko, Oyedolapo Bamigboye, and Moses O. Alfred

Abstract The surge in the world population and a paradigm shift in human diets into modern, and high protein-rich foodstuffs, have caused an increase in the production of poultry and poultry processing plant wastewater (PPW). Hence, the strategic management of PPW is crucial to ensuring environmental sustainability and compliance with environmental standards. Efficient treatment strategies of PPW can help reduce contaminants and its negative impact on water resources. Technologies such as membrane filtration, advanced oxidation processes (AOPs) like ozonation and photocatalysis, nutrient recovery technologies, etc. can be employed to treat PPW effectively. Conclusively, a good understanding of wastewater characteristics from indigenous poultry and poultry processing will be an immense advantage in implementing effective treatment strategies that will help ensure sustainable waste management in the poultry industry. Overall, this review will help in achieving Sustainable Development Goals (SDGs) 3, 12 and 15 on good health/wellbeing, responsible consumption/production and environmental sustainability respectively.

A. A. Sangoremi
Environmental Research Group Laboratory, Department of Chemistry, Federal University Otuoke, Yenagoa, Bayelsa State, Nigeria
e-mail: sangoremiaa@futuoke.edu.ng

O. K. Olayanju
Department of Civil Engineering, Redeemer's University, Ede, Osun State, Nigeria
e-mail: olayanjuo@run.edu.ng

D. T. Koko · O. Bamigboye (✉) · M. O. Alfred
African Centre of Excellence for Water and Environment Research (ACEWATER), Redeemer's University, Ede, Osun State, Nigeria
e-mail: bamigboye8679@run.edu.ng

Department of Chemical Sciences, Redeemer's University, Ede, Osun State, Nigeria

D. T. Koko
e-mail: kokod@run.edu.ng

M. O. Alfred
e-mail: alfredm@run.edu.ng

N. A. Oladoja and E. I. Unuabonah (eds.), *Strategic Management of Wastewater from Intensive Rural Industries*, Springer Water,
https://doi.org/10.1007/978-3-031-90314-4_9

Keywords Poultry · Wastewater · Treatment · Poultry processing plant wastewater · Sustainable development goals

1 Introduction

The poultry sector is a cornerstone in major supplies of meat protein to an increasing world population (Mottet & Tempio, 2017). It also serves as a major economic boost to a nation's internally generated revenue (IGR), as well as its export to other nations. Global chicken meat production was projected to reach 103.4 million metric tons before the end of 2023. In Africa, the practice of chicken farming is carried out by both the average and the poor alike, using different techniques including indigenous poultry, traditional method, free-range, local poultry, etc. (Sati et al., 2022). The indigenous chickens are recognized for their tendency to possess unique characteristics, such as environmental adaptability, high disease resistance, food scavenging and preservation ability, and high reproducibility, which ensures their survival against all odds (Ogbu, 2021). The birds in this system are managed by either free-range or semi-intensive management systems. Free-range poultry farming usually involves fixed or mobile facilities characterized by outdoor access for birds, organic feed, and lower production cost (Hayat et al., 2014; Petrović et al., 2024). While free-range systems may result in poorer feed conversion and lower body weights compared to intensive farming, they are associated with potential improvements in meat quality, including favorable fatty acid profiles and texture (Hayat et al., 2014; Muhammad et al., 2017).

With soaring poultry product demands, optimizing local poultry operations became essential. This resulted in the introduction of biogenetically engineered poultry breeds, development of intensive care system, use of drugs/chemicals, provision of processed feeds, and the development of other suitable conditions to enhance production. These improvement strategies have led to the increase in volume of wastewater generated from the poultry and poultry processing plants (Mukhtar, 2005). Water plays a vital role in the poultry and poultry processing industries, as it is used in feeding, slaughtering and processing plants. From the United States Department of Agriculture: National Agriculture Statistics Service (USDA:NASS) 2019 these processes has been factored to lead to the release of high volume of wastewater, which are challenging to treat, thereby requiring high costs of treatment (Rothrock Jr et al., 2019). The environmental and economic risk posed by the poultry processing wastewater (PPW) is the reason for the adoption of a series of methods for its management. The PPW have been characterized to contain high turbidity, high total dissolved solids, high biochemical oxygen demand, inorganics, and pharmaceuticals (Vaishnav et al., 2023). The presence of different organic fractions (e.g., carbohydrates, fats, oils, serum, blood, and plasma) have also been reported (Mohan et al., 2020b).

It has been observed that the majority of research on poultry and poultry processing operations consider waste management based on direct waste products, such as eggshells, feathers, bones, etc., without duly considering the economic and environmental constraints of generated wastewater. For example, the management of waste products, which include cracked egg/stale wastes and dead chicks/bird wastes (Onu et al., 2015), solid wastes (Balogun et al., 2017), fecal wastes, stale/broken eggs, and dead chicks/chickens (Akanni & Benson, 2014), mortality disposal (Moreki & Keaikitse, 2013), poultry litter (Waziri & Kaltungo, 2017), and mortality disposal (Adedayo, 2012), have been investigated and reported.

Considering the scarcity of clean drinking water and vulnerability of surface and groundwater to PPW, its proper management is essential for environmental sustainability and good public health. Therefore, this chapter is aimed at highlighting the peculiar features of PPW, the conventional and non-conventional approach adopted for its treatment and the economic and technological implications of the respective treatment options.

2 Overview of Poultry Operations

Indigenous poultry, also known as native or traditional poultry (e.g., Gallus domesticus), are raised in their native land without any technological interventions or variations (Anyona et al., 2023; Mappigau & Amar, 2020). Indigenous chicken breeds, such as the Kuroiler in Africa or the Kadaknath in India (Kassa et al., 2021; Kumar et al., 2022), are well-suited to local climates and conditions, and thrive in their natural habitat (Mathew and Mathew, 2023). The indigenous chicken is easy to distinguish from the commercial layers because of their uniqueness and morphology, like single-comb type with purported superior fertility, egg production, body mass, and plumage color (Chebo et al., 2023; Maharani et al., 2021; Winaya et al., 2023) (Fig. 1).

Indigenous poultry farming is common in developing countries because of the inadequate capital to fund large-scale poultry farms (Desta, 2021), and the significant contribution to poverty reduction. Unlike large-scale poultry farming, it requires limited financial resources, which enable resource-poor farmers to initiate business. The revenue generated from these ventures is a vital financial resource for rural households and individuals in low-income societies, facilitating business expansion (Desta, 2021). Furthermore, indigenous poultry and related processes create employment opportunities within local communities. This encompasses the production of poultry feeds, marketing and sales of products, as well as involvement in food processing and packaging. These avenues contribute substantially to societal growth, mitigating unemployment and poverty (Anyona et al., 2023; Desta, 2021). Indigenous poultry farming empowers families by diversifying their income sources, providing an alternative revenue stream (Subasinghe et al., 2021). By promoting income generation, employment, and improved nutrition, indigenous poultry farming effectively addresses key aspects of poverty. This poultry farming type grants individuals and communities the ability to break free from the cycle of poverty, fostering a sustainable and economically robust future.

Fig. 1 Some Indigenous Chicken; (A) Partridge color Hungarian rooster. (B) Yellow Hungarian rooster. (C) White Hungarian rooster. (D) Speckled Hungarian rooster. (E) Black Transylvanian naked neck rooster. (F) Speckled Transylvanian naked neck rooster. (http://www.geneconservation.hu/content/old-hungarian-farm-animals) (Lázár et al., 2021)

In rural sub-Saharan Africa, indigenous poultry has proven to be an excellent source of animal protein (Kpomasse et al., 2023). Indigenous poultry does not only provide nutritious meat and eggs but also enhance the overall health of families, leading to improved productivity, especially among children, whose physical and cognitive development relies on proper nutrition (Tandon et al., 2016). The diverse range of indigenous poultry breeds, which include native chickens, ducks, and turkeys, offers a source of lean protein, along with essential vitamins and minerals necessary for a balanced diet.

Additionally, their meat is low in fat while their eggs are densely packed with high-quality protein, vitamins, and minerals (Lordelo et al., 2020). Studies conducted in various countries have demonstrated the superior quality of products, like eggs from indigenous chicken breeds, when compared with those from commercial breeds (Lordelo et al., 2017, 2020). For instance, in Portugal, the eggs laid by traditional chicken breeds, particularly the Pedrês and Preta, often surpass the quality of commercial products in various aspects, as shown by comprehensive physical and chemical analyses (Lordelo et al., 2020). Eggs, from free ranged poultry, contain elevated levels of omega-3 fatty acids and vitamin D, when compared with those from intensively farmed poultry (Heflin et al., 2018; Sergin et al., 2022).

Indigenous poultry farming stands as a beacon for sustainable agricultural practices, emphasizing balanced farming system as its guiding principle. The scavenging nature of indigenous chickens, for instance, enables them to feed on insects and weeds, naturally promoting pest control within the ecosystem (Mathew and Mathew, 2023). Moreover, these chickens require small quantities of feeds, resulting in minimal waste when compared with commercial poultry production (Kpomasse et al., 2023). This sustainable approach fosters genetic diversity, as indigenous chickens exhibit resilience to climate stress, making them a viable option for farmers in climate-stressed rural communities (Kpomasse et al., 2023).

In addition, indigenous poultry farming advocates organic methods, steering clear of excessive antibiotics and chemicals. They exhibit hardiness and disease resistance among the birds, minimizing the need for antibiotics and vaccines (Kpomasse et al., 2023). This approach ensures that poultry products remain free of harmful residues, benefiting both individual health and environmental well-being (Gržinić et al., 2023). The incorporation of traditional knowledge, passed down through generations, further enriches this practice. This wisdom equips farmers with natural remedies for common poultry diseases, and fostering healthier birds without the need for extensive medication. The indigenous poultry farming often involves tasks that can be managed by women, providing them with opportunities for income generation and economic independence (Desta, 2021). The empowerment of women in these communities has far-reaching effects, bringing positive social changes within families and the wider society. These programs enhance farmers' skills in poultry management which improves the productivity and profitability of their poultry enterprises. Due to huge demand for animal protein, indigenous poultry have been sought after.

2.1 Wastewater Generation in Poultry and Poultry Processing Plants

The use of water in poultry farms is usually for drinking (by the) birds, sprinkling during the summer to keep the birds cool, and for cleaning sheds and equipment in between batch replacement. However, wastewater is generated during cleaning operations. Other identified sources of wastewater in poultry operations are runoff from poultry housing, feeding, and watering; from waste storage and management facilities. The wastewater is collected in a holding tank and utilized in gardening around the premises. Discharges from poultry operations typically have a high content of organic material, consequently a high biochemical oxygen demand (BOD) and chemical oxygen demand (COD), as well as nutrients and suspended solids (TSS).

According to EPA, to manage effectively the waste from poultry, it is important to minimize discharges to surface water and groundwater. To further reduce the impacts of runoffs from poultry operations, the following are recommended management techniques:

(i) During animal watering, overflowing from watering devices should be prevented, to reduce water spills and wastage. For effectiveness, watering devices should be calibrated, and well-maintained;
(ii) The use of vegetative filters should be encouraged as it helps to trap sediments;
(iii) Implementing surface water diversions to reroute uncontaminated runoff away from waste areas;
(iv) Implementation of buffer zones for surface water bodies, as appropriate to local conditions and requirements, and avoid land spreading of manure within these areas.
(v) A good drainage system should be maintained around the holding house. Drains for rainwater should be diverted from the holding house drain to avoid contamination of rainwater.
(vi) Wastewater from the cleaning of the holding pens must undergo treatment before discharge.

The influence of waste produced by native poultry and poultry processing facilities is considerably huge, making a substantial contribution to air and water pollution (Chaiwang et al., 2023; Colley, 2020). Based on EPA data, meat and poultry processing plants are identified as the second-largest industrial contributor of nitrogen released into water bodies (Hu et al., 2017). The various stages of poultry processing yield wastewater that is contaminated with a blend of biological and non-biological contaminants (Chaiwang et al., 2023). Poultry processing encompasses several steps, including slaughtering, defeathering, evisceration, washing, and packaging (Fig. 2). In the slaughtering process, poultry undergoes the removal of blood, feathers, and other organic matter, resulting in wastewater containing blood, proteins, and organic materials (Wang et al., 2024). Defeathering and evisceration processes involve the removal of feathers and internal organs, respectively, both contributing to wastewater containing feathers, intestinal contents, fats, and proteins (Jayathilakan et al., 2012). Washing poultry carcasses to eliminate contaminants, dirt, and bacteria introduces wastewater with detergents, disinfectants, and organic matter from the poultry surface. The packaging stage utilizes water, to clean packaging materials and equipment, generating wastewater with trace contaminants (Bayar et al., 2011).

The nutrient-rich wastewater discharged into aquatic environment leads to eutrophication, fostering excessive aquatic vegetation proliferation (Mohan et al., 2020a). The presence of ammonia in the wastewater results in reduced oxygen levels in water which adversely affects aquatic life (Mohan et al., 2020a). The elevated nutrient levels from poultry wastewater could induce changes in the composition of plants and aquatic animals in rivers, exerting a negative impact on the aquatic ecosystems. Poultry wastewater also harbors bacteria, viruses, and other pathogens, posing health risks to animals and humans (Gržinić et al., 2023). In addition, the release of odor from the wastewater create an unpleasant environment, affecting the quality of life for residents (Gržinić et al., 2023).

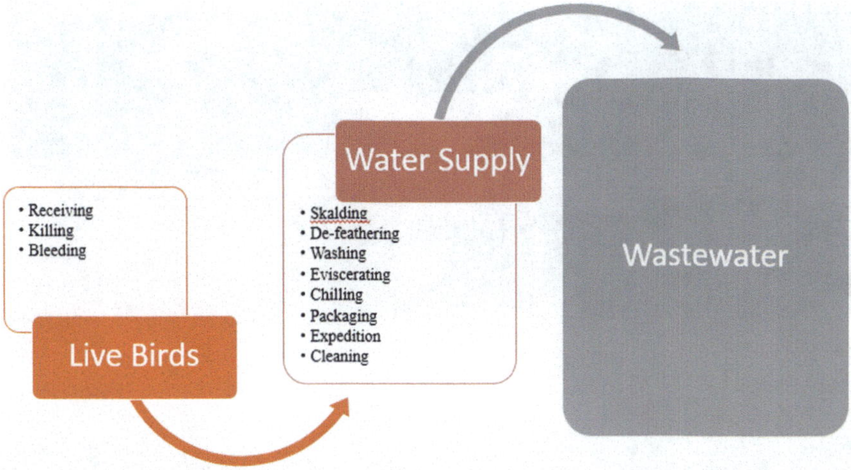

Fig. 2 Water consumption during poultry processing in the poultry slaughterhouse (Fatima et al., 2021)

Poultry processing is characterized by substantial water consumption, leading to the generation of large volume of wastewater. As illustrated in Fig. 3 (Fatima et al., 2021), the stages of poultry processing, begins with the killing of the birds, followed by scalding, de-feathering, evisceration, washing, chilling, and finally packaging. These processes are all highly water-intensive. Each of these steps requires significant water usage to ensure the cleanliness and safety of the poultry products. Due to the high organic load, the presence of pathogens, and the undesirable chemical constituents, wastewater from poultry processing facilities is deemed to be highly negatively impactful, without proper treatment. Evidence of this can be seen in Fig. 4, where the water from the river turned green due to the effluents from chicken farms.

2.2 Physicochemical Characteristics of PPW

Poultry processing wastewater (PPW) is generally characterized by its high levels of proteins, fats, oils, grease, carbohydrates, and nutrients, sanitizers, and antibiotic compounds (Table 1). The bulk physicochemical parameters used in defining PPW may include (but not limited to): total nitrogen (TN), biochemical oxygen demand (BOD, typically BODs), total phosphorus (TP), Chemical oxygen demand (COD), total suspended solid (TSS), total organic carbon (TOC), and other parameters such as color, odour, turbidity (Yeoh et al., 2022).

Research conducted by Yaakob et al. on characteristics of poultry slaughterhouse wastewater revealed a high concentration of BOD (1341–1821 mg/L), COD (3154.19–7719.3 mg/L), TSS (377.67–5462 mg/L). These values exceed the European Union 1991; 91/271/EEC (BOD $\leq$ 25 mg/l, COD $\leq$ 125 mg/l,) and World Bank

Fig. 3 Water pollution: River Wye turned green as a result of effluent from chicken farms in Wales, US (Colley, 2020)

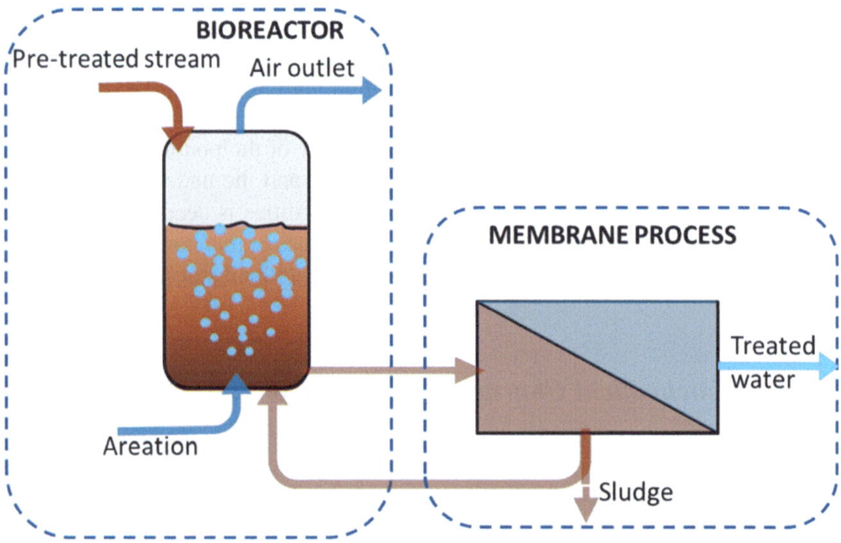

Fig. 4 Schematic representation of a membrane bioreactor (MBR) process (Poerio et al., 2019)

Table 1 Typical physicochemical parameters of PPW

Physicochemical parameters	Mean concentration (mg/L)			World bank standard (mg/L)
BOD	3000	1602	875	30.00
COD	5000	7719	5423	125.00
TOC	850	419	NR	
TN	450	361	258.5	10.00
TP	50	NR	NR	
TSS	3000	3438	589	50.00
K	50	NR	NR	
Colour (mg/L Pt scale)	300	NR	NR	
Turbidity (FAU)	275	NR	NR	
pH	6.5	8.02	6.6	
References	Bustillo-Lecompte & Mehrvar, 2015	Yaakob et al.	Aziz et al., 2018	Yeoh et al., 2022

All measurements in mg/l NR: not reported, TN–Total nitrogen, TOC–Total Organic Carbon, TP–Total Phosphorus

the standard limits for disposal of wastewater into the environments as seen in Table 1 (Preisner et al., 2020). The concentration ranges for total nitrogen and phosphate were (162.6–563.8 $\pm$ 215 mg/L), and (7.047–17.111 $\pm$ 4.25 mg/L), respectively, which are within the range required for microalgae growth, and the role of PPW in the occurrence of eutrophication phenomenon.

Similarly, Aziz et al. (2018) investigated the effects of sequential batch reactor (SBR) with different types of fibers on the treatment of poultry slaughterhouse wastewater. The results revealed the extent of treatment efficiency of the different reactors with and without fibers on biochemical oxygen demand (BOD), chemical oxygen demand (COD), ammonia–nitrogen (NH_3-N), phosphorus (P), nitrite (NO_2), nitrate (NO_3), total suspended solids (TSS), and oil-grease. It was inferred that the removal efficiency for the reactors with fibers was higher than that of the reactor without fibers for all pollutants. The treated effluent had 40 mg/L BOD_5 and 45 mg/L COD with an average removal efficiency of 96% and 93%, respectively, which meet the discharge limits stated in the Environmental Quality Act.

2.2.1 Designing Wastewater Treatment System for IPPPW

In order to gain insight into the qualities of the wastewater originating from indigenous poultry and poultry processing facilities (IPPP) and to enable the development of effective treatment strategies, there is a need to conduct information gathering.

The information to be gathered includes the wastewater volume, flow rate, temperature, pH, organic and inorganic pollutant concentrations, and any other specific contaminants of concern.

Analyzing the poultry wastewater composition will help to determine the appropriate treatment processes and the design parameters (Jebessa et al., 2022). Also, clearly defined treatment objectives, based on local regulations, environmental considerations, and the desired quality of the treated effluent must be put in place. Different treatment objectives may require specific treatment processes or additional treatment steps.

Some factors to consider during poultry wastewater treatment include space availability for the installation of the treatment system and the local environmental conditions is important. Oter factors such as soil type, topography, climate, and available utilities (electricity, water supply) (Fatima et al., 2021). In combination with the selection of the appropriate treatment processes is the issue of wastewater characteristics, treatment objectives, and available resources which might involve physical separation, biological treatment, chemical treatment, or a combination thereof, must be ensured. This is because each process has its design considerations, including size requirement, hydraulic retention times, and other operational parameters (Fatima et al., 2021; Jebessa et al., 2022). Furthermore, knowing the components required for the treatment system, such as tanks, reactors, filters, pumps, and pipes is important.

In poultry wastewater management it is also important to consider the materials and construction techniques suitable for the local conditions and the budget available for the project so as to achieve desirable in results (Jebessa et al., 2022). In addition, proper planning for the operation and maintenance of the poultry wastewater treatment system should be consideration. Factors such as the availability of skilled personnel, regular maintenance tasks, monitoring requirements, and the need for periodic system inspections and cleaning need to be considered (Bustillo-Lecompte & Mehrvar, 2015).

Furthermore, the health and safety of the operators and the surrounding environment should also be considered. These include measures to prevent the release of harmful gases, containment of odour, and proper handling and disposal of sludge or by-products generated during the wastewater treatment process (Yeoh et al., 2022).

In addition, adequate familiarization with local regulations and permit requirements related to wastewater treatment are necessary. It should be ensured that the design of the homegrown wastewater treatment system complies with the applicable standards and guidelines (Yeoh et al., 2022). This can be facilitated through deliberate efforts at educating indigenous poultry farmers on the existing local regulations and the need to comply. Some of the existing standard limits for selected contaminants in poultry wastewater are presented in Table 2. In addition, seeking expert advice and consulting with wastewater treatment professionals, engineers, or local authorities to ensure that the homegrown poultry wastewater treatment system is appropriately designed and meets the necessary standards are required (Yeoh et al., 2022).

Table 2 Global agencies' standards for dischargeable PPW effluents (Yeoh et al., 2022)

	World Bank	European Union	United State of America	Canadian	Australian
BOD_5 (mg/g)	30.00	25.00	26.00	5.00–30.00	6.00–10.00
COD (mg/L)	125.00	125.00	NR	NR	3 * BOD_5
TSS (mg/L)	50.00	35.00	30.00	5.00–30.00	10.00–15.00
TN (mg/L)	10.00	10.00	8.00	1.00	50.00

NR: Not Reported

2.3 Management of Poultry and Poultry Processing Plants Wastewater

The management of wastewater from indigenous poultry plants does pose unique challenges due to careless practices and the scale of operations. Emerging technologies offer innovative solutions to these challenges, enhancing the efficiency and sustainability of wastewater treatment processes. This section explores cutting-edge advancements in wastewater management tailored for Indigenous poultry operations. By integrating modern techniques such as biological treatment, advanced filtration, and resource recovery, these technologies aim to reduce environmental impact and improve water reuse. The adoption of these emerging solutions holds the potential to transform wastewater management in indigenous poultry farms, promoting environmental stewardship and operational efficiency.

The treatment processes for poultry wastewater often align with those used in municipal wastewater treatment, which involves primary, secondary, and tertiary treatment stages. There are several wastewater treatment systems after preliminary treatment, which can be split into four main groups: physicochemical treatment; biological treatment; Advanced oxidation processes (AOPs); and combined processes (Bustillo-Lecompte and Mehrvar, 2015).

Akpenpuun et al., developed a gravity sand filter technique for the management of poultry waste in Ilorin, Kwara State Nigeria. The setup which was run for 12 days each had two major configurations of single media and dual media. The single media was made up of silica sand (average particle size diameter of 0.5 mm and depth 150 mm) and silica gravel (average particle size diameter of 5–50 mm and depth 130 mm) while the dual media consist of silica sand (average particle size diameter of 0.5 mm and depth 50 mm), coal (average particle size of 1.0 mm and depth 100 mm) and silica gravel (average particle size diameter of 5–50 mm and depth 130 mm). The study investigated and evaluated the effect of time of operation on the BOD_5, TSS, TDS, turbidity, filtration rate and head loss in each of the gravity sand filter configurations. The performance evaluation of the technique showed that the configuration with the dual media operated on higher filtration rate with an average filtration rate of 45.50 mm/hr compared to the single media which operated at an average filtration rate of 39.60 mm/hr, during the 12 days operating period. It was observed that there was a lower head loss in dual media configuration than in the single media configuration. Overall, effluent from the dual media configuration shows better effluent characteristic with BOD removal of 53%, TSS removal of 50%, TDS removal of 42% and turbidity removal of 45% when compared with the single media configuration, which gave an effluent quality with BOD removal of 42%, TSS removal of 46%, TDS removal of 40% and turbidity removal of 40%. It was concluded that the sand filter could be said to be effective in the treatment of poultry wastewater.

There are often challenges with the operation of IPPP wastewater treatment facilities one of which is that the wastewater treatment facilities are often designed and operated by individuals with little knowledge or experience in wastewater treatment. Limited expertise from the operators of wastewater treatment facilities can lead to difficulties in understanding its functionalities, troubleshooting and operational issues, as well as how to optimize the treatment facility's performance (Fatima et al., 2021). Furthermore, inadequate monitoring and maintenance can result in system malfunctions and reduced treatment efficiency after several cycles of use. Lack of routine inspections, cleaning, and calibration of equipment can lead to clogging, decreased treatment capacity and increased treatment capacity, and increased energy consumption. Also, homegrown systems may have limited access to resources such as funding, equipment, and skilled labour (Yeoh et al., 2022). This can hinder the ability to invest in necessary upgrades and appropriately remunerate high-quality personnel for system operation and maintenance.

Indigenous wastewater from poultry and poultry processing farms can exhibit significant variability in terms of volume, composition, and pollutants' concentrations. Dealing with such variability can pose challenges with maintaining consistent treatment performance and meeting regulatory requirements (Fatima et al., 2021). Additionally, poultry wastewater treatment systems can generate odor, attracting pests and causing nuisances. Controlling odor and pests can be challenging, especially in home-grown systems that lack specialized odor control measures and pest management strategies (Yeoh et al., 2022). Also, homegrown systems often generate sludge as a by-product of the treatment process. Proper sludge removal dewatering, and appropriate disposal or reuse, can be challenging without access to specialized equipment and disposal facilities together with compliance with regulatory standards may pose a serious challenge to home-grown system (Fatima et al., 2021).

Furthermore, as the need arises or as regulatory requirements change, home-grown systems may require upgrades or expansion. However, limited resources and technical expertise can make these modifications challenging, potentially leading to system inefficiencies or non-compliance (Bustillo-Lecompte & Mehrvar, 2015). Also, home-grown poultry wastewater treatment systems can be vulnerable to environmental factors such as extreme weather conditions, flooding, or changes in groundwater levels. These factors can impact the PPW treatment system's performance and will require additional measures for protection and resilience (Bustillo-Lecompte & Mehrvar, 2015). It should be noted that lack of 'succession' planning and knowledge transfer often lead to disruption in system's operation and maintenance as emerging pollutants are discovered from wastewater. Hence, numerous emerging technologies are now available for handling wastewater generated by poultry and poultry processing facilities. These innovations focus on minimizing the environmental repercussions of wastewater release and advocate for sustainable wastewater management practices (Fatima et al., 2021). Some of these emerging technologies include:

2.3.1 Membrane Bioreactors (MBRs)

Membrane bioreactors (MBRs) contain membrane filtration with biological treatment to treat wastewater effectively as illustrated in Fig. 4. They use fine pores in membranes with definite sizes to separate solids, microorganisms, and other contaminants from the water. MBRs offer high-quality effluents, compact footprint, and high pathogen removal efficiency, compared to the conventional treatment methods (Avula et al., 2009). Fatima et al., (2021), reported high efficiency in removing pollutants and valuable nutrients while treating poultry and poultry processing wastewater using membrane technologies. The treated effluent was reportedly safe for discharge into the environment, and possibly for further reuse in poultry processes.

2.3.2 Anaerobic Digestion

This is a biological process that decomposes organic matter in the absence of free oxygen, yielding biogas and nutrient-rich digestate. Employing anaerobic digesters for the treatment of poultry wastewater greatly eliminates the organic content fraction of the wastewater and also generates abundant renewable biogas energy. This approach aids in curbing odor emissions and diminishes the volume of sludge produced (Yeoh et al., 2022). Constructed wetlands, an engineering system crafted to emulate the natural treatment mechanisms observed in wetlands, are often incorporated for effective poultry wastewater treatment. They harness wetland vegetation, soils, and microbes to eliminate contaminants from wastewater. These systems efficiently extract nutrients, organic matter, and pathogens from poultry wastewater. Moreover, constructed wetlands offer supplementary advantages like habitat establishment and aesthetic enhancements (Fatima et al., 2021).

2.3.3 Advance Oxidation Processes (AOPs)

AOPs involve the use of oxidizing agents or catalysts to degrade contaminants in wastewater. Technologies such as the use of ozone (O_3), ultraviolet (UV) radiation, and hydrogen peroxide (H_2O_2) are employed to break down organic compounds and destroy pathogens in poultry wastewater. AOPs are effective in removing persistent organic pollutants and for wastewater disinfection. (Adesina et al., 2023; Anjorin et al.; Yeoh et al., 2022), investigated the use of AOPs over other conventional methods in the treatment of poultry and poultry processing wastewater (PPPW). It was reported that AOPs performed excellently due to high treatment efficiency of

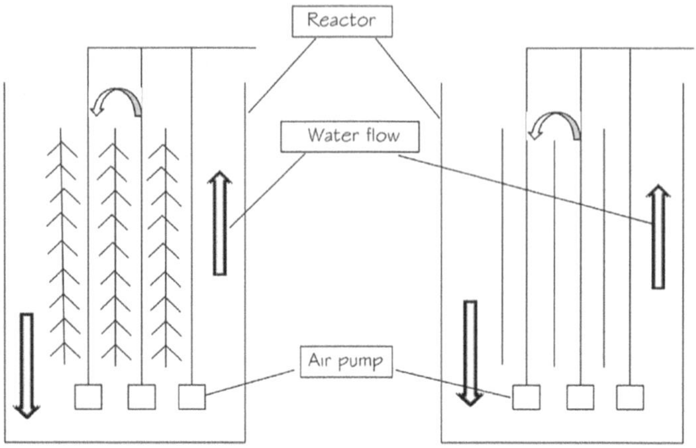

Fig. 5 The schematic diagram of sequential batch reactor (SBR) (Abdul Aziz et al., 2018)

96% pollutants removal, low energy consumption, and absence of foul odour in the treated PPPW.

2.3.4 Nutrient Recovery Technologies (NRTs)

The nutrient-rich composition of poultry wastewater, notably containing nitrogen and phosphorus, presents an opportunity for recovery and reuse. Utilizing sophisticated methods like struvite precipitation or ion exchange allows the extraction and capture of these valuable nutrients from the wastewater These recovered nutrients are used as fertilizers or sold as valuable products. Aziz et al., (2018), investigated the possibility of nutrient recovery from poultry and poultry processing plants wastewater using Sequential Batch Reactor (SBR) as illustrated in Fig. 5. The recovery efficiency of 96 to 93% was reported for phosphorous and nitrogen, respectively.

Struvite precipitation is another effective method for recovering nitrogen and phosphorus from poultry wastewater, with optimal conditions including pH 9.0, temperature 30 °C, and PO_4^{3-}/Mg^{2+} ratio of 1:3 (Sreyvich et al., 2020). Adjusting the Mg:N:P ratio to 1.5:1:1 can achieve up to 97.4% NH_4-N removal efficiency (Yilmazel & Demirer, 2011).

2.3.5 Innovative Filtration Systems (IFSs)

Various innovative filtration systems, such as rotating biological contactors, sand filters, and microfiltration, ultrafiltration and nanofiltration membranes, have been used to treat poultry wastewater (Mkilima et al., 2022). These systems provide effective removal of suspended solids, microorganisms, and other contaminants, producing high-quality effluents (Bayar et al., 2011).

It is important to note that the selection of a wastewater management technology should be based on specific requirements and characteristics of poultry wastewater, as well as local regulation and infrastructure. Additionally, a comprehensive approach that combines multiple technologies and best management practices is often necessary for effective wastewater management in poultry and poultry processing plants.

2.4 Odour and Pests Management in IPPPW

One of the challenges of poultry operations is the generation of pungent odour. Therefore, adequate ventilation, which ensures proper ventilation in the treatment system is required to minimize the accumulation of odiferous gases. Proper airflow helps disperse odor and prevent their concentration in specific areas. Strategically employing fans or vents to enhance air circulation is crucial. Implementing targeted odor control measures tailored to the traits of the wastewater and treatment processes is equally vital (Bustillo-Lecompte & Mehrvar, 2015). Common methods include the use of activated carbon filters, biofilters, or chemical scrubbers to remove compounds with foul odours. These systems can be installed in key areas where odors are likely to be generated, such as vents or outlets. Further, implementation of regular system inspections of the treatment system, to identify potential odour sources or areas where pest infestation may occur is necessary. Regular monitoring allows for timely identification and mitigation of odor or pest-related problems.

Implementation of proper sludge management, which include, regular sludge removal, dewatering, and appropriate disposal or reuse is also important. The coverage of sludge storage areas to minimize odor and prevent pests from accessing the sludge is needed. Implementation of good housekeeping practices, in and around the wastewater treatment system, which include prompt cleaning of spills, removal of debris, and ensuring that the containers, tanks, and equipment are properly sealed to prevent odors and prevent pests from accessing the sludge is a requirement that must be fulfilled (Fatima et al., 2021).

2.5 Case Studies of IPPW

Oduwaiye et al. carried out a study to assess the waste management methods used by commercial poultry farmers in Kogi and Kwara States, in Nigeria. The study used a questionnaire to obtain data from registered and unregistered poultry farmers from both states. The data obtained were analyzed using both descriptive and inferential statistics. The results obtained from the analysis show that mean age of the respondents was 46.7 years, farm size was 991 birds and poultry farming experience was 11.24 years. Also, majority of the respondents were male (82.0%), married (88.8%), with tertiary education (94.0%). The average age of farm was 9.24 years. The most frequent waste management methods used by poultry farmers were open dumping around the farm (mean = 2.23); giving out wastes free to the public as farmyard manure (mean = 2.16); dumping wastes in nearby bushes (mean = 2.09); burying dead birds in pit near the farm (mean = 2.08), and throwing dead birds into nearby bush (mean = 2.01) and burning dead birds in pit or heap near the farm (mean = 2.01). Burning technique was the most (mean = 1.91) popular method of treating wastes against odour and flies by the respondents. The major constraints faced by farmers in waste management practices were: lack of awareness on how to use the wastes productively (mean-4.06); No agricultural land nearby where the wastes can be applied (mean = 3.69); excessive odour from waste (mean = 3.66); high cost of chemical treatment (mean = 3.56); high transportation cost (mean = 3.24) and high cost of private waste management agencies (mean = 3.01). In conclusion, the study established that poultry waste management practices in the study area were confronted by high (mean = 3.39) level of constraints. It was then recommended that the poultry farmers' association should have training and retraining programs for the farmers regularly; also, experts should be invited to train them in the areas of waste management and general environmental health.

As a means of managing wastewater from poultry processes, different researchers have documented different approaches for the management and treatment. The goal is always to have an environment that is free from pollution that may arise from improper management and indiscriminate discharge of wastewater from poultry processes.

3 Cost Implications of Managing Wastewater from Poultry and Poultry Processing Plants

The interest in understanding the financial impacts on the management of wastewater from poultry results from the upsurge in concern from environmental monitoring bodies to ensure proper discharge of wastewater from these plants to avoid hazards to humans. Humans have been treating drinking water for a long time, it was not until around 4000 BC that methods for adjusting physicochemical properties were established. Standard techniques for the management of wastewater began to sprout because of scientific discoveries and more understanding of hazards as well as the economic implications (Lofrano & Brown, 2010).

The cost implication of any treatment technique for wastewater reclamation is dependent on the complexity of the water matrix and the authenticity of the technique, bearing in mind energy consumption during the treatment process. For most processing industries in urban cities, their wastewater is channeled into large surface water bodies that sometimes cost more than the recyclability of the wastewater. Poultry and poultry processing plants are known to generate large quantities of wastewater that are treated using conventional and other improved techniques.

In another report, Rossi and et al., (1979) showed the cost effects of using combined treatment techniques for wastewater from domestic and poultry processing plants, considering activated sludge, filtration, and aerated lagoons as the treatment techniques. They observed that depending on the targeted population as well as the size of the wastewater treatment plant, the cost implication increases or decreases in correspondence with each variable (Rossi et al., 1979).

In 1972 a cost report was also compiled by Jones and colleague on the recyclability of poultry wastewater from a plant in Georgia involved in the processing of poultry products, considering the volume of water inputted in rendering plants as well. The overall cost implication expended on the treatment of wastewater from the poultry plants varied from $0.04 to $0.75 million, based on the volume of water generated by the plants (Jones & Kerns, 1974).

In the new era, wastewater management from indigenous poultry and poultry plants is promising. According to Mijinyawa and Lawal, proper management and recycling of poultry process wastewater is more economical, compared to continuous water supply of fresh water via tankers to a chicken abattoir in Ibadan, Oyo State, Nigeria. The wastewater treatment system comprises a combination of stepwise conventional treatment techniques designed to efficiently eliminate chemical and biological contaminants. The management of wastewater through the extraction, purification, and utilization of individual components contained in the poultry wastewater should be to maximize gain. According to findings by Mijinyawa and Lawal (2008), the cost of treating water at the wastewater treatment plant is ₦0.15 per litre (₦ is naira, the Nigerian currency), compared to ₦0.50 per litre from tanker suppliers. This represents a daily saving of ₦120,506.95, while also limiting the discharge of contaminated water into the environment (Mijinyawa & Lawal, 2008). The financial impact of recovering protein from poultry processing plants using three combined pressurized filtration apparatus was evaluated (Białas et al., 2015). About 84 and 70% of soluble protein and water were successfully recovered, respectively, with the reclaimed water saving cost of € 10,850/month towards the utilization of the recycled water. This also reduced the volumes of pure water utilized at the rate of 7000 m^3 per month, which in turn increased the monthly savings by an additional amount of € 6,166 for a month. The protein, if employed for the manufacture of food, could raise an amount of € 33,000 per month, using an assumed product price rate of t € 300/ton (Białas et al., 2015).

4 Guidelines for Wastewater Management in Poultry and Poultry Processing Plant

Managing wastewater in poultry and poultry processing plants is crucial for environmental sustainability. Lack of proper waste management can have negative impacts on the environment, which can include pollution and health hazards. Wastes from poultry processes contains high levels of nutrients, such as nitrogen and phosphorus, which can seep into water bodies if not adequately managed. This can lead to water pollution, causing harmful algal blooms and fish kills. The negative effects of improper waste management can also lead to air pollution, which pose risks to human health as well. In addition, poultry waste and wastewater contain pathogens that can pollute crops, water, and soil, spreading illnesses like *E. Coli* and salmonella. Therefore, proper wastewater management not only protects the environment but also improves farm efficiency and reduces costs associated with waste and wastewater disposal.

Implementing effective poultry wastewater management strategies is crucial for maintaining a sustainable and environmentally friendly poultry industry and therefore to this, the following best practices should be adhered to.

a. **Education and training**: this plays a vital role in ensuring effective poultry wastewater handling and management. Proper handling techniques must be taught to all personnel involved in waste management to minimize contamination risks and promote optimal handling practices.
b. **Compliance with regulations and guidelines**: this is important to make sure that all poultry waste management practices adhere to environmental regulations and health and safety standards. Regulations are designed and provided to protect the environment and prevent any potential negative impacts on human health.
c. **Continuous monitoring and evaluation**: It is highly important to measure the effectiveness of implemented waste management strategies. Routine inspections by trained professionals should be conducted to identify any issues or areas for improvement. The monitoring and evaluation include data collection and analysis which will provide valuable insights into the efficiency of waste management practices and based on the data collected and analyzed, necessary adjustments and improvements can be made to management practices. This will invariably allow for the identification of emerging trends and challenges, allowing for proactive measures to be implemented.
d. The treatment processes for poultry wastewater often align with those used in municipal wastewater treatment, involving primary, secondary, and tertiary treatment stages. There are several wastewater treatment systems after preliminary treatment, which can be split into four main groups: physicochemical treatment; biological treatment; Advanced oxidation processes (AOPs); and combined processes (Bustillo-Lecompte and Mehrvar, 2015). Wastewater from Poultry slaughterhouse wastewater (PSW) usually poses environmental challenges due to its composition, which includes proteins, fats, carbohydrates, blood, feathers, meat particles, residual acids, metals, and chemicals. Generally, similar methods in treating wastewater from meat processing is applicable with minor difference.

Conventional treatment methods usually employed for the treatment of poultry slaughter house wastewater include;

1. **Preliminary Treatment**: this treatment process basically removes suspended solids from PSW; screeners, sieves, and strainers are used most frequently in the preliminary treatment process. In preliminary treatment, 60% of suspended solids and 30% of BOD are removed from PSW. Other preliminary methods are catch basins, flotation, equalization, and settlers

2. **Primary Treatment**: as soon as the preliminary treatment is completed, the PSW undergoes primary treatment, which is a physicochemical process that removes BOD, COD, oil, grease, fats, and residual TSS. The conventional primary methods are dissolved air flotation (DAF), electrocoagulation, coagulation-flocculation, and sedimentation (Fatima et al., 2021).
3. **Secondary Treatment**: Secondary treatment is used to further treat the contaminants in wastewater that are not eliminated by the primary treatment. Removing organic compounds to lower the BOD level is the primary objective of secondary treatment. PSW are treated in aerobic and anaerobic digesters as part of the biological process utilized for secondary treatment (Bustillo-Lecompte & Mehrvar, 2015). In aerobic and anaerobic treatments, organic matter is degraded into simple compounds with the help of microorganisms.

Some of the aerobic and anaerobic treatment methods that have been developed include.

a. Submerged Fibers in an Attached Growth Sequential Batch Reactor (SBR): as discussed by Aziz et al., (2018), the SBR with fibers has a removal efficiency of 96% and 93% for COD and BOD_5 respectively.
b. Aerobic Moving Bed Biofilm Reactor (MBBR): the MBBR was employed by Baddour et al., for the treatment of PSW, the result obtained showed that with increase in the detention time of the wastewater in the reactor, there was significant reduction in the value of total dissolved solid and COD.
c. Up-flow Anaerobic Sludge Blanket (UASB) Reactor: according to Chavez et al., removal of organic material from poultry slaughter wastewater as determined by changes in biological oxygen demand was investigated by adding three different types of inoculum combining cow manure, yeast extract or hydraulic residence time as variables with response vector of reduction of BOD_5. In 3-l reactors, a 95% removal of BOD_5 from poultry slaughter wastewater was obtained with organic loading rates up to 31 kg BOD_5 $m^{-3}d^{-1}$ without loss of stability. This 95% removal was obtained between 25 and 39 °C with a hydraulic residence time between 3.5 and 4.5 h.

The above discussed conventional treatment methods often do not efficiently recover valuable nutrients and enable the wastewater reuse. However, membrane technologies have been described as efficient for the effective management of poultry slaughterhouse wastewater. Membrane technologies processes are pressure driven and they include Microfiltration (MF), ultrafiltration (UF), nanofiltration (NF), and reverse osmosis (RO). They can either be used as stand-alone processes or integrated with other methods. Membrane technology efficiently removes pollutants from PSW, purifying water for discharge or even reuse in poultry processing (Fatima et al., 2021). The benefits of membrane technology include high efficiency in removal of pollutants; nutrient recovery and high-quality water.

In summary, poultry waste and wastewater management is of utmost importance to protect the environment, human health, and the sustainability of the poultry industry.

4.1 Future Perspectives and Recommendations

Awareness of the management of poultry and wastewater is very important and will make some of the United Nations sustainable development goals (SDGs), geared towards providing good health and well-being for all (SDG 3), clean water and sanitation for all (SDG 6), and life below water and land (SDG 14 and SDG 15), to be achieved. It is recommended to enhance pretreatment processes by implementing good screening and filtration systems, to remove contaminants before biological treatment. Efficient sludge management strategies, including dewatering and composting, are essential for sustainable operations. Ensuring regulatory compliance through regular monitoring and stakeholder engagement, along with investing in research and development, can drive continuous improvement in wastewater management. Training and capacity-building programs for staff are also crucial to maintain effective and sustainable wastewater treatment practices.

5 Conclusion

This chapter has explored innovative approaches to poultry processing wastewater (PPW) treatment. It has comprehensively examined the various aspects of managing wastewater generated from indigenous poultry and poultry processing plants. Effective management practices are critical to maintaining environmental health and ensuring compliance with regulatory standards. The discussion highlighted various treatment technologies such as membrane filtration, advanced oxidation processes (AOPs), and nutrient recovery systems, emphasizing their efficiency and cost-effectiveness in handling poultry wastewater. By implementing these advanced treatment strategies, it is possible to significantly reduce the environmental impact of poultry processing operations, thus contributing to sustainable agricultural practices. Also, the management of poultry processing wastewater will contribute to the actualization of some of the United Nations Sustainable Development Goals; this directly contributes to the protection of the environment, human health, and the sustainability of the poultry industry. This is because effective wastewater treatment not only reduces environmental harm but also contributes to sustainable practices and cost savings for poultry processing plants. In summary, adopting innovative technologies, optimizing membrane-based treatments, and promoting sustainable practices will shape the future of poultry wastewater management.

References

Adedayo, V. (2012). Poultry waste management techniques in urban agriculture and its implications: A case of metropolitan Lagos, Nigeria. *Asian Journal of Agricultural Sciences, 4*, 258–263.

Adesina, M. O., Block, I., Günter, C., Unuabonah, E. I., & Taubert, A. (2023). Efficient removal of tetracycline and bisphenol a from water with a new hybrid clay/TiO_2 composite. *ACS Omega, 8*, 21594–21604.

Akanni, K., Benson, O. (2014). Poultry waste management strategies and environmental implications on human health in ogun state (Nigeria).

Anjorin, E. O., Alfred, M. O., Sotunde, B., Nnamani, E. A., Bayode, A. A., Unuabonah, E. I., Helmreich, B., Omorogie, M. O. Overview of the mechanism of degradation of pharmaceuticals by persulfate/peroxysulfate catalysts. ChemBioEng Reviews, e202300079.

Anyona, D. N., Musyoka, M. M., Ogolla, K. O., Chemuliti, J. K., Nyamongo, I. K., & Bukachi, S. A. (2023). Characterization of indigenous chicken production and related constraints: Insights from smallholder households in rural Kenya. *Scientific African, 20*, Article e01717.

Avula, R. Y., Nelson, H. M., & Singh, R. K. (2009). Recycling of poultry process wastewater by ultrafiltration. *Innovative Food Science and Emerging Technologies, 10*, 1–8.

Aziz, H. A., Puat, N. N. A., Alazaiza, M. Y., & Hung, Y.-T. (2018). Poultry slaughterhouse wastewater treatment using submerged fibers in an attached growth sequential batch reactor. *International Journal of Environmental Research and Public Health, 15*, 1734.

Balogun, O. L., Ayo-Bello, T. A., Afodu, O. J., Akinwole, O. T., & Ndubuisi-Ogbonna, L. C. (2017). Determinants of waste management techniques among the poultry farmers in Ikenne Local Government Area of Ogun State. *Nigeria International Journal Livest Resource, 7*, 41–51.

Bayar, S., Yıldız, Y. Ş, Yılmaz, A. E., & İrdemez, Ş. (2011). The effect of stirring speed and current density on removal efficiency of poultry slaughterhouse wastewater by electrocoagulation method. *Desalination, 280*, 103–107.

Białas, W., Stangierski, J., & Konieczny, P. (2015). Protein and water recovery from poultry processing wastewater integrating microfiltration, ultrafiltration and vacuum membrane distillation. *International Journal of Environmental Science Technology, 12*, 1875–1888.

Bustillo-Lecompte, C. F., & Mehrvar, M. (2015). Slaughterhouse wastewater characteristics, treatment, and management in the meat processing industry: A review on trends and advances. *Journal of Environmental Management, 161*, 287–302.

Chaiwang, N., Marupanthorn, K., Krutthai, N., Wattanakul, W., Jaturasitha, S., Arjin, C., Sringarm, K., & Setthaya, P. (2023). Assessment of nucleic acid content, amino acid profile, carcass, and meat quality of Thai native chicken. *Poultry Science, 102*, Article 103067.

Chebo, C., Melesse, A., & Betsha, S. (2023). Morphological descriptors of indigenous chickens: As a selection guide. *Journal of Applied Animal Research, 51*, 729–742.

Colley, C. (2020). River pollution leads to Welsh demand for halt to intensive chicken farms. The Guardian. United Kingdom. https://www.theguardian.com/environment/2020/oct/05/river-pollution-leads-towelsh-demand-for-halt-to-intensive-poultry-units.

Desta, T. T. (2021). Indigenous village chicken production: A tool for poverty alleviation, the empowerment of women, and rural development. *Tropical Animal Health and Production, 53*, 1.

Fatima, F., Du, H., & Kommalapati, R. R. (2021). Treatment of poultry slaughterhouse wastewater with membrane technologies: A review. *Water, 13*, 1905.

Gržinić, G., Piotrowicz-Cieślak, A., Klimkowicz-Pawlas, A., Górny, R. L., Ławniczek-Wałczyk, A., Piechowicz, L., Olkowska, E., Potrykus, M., Tankiewicz, M., & Krupka, M. (2023). Intensive poultry farming: A review of the impact on the environment and human health. *Science of the Total Environment, 858*, Article 160014.

Hayat, A., Basheer, A., Zahoor, I., Mahmud, A. (2014). Free-range rearing system and its impact on production and consumption of poultry: a review. In.

Heflin, L. E., Malheiros, R., Anderson, K. E., Johnson, L. K., & Raatz, S. K. (2018). Mineral content of eggs differs with hen strain, age, and rearing environment. *Poultry Science, 97*, 1605–1613.

Hu, Y., Cheng, H., & Tao, S. (2017). Environmental and human health challenges of industrial livestock and poultry farming in China and their mitigation. *Environment International, 107*, 111–130.

Jayathilakan, K., Sultana, K., Radhakrishna, K., & Bawa, A. (2012). Utilization of byproducts and waste materials from meat, poultry and fish processing industries: A review. *Journal of Food Science and Technology, 49*, 278–293.

Jebessa, E., Guo, L., Chen, X., Bello, S. F., Cai, B., Girma, M., Hanotte, O., & Nie, Q. (2022). Influence of Eimeria maxima coccidia infection on gut microbiome diversity and composition of the jejunum and cecum of indigenous chicken. *Frontiers in Immunology, 13*, Article 994224.

Jones, H. B., & Kerns, W. R. (1974). Waste water treatment costs in poultry rendering plants in Georgia. *Journal of Agricultural Applied Economics, 6*, 65–71.

Kassa, B., Tadesse, Y., Esatu, W., & Dessie, T. (2021). On-farm comparative evaluation of production performance of tropically adapted exotic chicken breeds in western Amhara. *Ethiopia Journal of Applied Poultry Research, 30*, Article 100194.

Kpomasse, C. C., Kouame, Y. A. E., N'nanle, O., Houndonougbo, F. M., Tona, K., Oke, O. E. (2023). The productivity and resilience of the indigenous chickens in the tropical environments: improvement and future perspectives. *Journal of Applied Animal Research, 51*, 456–469.

Kumar, M., Dahiya, S., Ratwan, P., Sheoran, N., Kumar, S., & Kumar, N. (2022). Assessment of egg quality and biochemical parameters of Aseel and Kadaknath indigenous chicken breeds of India under backyard poultry farming. *Poultry Science, 101*, Article 101589.

Lázár, B., Molnár, M., Sztán, N., Végi, B., Drobnyák, Á., Tóth, R., Szabadi, N. T., McGrew, M. J., Gócza, E., & Várkonyi, E. P. (2021). Successful cryopreservation and regeneration of a partridge colored Hungarian native chicken breed using primordial germ cells. *Poultry Science, 100*, Article 101207.

Lofrano, G., & Brown, J. (2010). Wastewater management through the ages: A history of mankind. *Science of the Total Environment, 408*, 5254–5264.

Lordelo, M., Cid, J., Cordovil, C. M., Alves, S. P., Bessa, R. J., & Carolino, I. (2020). A comparison between the quality of eggs from indigenous chicken breeds and that from commercial layers. *Poultry Science, 99*, 1768–1776.

Lordelo, M., Fernandes, E., Bessa, R., & Alves, S. (2017). Quality of eggs from different laying hen production systems, from indigenous breeds and specialty eggs. *Poultry Science, 96*, 1485–1491.

Maharani, D., Mustofa, F., Sari, A. P. Z., Fathoni, A., Sasongko, H., & Hariyono, D. N. H. (2021). Phenotypic characterization and principal component analyses of indigenous chicken breeds in Indonesia. *Veterinary World, 14*, 1665.

Mappigau, P., & Amar, M. Y. (2020). Knowledge of technology poultry production and entrepreneurship for development of indigenous chicken farms in Indonesia. *Journal of Business and Management Review, 1*, 373–389.

Mathew, E., Mathew, L. (2023). Conservation of landraces and indigenous breeds: An investment for the future, In: Conservation and sustainable utilization of bioresources. Springer, pp. 291–321.

Mijinyawa, Y., & Lawal, N. S. (2008). Treatment efficiency and economic benefit of Zartech poultry slaughter house waste water treatment plant, Ibadan, Nigeria. *Scientific Research Essay, 3*, 219–223.

Mkilima, T., Bazarbayeva, T., Assel, K., Nurmukhanbetova, N. N., Ostretsova, I. B., Khamitova, A. S., Makhanova, S., & Sergazina, S. J. W. (2022). Pore size in the removal of phosphorus and nitrogen from poultry slaughterhouse wastewater using polymeric nanofiltration membranes. *Water, 14*, 2929.

Mohan, T., Sheik Farid, N. S., KV, S., K, R. (2020b). Sustainable biological system for the removal of high strength ammoniacal nitrogen and organic pollutants in poultry waste processing industrial effluent. *Journal of the Air Waste Management Association, 70*, 1236–1243.

Moreki, J. C., & Keaikitse, T. (2013). Poultry waste management practices in selected poultry operations around Gaborone, Botswana. *International Journal of Current Microbiology Applied Sciences, 2*, 240–248.

Mottet, A., & Tempio, G. (2017). Global poultry production: Current state and future outlook and challenges. *World's Poultry Science Journal, 73*, 245–256.

Muhammad, I. K., Faisal, F. K., Majid, M., Mohammad, U., Ali, S. M., Nikolai, P. T., Igor, P., Alexey, G., & Alexey, L. (2017). Free range poultry husbandry and physiochemical quality of meat: A review. *International Journal of Research in Ayurveda and Pharmacy, 8*, 74–79.

Mukhtar, S. (2005). *Poultry production: Manure and wastewater management* (p. 744). Marcel Decker, Inc.

Ogbu, C. C. (2021). Utilization and conservation of landrace chickens of Nigeria: Physical and performance characteristics, issues and concerns. Landraces-Traditional Variety Natural Breed

Onu, D., Offor, E., Okpara, B. (2015). Poultry wastes management strategies and environmental implications in Abia State. *International Research Journal of Agricultural Science Soil Science*

Petrović, M., Bogosavljević- Bošković, S., Bogdanović, V., Đoković, R., Rakonjac, S., Petrović, M. (2024). Rearing systems and production in organic poultry farming. In Proceedings/2nd International Symposium on Biotechnology.

Poerio, T., Piacentini, E., & Mazzei, R. (2019). Membrane processes for microplastic removal. *Molecules, 24*, 4148.

Preisner, M., Neverova-Dziopak, E., Kowalewski, Z. J. E. M. (2020). An analytical review of different approaches to wastewater discharge standards with particular emphasis on nutrients. *Environmental Management, 66*, 694–708.

Rossi, D., Young, C. E., & Epp, D. J. (1979). The cost impact of joint treatment of domestic and poultry processing wastewaters. *Land Economics, 55*, 444–459.

Rothrock, M. J., Jr., Gibson, K. E., Micciche, A. C., & Ricke, S. C. (2019). Pastured poultry production in the united states: Strategies to balance system sustainability and environmental impact. *Frontiers in Sustainable Food Systems, 3*, 74.

Sati, N. M., Luka, P. D., Mwiine, F. N., Fagbamila, I. O., Weka, R. P., Muhammad, M., Erume, J. (2022). Perceptions and practices of farmers of indigenous poultry towards Salmonella infections in North-Central Nigeria. *Open Veterinary Journal, 12*, 567–577.

Sergin, S., Jambunathan, V., Garg, E., Rowntree, J. E., & Fenton, J. I. (2022). Fatty acid and antioxidant profile of eggs from pasture-raised hens fed a corn-and soy-free diet and supplemented with grass-fed beef suet and liver. *Foods, 11*, 3404.

Sreyvich, S., Petrus, H. T. B. M., Purnomo, C. W. (2020). Nutrient recovery from slaughterhouse wastewater. *IOP Conference Series: Materials Science and Engineering, 778*.

Subasinghe, R., Siriwardena, S., Byrd, K., Chan, C., Dizyee, K., Shikuku, K., Tran, N., Adegoke, A., Adeleke, L., Anastasiou, K. (2021). Nigeria fish futures. Aquaculture in Nigeria: Increasing income, diversifying diets and empowering women. Report of the scoping study.

Tandon, P. S., Tovar, A., Jayasuriya, A. T., Welker, E., Schober, D. J., Copeland, K., Dev, D. A., Murriel, A. L., Amso, D., & Ward, D. S. (2016). The relationship between physical activity and diet and young children's cognitive development: A systematic review. *Preventive Medicine Reports, 3*, 379–390.

Vaishnav, S., Saini, T., Chauhan, A., Gaur, G. K., Tiwari, R., Dutt, T., Tarafdar, A. (2023). Livestock and poultry farm wastewater treatment and its valorization for generating value-added products: Recent updates and way forward. *Bioresource Technology, 129170*

Wang, S., Wei, Z., & Wang, L. (2024). Improving slaughterhouse byproducts utilization via anaerobic digestion, composting, and rendering. *Renewable and Sustainable Energy Reviews, 189*, Article 113881.

Waziri, M. I., Kaltungo, B. Y. (2017). Poultry litter selection, management and utilization in the tropics. *Poultry science, 191*

Winaya, A., Fahmiady, D. I., Suyatno, S., Malik, A., Mahmud, A., Jaganathan, R. (2023). Morphometric diversity and genetic relationship of "Bangkok" chicken (Thai Game Fowl) in East Java, Indonesia. *Jordan Journal of Biological Sciences, 16*.

Yeoh, J. X., Md. Jamil, S. N. A., Syukri, F., Koyama, M., Nourouzi Mobarekeh, M. (2022). Comparison between conventional treatment processes and advanced oxidation processes in treating slaughterhouse wastewater: A review. *Water 14*, 3778.

Yilmazel, Y. D., & Demirer, G. N. (2011). Removal and recovery of nutrients as struvite from anaerobic digestion residues of poultry manure. *Environmental TecHnology, 32*, 783–794.

Strategic Management of Wastewater from Ranches and Cattle Lots and Their Value-Added Products

R. O. A. Adelagun, A. H. Labulo, C. O. Obadimu, and A. D. Omoyeni

Abstract Ranch and cattle lot operations are intensive animal production systems, generate large amounts of manure, solid wastes and wastewater. These wastes contain high concentrations of organic matter, nutrients, antibiotics, hormones, contaminants and pathogens. The importance of effective manure and wastewater management systems in cattle farms is of great concern because of the risks to the environment and public health. This chapter highlights the characteristics of the wastewater from the cattle ranch and feedlot operations, environmental impacts of the waste effluents of the operation as well as the description of the different practices including physical, biological, AI- assisted, chemical as well as integrated practices adopted in the management of the wastewater. The challenges, constraints and benefits of resource recovery and reuse of the wastewater is highlighted for stakeholders. The potential value-added products from the cattle lot wastewater are enumerated for consideration.

Keywords Ranches · Cattle lot · Wastewater · Nutrient recovery · Valorization · Eutrophication

R. O. A. Adelagun (✉) · A. D. Omoyeni
Department of Chemical Sciences, Federal University Wukari, Taraba State, Katsina Ala Way, Wukari, Nigeria
e-mail: jemiruth2009@yahoo.com

A. H. Labulo
Department of Chemistry, Federal University of Lafia, Lafia, Nassarawa State, Nigeria
e-mail: ayomide.labulo@science.fulafia.edu.ng

C. O. Obadimu
Department of Chemical Sciences, Akwa Ibom State University, Akwa Ibom State, Uyo, Nigeria
e-mail: clementobadimu@aksu.edu.ng

N. A. Oladoja and E. I. Unuabonah (eds.), *Strategic Management of Wastewater from Intensive Rural Industries*, Springer Water,
https://doi.org/10.1007/978-3-031-90314-4_10

1 Introduction

A ranch is a property used for raising livestock, particularly cattle often located on public lands, while cattle lots are areas where cattle are confined, typically for feeding and fattening (Novikova & Startiene, 2018). Cattle ranching and cattle lot operations differ in management systems and performance outcomes. Ranching is typically practiced in open-range settings, which allows for natural grazing and movement, while feedlot operations, such as deep-bedded confinement buildings, are more controlled environments (Fig. 1). The specific farming systems within these operations, such as meat production in urban areas and milk production in semi-urban areas, can further differentiate the two.

Ranch and cattle lot operations are intensive animal production systems and generate large amounts of manure, wastewater, and other wastes that have constituted a major environmental and public health concern. These wastes contain high concentrations of organic matter, nutrients, pathogens, antibiotics, hormones, and other contaminants that can pose serious risks to water, soil, air, and human and animal health. The organic matter depletes the dissolved oxygen, cause eutrophication and hypoxia in water bodies. The nutrients, such as nitrogen and phosphorus stimulate the growth of algae and cyanobacteria, and produce harmful algal blooms and toxins (Adelagun, 2016). The pathogens, such as bacteria, viruses, and parasites cause infections and diseases in humans and animals. The antibiotics, such as tetracycline, sulfonamides, and macrolides induce antibiotic resistance and reduce the effectiveness of medical treatments. The hormones, such as estrogens, androgens, and progestin disrupt the endocrine system and affect the reproduction and development of humans and wildlife (Gu et al., 2020).

The types and amounts of manure and wastewater produced by ranch and cattle lot operations depend on the animal type, size, weight, age, breed, diet, health, and productivity, as well as the housing, bedding, feeding, watering, and waste management systems (USEPA, 2004). According to the Food and Agricultural Organization (FAO) (2020), the average daily manure production per herd of cattle ranges from 10 to 20 kg for beef cattle, and from 15 to 25 kg for dairy cattle, depending on the

Fig. 1 Cattle lot and ranch . *Source* www.pixabay.com/images

animal category and the moisture content of the manure. The average daily wastewater production per herd of cattle ranges from 20 to 40 L for beef cattle, and from 40 to 80 L for dairy cattle, depending on the water consumption and the wash water usage (USEPA, 2004).

The waste from ranch and cattle lot operations can be classified into three main physical states: solid; semi-solid; and liquid (USEPA, 2004). The solid waste has a moisture content of less than 20%, and it is usually collected from dry lots or pasture systems, where the cattle are kept on bare ground or vegetated land, without any bedding material. Semi-solid manure has a moisture content of 20% to 40%, and it is usually collected from bedded lot or open lot systems, where the cattle are kept on a concrete or earthen pad, with some bedding material, such as straw, sawdust, or wood shavings. Liquid manure has a moisture content of more than 40%, and it is usually collected from confinement or feedlot systems, where the cattle are kept in stalls or pens, with little or no bedding material, and the manure is flushed or scraped into a storage or treatment facility.

The wastewater from ranch and cattle lot operations have been classified into three subcategories viz: runoff, leachate, and wash water (USEPA, 2004). Runoff is the water that flows over the surface of the cattle lot during rainfall or snowmelt events, carrying manure, bedding, feed, and soil particles. Leachate is the water that percolates through the manure pile or the soil under the cattle lot, dissolving and transporting soluble and particulate constituents, while wash water is the water that is used to clean the cattle, the equipment, and the facilities, containing manure, urine, soap, and disinfectants.

2 Wastewater from Ranch and Cattle Lot

Research on water use and wastewater generation in cattle ranching and cattle lots has revealed several key findings. The global water supply estimated to be consumed by cattle ranching and feedlots industries is approximately 8% (Schlink et al., 2010). This includes water used for various purposes, such as cattle consumption, irrigation of feed crops, watering pastures, and processing animals for marketing (Schlink et al., 2010). Ziara (2015) found that the main sources of wastewater in beef packing plants are viscera processing and overnight cleaning.

The properties that reflect the quality, quantity, and potential impacts of wastewater of the cattle lot wastewater can be classified as physical, chemical, and biological. The color of the wastewater usually varies from yellow to brown to black, depending on the degree of decomposition and oxidation of the organic matter constituents. The odor varies from earthy to pungent to foul, depending on the type and concentration of the volatile compounds produced by the microbial activity. As seen in Table 1, which was adapted from USEPA, 2004 and Bolan et al., 2010, cattle lot wastewater is generally characterized by a high content of organic matter, nutrients, solids, pathogens, and contaminants, which indicates a high pollution potential and a low quality for reuse or discharge.

Table 1 Characteristics of Cattle Lot Wastewater (Vaishnav et al., 2023; Sweeten, 2014)

Indicator	Unit	Range
Color	-	Brown, black, green
Odor	-	Foul, pungent, rotten
Turbidity	NTU	100 - 10,000
Temperature	°C	10 - 40
Density	kg/m3	980 - 1,050
Viscosity	Pa.s	0.001 - 0.01
Conductivity	mS/cm	1 - 20
pH	-	6 - 9
Alkalinity	mg/L as CaCO3	500 - 5,000
Acidity	mg/L as CaCO3	100 - 1,000
Hardness	mg/L as CaCO3	100 - 1,000
Salinity	mg/L	500 - 5,000
Dissolved oxygen	mg/L	0 - 10
Biochemical oxygen demand	mg/L	1,000 - 10,000
Chemical oxygen demand	mg/L	2,000 - 20,000
Total organic carbon	mg/L	500 - 5,000
Total nitrogen	mg/L	50 - 500
Ammonia nitrogen	mg/L	10 - 200
Nitrate nitrogen	mg/L	0 - 50
Nitrite nitrogen	mg/L	0 - 10
Total phosphorus	mg/L	10 - 100
Orthophosphate	mg/L	5 - 50
Total suspended solids	mg/L	500 - 5,000
Total dissolved solids	mg/L	500 - 5,000
Volatile suspended solids	mg/L	200 - 2,000
Fixed suspended solids	mg/L	300 - 3,000
Total solids	mg/L	1,000 - 10,000
Volatile solids	mg/L	400 - 4,000
Fixed solids	mg/L	600 - 6,000
Total kjeldahl nitrogen	mg/L	50 - 500
Organic nitrogen	mg/L	10 - 100
Total coliforms	MPN/100 mL	106 - 109
Fecal coliforms	MPN/100 mL	105 - 108
Escherichia coli	MPN/100 mL	104 - 107
Fecal streptococci	MPN/100 mL	104 - 107

(continued)

Table 1 (continued)

Indicator	Unit	Range
Enterococci	MPN/100 mL	104 - 107
Clostridium perfringens	MPN/100 mL	103 - 106
Salmonella spp.	MPN/100 mL	102 - 105
Campylobacter spp.	MPN/100 mL	102 - 105
Cryptosporidium spp.	oocysts/100 L	10 - 1,000
Giardia spp.	cysts/100 L	10 - 1,000
Helminth eggs	eggs/L	1 - 100
Antibiotics	mg/L	0.001 - 1
Hormones	ng/L	0.01 - 100
Heavy metals	mg/L	0.001 - 10
Other organic or inorganic contaminants	mg/L	0.001 - 10

The characteristics of the ranch and cattle lot wastewater vary and depend on several factors, such as the type, size, and management of the operation, the feed and water quality, the climate and season, and the treatment and disposal methods. The influence of these factors on the characteristics of the cattle lot wastewater are enunciated below.

Wastewater from cattle ranches and cattle lots is characterized by high levels of organic matter, ammonia nitrogen, and biodegradability (Jin et al., 2011). They are usually characterized by high biochemical oxygen demand (BOD and COD) chemical oxygen demand (COD), suspended solids (SS), total nitrogen (T-N), and total phosphorus (T-P) in the effluent (Ming et al., 2007; Park et al., 2013). The characteristics of wastewater from cattle farms, as reported by different Authors are presented in Table 2.

2.1 Type, Size, and Management of the Operation

The type, size, and management of the cattle operation influence the amount and composition of the manure and wastewater produced, as well as the potential for environmental and public health impacts. For example, beef cattle produce more manure and wastewater than dairy cattle, due to their higher body weight and lower water efficiency. Larger and more intensive cattle operations tend to generate more concentrated and contaminated wastewater than smaller and less intensive cattle operations, due to the higher animal density and lower waste removal frequency (Bolan et al., 2021). Different cattle management systems, such as dry lot, bedded lot, open lot, confinement, or feedlot, have different effects on the characteristics of the wastewater, such as the moisture content, the organic matter content, the nutrient content, and the pathogen content (Vanderholm et al., 2022).

Table 2 Features of wastewater from cattle lot

Parameters	Ungureanu et al., 2019)	(Chandrasasi et al., 2021)	(Knight et al., 2000)	(Sweeten & Wolfe, 1994)	(Sweeten & Wolfe, 1994)
pH	7.8	8.4		7.6	7.6
Conductivity (mS/cm)	3.84			2.82	4.3
Turbidity (NTU)	544				
Suspended Solids (mg/L)	1216	46.50	291	839	2179
COD (mg O_2/L)	1512.6	102.0		1480	6086
BOD_5 (mg O_2/L)	204.6	64.3	137		
Total N (mg/L)	234.2	65.9*	103 (5.1*)	171.7	305
Total P (mg/L)	56.4			53	58
Total coliforms (CFU	700,000				

* N-NH_3

2.2 *Feed and Water Quality*

The feed and water quality affect the characteristics of the cattle lot wastewater, as they determine the amount and composition of the cattle's manure and urine excreted by the cattle and the amount and quality of the wash water used. For example, the type and amount of feed influence the manure's digestibility, nutrient content, and energy value, as well as the emission of greenhouse gases, such as methane and nitrous oxide, from the cattle and the wastewater (Gerber et al., 2013). The quality and quantity of water influence the water consumption, urine production, and water efficiency of the cattle, as well as the water demand, water pollution, and water recycling of the cattle operation (Schlink et al., 2010).

2.3 *Climate and Season*

The climate and season affect the characteristics of the cattle lot wastewater, as they influence the temperature, precipitation, evaporation, and runoff patterns of the cattle lot, as well as the biological and chemical processes occurring in the manure and wastewater. For example, higher temperatures increase the wastewater's evaporation and volatilization rates, resulting in higher concentrations of solids, nutrients, and ammonia, as well as lower pH and dissolved oxygen levels. Higher precipitation increases the runoff and leachate volumes and loads of the wastewater, resulting in higher risks of surface and groundwater contamination, as well as higher demands for storage and treatment facilities (Bolan et al., 2021).

2.4 *Treatment and Disposal Methods*

The treatment and disposal methods affect the characteristics of the cattle lot wastewater, as they alter the wastewater's physical, chemical, and biological properties, as well as the potential for environmental and public health impact. For example, different treatment methods, such as sedimentation, filtration, aeration, coagulation, flocculation, precipitation, oxidation, reduction, disinfection, or digestion, can remove or reduce the solids, nutrients, pathogens, or contaminants in the wastewater, improving its quality and suitability for reuse or discharge (Oladoja et al., 2015). Different disposal methods, such as land application, irrigation, injection, or discharge, affect the fate and transport of the wastewater and its constituents in the soil, water, or air, influencing its impacts on the ecosystem and human health (Bolan et al., 2021).

Sweeten (1994) highlighted the importance of effective manure and wastewater management systems in open lot dairy operations, with two-stage anaerobic lagoon systems showing high removal efficiencies for solids and nutrients. Chandrasasi (2021) emphasized the need for wastewater treatment plants in cattle farms to prevent surface water and groundwater pollution, with the Multi Soil Layering (MSL) method and anaerobic wasps nest bio-filter showing promise as treatment alternatives.

3 Environmental Impact of Ranch and Cattle Lot Operations

The livestock industry wastes have significant negative impacts on the environment, especially on the water, soil, and air quality. The impacts have been found to vary with the type, quantity, and quality of the wastes, as well as the management practices, climatic conditions, and environmental regulations (Gaur et al., 2020). Some of the major environmental impacts of the livestock industry wastes include:

Water Usage and Pollution: Cattle ranching requires significant water resources for various purposes such as cattle consumption, irrigation of feed crops, and processing activities. This can lead to water scarcity in regions with intensive livestock production (Schlink et al., 2010). Water pollution is one of the most serious and widespread environmental impacts of the livestock industry wastes. Runoff from cattle ranches can contain pollutants such as antibiotics, hormones, and pathogens from manure, which can contaminate water sources and affect aquatic ecosystems (Schlink et al., 2010). Water pollution occurs when the wastes or their constituents are transported by runoff, leaching, or overflow to the surface water or groundwater resources. The pollutants include organic matter, nutrients, pathogens, antibiotics, hormones, metals, and other contaminants that can impair the water quality and affect the aquatic ecosystems and human health. Hormones can contaminate the water resources and affect the endocrine system and the reproductive function of the aquatic life and humans. Hormones can also cause the feminization or masculinization of fish and amphibians, which can alter the sex ratio and the population dynamics of the aquatic species. Antibiotics also contribute to the development and spread of antibiotic-resistant bacteria, which pose serious threat to human and animal health.

Soil Pollution: Soil pollution is another major environmental impact of the livestock industry wastes. For instance, excessive organic matter cause soil compaction, anaerobic conditions, and nutrient imbalance, which reduce soil aeration, drainage, and productivity. Nutrient accumulation in soil cause soil salinization, acidification, and toxicity, which affect the soil pH, electrical conductivity, and cation exchange capacity. Hormones contaminate the soil and affect the endocrine system, the reproductive function of the soil organisms, and humans. Hormones also cause the feminization or masculinization of soil invertebrates, such as earthworms and nematodes, which ultimately alter the soil structure and functions.

Air Pollution: Air pollution occurs when the wastes or their constituents are emitted to the atmosphere as gases, vapours, aerosols, or particulates. These include ammonia, hydrogen sulfide, methane, nitrous oxide, carbon dioxide and volatile organic compounds that impair the air quality. Hydrogen sulfide volatilizes from anaerobic waste storage or treatment facilities and is released during the agitation or pumping of the wastes. It causes corrosion of metal structures, as well as olfactory fatigue, headache, nausea, and death in humans and animals (Li, et al., 2021). Volatile organic compounds cause odors, smog, and photochemical reactions in the atmosphere, irritation, allergy and cancer in humans and animals (Li, et al., 2021). Odor (foul smell) is emitted from animal housing and waste storage facilities, and causes nuisance, annoyance and stress in humans and animals, as well as reduce the property value and quality of life. Cattle ranching is a significant contributor to greenhouse gas emissions, particularly methane produced by cattle during digestion and manure decomposition. These emissions contribute to climate change and global warming (Cooprider et al., 2011). Beef cattle farming proved to be responsible for intensive greenhouse gas emissions (22.0 ± 3.9 kg CO_2 eq$^-$kg LW^{-1}), especially when it occurs predominantly on pasture (Costantini et al., 2021).

4 Management of Wastewater from Ranch and Cattle Lot

Cattle lot generates a large amount of wastewater usually collected and stored in lagoons, ponds, tanks, or pits, before being applied to land as fertilizer, or discharged to surface water or groundwater. The various methods used for cattle ranch and feed lot wastewater management practices are,

Physical**Practices.**

Physical practices are the application of physical methods to separate, filter, or treat the cattle lot wastewater, based on the physical properties of the wastewater components, such as size, density, or charge. Physical practices are classified into three main types: solid–liquid separation, advanced filtration, and disinfection (Vaishnav et al., 2023). Solid–liquid separation is the process of removing the solid particles, such as manure, bedding, and feed, from the liquid phase of the wastewater, using gravity, centrifugal force, or mechanical force. Solid–liquid separation reduce the organic load, the suspended solids, and the turbidity of the wastewater, to produce a solid fraction, which can be used as fertilizer or a feedstock for other processes, such as composting or anaerobic digestion. The liquid fraction can be further treated by other methods, such as biological or chemical practices (Vaishnav et al., 2023) for other uses or safe discharge. Advanced filtration techniques, via the applications of membrane in aerobic and anaerobic conditions have been commonly utilized in the management of cattle lot wastewater, with the membrane being used as the liquid/solid separator (Visvanathan et al., 2000; Iorhemen et al., 2016; Meng et al., 2009). A good example of membrane technology successfully applied is the membrane bioreactor (MBR) system, which involves a semi-permeable membrane, such as microfiltration or ultrafiltration integrated with a relevant biological process. The solid–liquid separation process is done through the membrane. Among the several versions of the MBR, the most commonly applied are the gravity-driven and the pressure-driven methods (Hu et al., 2020). The membrane technology is becoming more economically competitive with other conventional technologies due to the significant improvement in the treatment cost. Advance disinfection process has been utilized in the treatment of cattle lot wastewater by combining ozone with the coagulation-flocculation process. Due to its oxidative and disinfectant properties of ozone, the physicochemical and microbial pollutants in the wastewater was considerably reduced, making it effective for reuse (Hung e al., 2012).

- **Biological Practices**

Biological practices are the application of biological methods to treat or modify the cattle lot wastewater, based on the biological activities or reactions of the microorganisms, such as bacteria, fungi, algae, or plants that can degrade or transform the organic or inorganic pollutants, and produce biomass or bio-products. Biological practices is classified into three main types: aerobic treatment, anaerobic treatment, and phytoremediation (Vaishnav et al., 2023). Aerobic treatment is the process of supplying oxygen to the wastewater, using an aerator, a blower, or a compressor, to enhance the aerobic degradation of the organic matter by the microorganisms, and produce carbon dioxide, water, and biomass. Aerobic treatment reduces the biochemical oxygen demand (BOD) and the chemical oxygen demand (COD) of the wastewater, to produce sludge, which can be used as fertilizer or a feedstock for other processes, such as composting or anaerobic digestion (Vaishnav et al., 2023).

Anaerobic treatment is the process of excluding oxygen from the wastewater, using a closed reactor or a digester, to enhance the anaerobic degradation of the organic matter by the microorganisms, and produce biogas, mainly composed of methane (CH_4) and carbon dioxide (CO_2), and digestate, which is the solid and liquid residue after anaerobic digestion. Anaerobic treatment reduces the organic load, the pathogens, and the odor of the wastewater, and generate renewable energy and fertilizer (Kariyama et al., 2023).

Phytoremediation is another biological technique that uses plants, such as grasses, crops, or trees, to treat or modify the cattle lot wastewater, based on the biological activities or reactions of the plants, such as uptake, transpiration, degradation, or volatilization. Phytoremediation can reduce the pollutants, such as nitrogen, phosphorus, metals, and organic compounds, from the wastewater, and produce biomass or bio-products, such as food, feed, fiber, or fuel (Wang et al., 2021).

- **AI-Based Practices**

AI-based practices are the application of artificial intelligence (AI) methods to monitor, control, or optimize the cattle lot wastewater management, based on the data analysis, machine learning, or decision making of the AI algorithms, such as neural networks, fuzzy logic, or genetic algorithms. AI-based practices can improve the performance, the efficiency, and the reliability of the wastewater management, and reduce the costs, the risks, and the environmental impact (Rane et al., 2023). AI-based practices include:

i) **AI-based monitoring:** AI-based monitoring is the process of using AI algorithms to collect, process, and analyze the data from the sensors, the cameras, or the online platforms, which can measure the parameters, such as the flow rate, the pH, the temperature, the dissolved oxygen, the BOD, the COD, the nutrients, the metals, the pathogens, the antibiotics, or the hormones, of the wastewater or the treatment process. AI-based monitoring can provide real-time and accurate information, and detect the anomalies, the faults, or the changes of the wastewater or the treatment process (Rane et al., 2023)

ii) **AI-based control:** AI-based control is the process of using AI algorithms to adjust the operating parameters or conditions, such as the applied voltage, the pH, the temperature, the HRT, the transmembrane pressure, the cross-flow velocity, the cleaning frequency, the coagulant or flocculant dosage, the oxygen supply, or the recycle ratio, of the wastewater or the treatment process, based on the feedback from the AI-based monitoring or the predefined objectives or constraints. AI-based control can optimize the performance, the efficiency, and the reliability of the wastewater or the treatment process, and reduce the energy consumption and the environmental impact (Rane et al., 2023).

iii) **AI-based optimization:** AI-based optimization is the process of using AI algorithms to find the optimal solution or configuration, such as the membrane type, the membrane configuration, the coagulant or flocculants type, the coagulant or flocculants dosage, the plant species, the plant density, or the integration scheme, of the wastewater or the treatment process, based on the data from the AI-based monitoring or the simulation models, and the objective function or the constraints, such as the water quality, the water quantity, the product yield, the product quality, the energy consumption, the environmental impact, or the economic cost. AI-based optimization can enhance the performance, efficiency, and reliability of the wastewater or the treatment process, and reduce the costs, risks, and environmental impact (Nam et al., 2023).

- **Chemical Practices**

Chemical practices refer to the use of chemical agents or processes to treat and manage Cattle Lot Wastewater (CLW). Chemical practice is via chemical precipitation and coagulation treatment. Chemical precipitation involves the addition of coagulants and/or flocculants to CLW, to induce the formation and settling of solid particles that can remove pollutants from the liquid phase. Chemical precipitation is one of the most widely used methods for the treatment of CLW, especially for the removal of nutrients, such as nitrogen (N) and phosphorus (P), that are responsible for eutrophication and algal blooms. Chemical precipitation can also remove other pollutants, such as metals, pathogens, and organic matter, by co-precipitation or adsorption mechanisms (Akinnawo et al., 2023). The principle of chemical precipitation is to add a chemical agent, such as lime, alum, iron salts, or polymers, to CLW, to adjust the pH and/or form insoluble compounds with the pollutants. The resulting solid particles, called flocs, can then be separated from the liquid phase by sedimentation, filtration, or centrifugation. The separated solid phase, called sludge, can be further processed or disposed of, while the treated liquid phase, called effluent, can be reused or discharged. The performance of chemical precipitation depends on several factors, such as the type and dose of the chemical agent, the pH and temperature of the CLW, the mixing and settling conditions, and the characteristics and concentrations of the pollutants. The optimal conditions for chemical precipitation vary depending on the target pollutants and the desired removal efficiency. Generally, higher doses of chemical agents and lower pH values can enhance the removal of N and P, while higher pH values can enhance the removal of metals and pathogens.

Chemical precipitation has some advantages, such as simplicity, rapidity, reliability, and versatility. It can achieve high removal efficiencies of N and P, up to 90% and 99%, respectively, as well as significant reductions of other pollutants, such as metals, pathogens, and organic matter (An et al., 2023). It can also be easily integrated with other treatment methods, such as biological or electrochemical processes, to improve the overall performance and quality of CLW. However, chemical precipitation also has some disadvantages, such as high cost, low selectivity, secondary pollution, and sludge disposal. The cost of chemical precipitation mainly depends on the type and amount of the chemical agent, which can be expensive and variable depending on the market and availability. The selectivity of chemical precipitation is low, meaning that it can remove not only the target pollutants, but also some beneficial components, such as alkalinity and trace elements, that are essential for the biological processes and the soil fertility. The secondary pollution of chemical precipitation is related to the formation of by-products, such as ammonia, nitrate, sulfate, and chloride that can increase the salinity and toxicity of the CLW. The sludge disposal of chemical precipitation is another challenge, as the sludge can contain high amounts of water, organic matter, and pollutants, that can pose environmental and health risks if not properly handled.

5 Integrated Practices

Integrated practices refer to the use of multiple methods or processes to treat and manage CLW that can achieve synergistic effects and benefits for the environment, the economy, and society. This is a combination of different approaches, from a systemic perspective, to account for all the contributions derived from both economic and environmental systems, which represent a holistic picture of ecological sustainability on larger and spatial scales and times (Björklund et al., 2001). Cattle lot wastewater treatment strategies aim at raising resource efficiency, whilst minimizing the final amount of waste delivered to the environment. These targets can be attained through appropriate technological options that emulate natural cycles and dynamics. Integrated practices can be classified into resource recovery and reuse, and ecological engineering. Resource recovery and reuse involves the extraction and utilization of valuable components, such as nutrients, water, energy, and organic matter from CLW to produce useful products, such as fertilizers, irrigation water, biogas, and compost. Ecological engineering involves the use of natural or engineered systems, such as wetlands, ponds, lagoons, and constructed ecosystems to treat and manage CLW by mimicking or enhancing the natural processes such as biodegradation, nitrification, denitrification, and phytoremediation. We describe the principles, mechanisms, and performance of each category in the following sections.

6 Resource Recovery and Reuse of Ranch and Cattle Lot Wastewater

Resource recovery and reuse is a promising strategy for the management of CLW that can transform the waste into a resource, and create a circular economy for the livestock industry and the society. Resource recovery and reuse can reduce the environmental impacts and the costs of CLW treatment and disposal, as well as generate additional income and benefits for the farmers and the consumers (Hossain et al., 2022). The principle of resource recovery and reuse is to apply various methods or processes to separate and concentrate the valuable components from CLW, and to convert them into useful products, such as fertilizers, irrigation water, biogas, and compost. The methods or processes can include physical, chemical, biological, or electrochemical techniques, such as filtration, precipitation, adsorption, ion exchange, membrane separation, anaerobic digestion, composting, and electrolysis. The products can be used for various purposes, such as crop production, soil improvement, energy generation, and greenhouse gas mitigation (Kesari et al., 2021). The performance of resource recovery and reuse depends on several factors, such as the type and quality of the products, the efficiency and cost of the methods or processes, the market and demand of the products, and the regulations and policies of the products. The optimal conditions for resource recovery and reuse vary depending on the objectives and constraints of the stakeholders, such as the farmers, the consumers, the regulators, and the society. Generally, higher quality and quantity of the products can increase the economic and environmental benefits of resource recovery and reuse, but also require higher efficiency and cost of the methods or processes. Higher market and demand of the products can increase the profitability and feasibility of resource recovery and reuse, but also require higher standards and compliance of the products (Elbana et al., 2019). Resource recovery and reuse has some advantages, such as sustainability, profitability, and social responsibility. It can achieve sustainability, by reducing the waste generation and the resource consumption, and by enhancing the resource efficiency and the environmental protection. It achieves profitability, by creating new revenue streams and reducing operational costs, and by improving the productivity and the competitiveness of the livestock industry. It achieves social responsibility, by providing safe and affordable products and services, and by contributing to the food security and the climate action of the society (Fito & Van Hulle, 2021).

Resource recovery and reuse also has some challenges, such as technical complexity, market uncertainty, and regulatory barriers. The technical complexity of resource recovery and reuse is related to the design and operation of the methods or processes, which can be challenging and require skilled personnel and maintenance. The market uncertainty of resource recovery and reuse is related to the supply and demand of the products, which can be variable and influenced by the quality, price, availability, and preference of the products. The regulatory barriers of resource recovery and reuse are related to the standards and policies of the products, which can be strict and inconsistent across different regions and countries.

7 Ecological Engineering Practice of Cattle Lot and Cattle Ranch Wastewater Management

Ecological engineering is another strategy for the management of CLW that can utilize the natural or engineered systems, such as wetlands, ponds, lagoons, and constructed ecosystems, to treat and manage CLW, by mimicking or enhancing the natural processes, such as biodegradation, nitrification, denitrification, and phytoremediation. Ecological engineering can improve the quality and the diversity of the ecosystems, as well as provide multiple ecosystem services, such as habitat provision, carbon sequestration, and aesthetic value (Winchell et al., 2022). The principle of ecological engineering is to design and construct the systems, such as wetlands, ponds, lagoons, and constructed ecosystems that can receive and treat CLW, by using the physical, chemical, and biological interactions among the water, soil, plants, microorganisms, and animals. The systems can have different configurations, such as surface flow, subsurface flow, horizontal flow, or vertical flow, depending on the hydraulic and hydrologic characteristics of the systems. The systems can also have different components, such as vegetation, substrate, aeration, or biofilm, depending on the treatment and performance objectives of the systems (Dagnaisser et al., 2022).

The performance of ecological engineering depends on several factors, such as the type and design of the systems, the loading and retention time of the CLW, the climate and season of the systems, and the characteristics and concentrations of the pollutants. The optimal conditions for ecological engineering vary depending on the objectives and constraints of the stakeholders, such as the farmers, the regulators, and the society. Generally, higher loading and retention time of the CLW can increase the removal of pollutants, but also increase the land requirement and the maintenance of the systems. Higher climate and season variability of the systems can affect the removal of pollutants, especially for temperature-sensitive processes, such as nitrification and denitrification. Higher diversity and complexity of the systems can improve the removal of pollutants, but also increase the design and construction costs of the systems (Al Ali et al., 2021).

Ecological engineering has some advantages, such as sustainability, resilience, and multifunctionality. It can achieve sustainability, by using natural or renewable resources, such as solar energy, water cycle, and biomass, and by minimizing external inputs, such as chemicals, energy, and labour. It can achieve resilience, by adapting to changing conditions, such as climate, season, and loading, and by recovering from the disturbances, such as floods, droughts, and shocks. It can achieve multifunctionality, by providing multiple benefits and services, such as water quality improvement, nutrient recovery, biodiversity conservation, landscape enhancement, and recreation opportunities (Long et al., 2022).

8 Strategies for Mitigating the Environmental Impact of Cattle Lot and Ranch Wastes–Current Practices

Ranch wastewater treatment is the process of removing or reducing the pollutants in the livestock wastewater to meet the desired quality standards for disposal or reuse. Cattle lot and ranch wastewater treatment can involve physical, chemical, biological, or thermal methods, or a combination of them. The choice of the treatment method depends on the characteristics, quantity, and quality of the wastewater, as well as the availability, feasibility, and cost of the technology (Al-Hazmi et al., 2023). Some of the common livestock wastewater treatment practices are described below.

i) **Composting:** Composting is a biological method of treating the solid wastes from ranch and cattle lot operations by aerobic decomposition of the organic matter. Composting can reduce the volume, weight, and odor of the solid wastes, and produce a stable and sanitized product that can be used as a soil amendment or fertilizer. Composting can be done in piles, windrows, bins, or vessels, with or without mechanical mixing or aeration. Composting requires proper control of the moisture, temperature, oxygen, and carbon-to-nitrogen ratio of the composting materials, to ensure the optimal microbial activity and decomposition (Ibekwe et al., 2023).

ii) **Aerobic digestion:** Aerobic digestion is a biological method of treating the liquid wastes from ranch and cattle lot operations by aerobic decomposition of the organic matter. Aerobic digestion can reduce the organic load, nutrient content, and pathogen level of the liquid wastes, and produce a clarified effluent that can be reused or discharged. Aerobic digestion can be done in tanks, ponds, or lagoons, with or without mechanical mixing or aeration. Aerobic digestion requires proper control of the pH, temperature, oxygen, and retention time of the liquid wastes, to ensure the optimal microbial activity and decomposition.

iii) **Activated sludge**: Activated sludge is a biological method of treating the liquid wastes from ranch and cattle lot operations by aerobic decomposition of the organic matter in a suspended growth system. Activated sludge can reduce the organic load, nutrient content, and pathogen level of the liquid wastes, and produce a clarified effluent that can be reused or discharged. Activated sludge can be done in tanks, reactors, or basins, with mechanical mixing and aeration. Activated sludge requires proper control of the pH, temperature, oxygen, retention time, and sludge concentration of the liquid wastes, to ensure the optimal microbial activity and decomposition (Thao et al., 2023).

iv) **Anaerobic digestion:** Anaerobic digestion is a biological method of treating the solid or liquid wastes from ranch and cattle lot operations by anaerobic decomposition of the organic matter. Anaerobic digestion can reduce the volume, weight, and odor of the solid or liquid wastes, and produce a stabilized and sanitized product that can be used as a soil amendment or fertilizer, as well as a biogas that can be used as a renewable energy source. Anaerobic digestion can be done in tanks, digesters, or lagoons, with or without mechanical mixing or heating. Anaerobic digestion requires proper control of the pH, temperature, retention time, and organic loading rate of the solid or liquid wastes, to ensure the optimal microbial activity and decomposition.

v) **Anaerobic lagoons**: Anaerobic lagoons are a biological method of treating the liquid wastes from ranch and cattle lot operations by anaerobic decomposition of the organic matter in a natural or artificial pond. Anaerobic lagoons can reduce the organic load, nutrient content, and pathogen level of the liquid wastes, and produce a clarified effluent that can be reused or discharged, as well as a biogas that can be used as a renewable energy source. Anaerobic lagoons require proper design and maintenance of the size, depth, shape, and lining of the pond, as well as the influent and effluent flow rates, to ensure the optimal hydraulic and organic loading and retention of the liquid wastes.

vi) **Anaerobic filters**: Anaerobic filters are a biological method of treating the liquid wastes from ranch and cattle lot operations by anaerobic decomposition of the organic matter in an attached growth system. Anaerobic filters can reduce the organic load, nutrient content, and pathogen level of the liquid wastes, and produce a clarified effluent that can be reused or discharged, as well as a biogas that can be used as a renewable energy source. Anaerobic filters can be done in tanks, reactors, or columns, with a packing material that provides a surface for the attachment and growth of the microorganisms. Anaerobic filters require proper control of the pH, temperature, retention time, and organic loading rate of the liquid wastes, to ensure the optimal microbial activity and decomposition.

9 Challenges of Current Practices

The current practices of ranch wastewater treatment face various challenges, such as technical, economic, social, and environmental challenges. Some of the examples of the challenges are:

a) **Technical challenges:** Technical challenges are the difficulties or limitations of the design, operation, and maintenance of the wastewater treatment systems, such as the complexity, reliability, efficiency, and adaptability of the systems. Technical challenges can affect the performance, quality, and safety of the wastewater treatment processes and products. Some of the technical challenges are:
- The variability and uncertainty of the wastewater characteristics, quantity, and quality, which can affect the selection, optimization, and control of the treatment methods and parameters.
- The lack of adequate and appropriate technologies, equipment, and infrastructure, especially for the small-scale and remote operations, which can affect the availability, accessibility, and affordability of the wastewater treatment systems.
- The lack of sufficient and skilled labor, training, and education, especially for the complex and advanced systems, which can affect the operation, maintenance, and troubleshooting of the wastewater treatment systems.
- The lack of sufficient and standardized monitoring, testing, and evaluation methods and tools, especially for the emerging and innovative systems, which can affect the verification, validation, and regulation of the wastewater treatment processes and products.

b) **Economic challenges:** Economic challenges are the difficulties or limitations of the financial and market aspects of the wastewater treatment systems, such as the cost, benefit, and profitability of the systems. Economic challenges can affect the feasibility, viability, and sustainability of the wastewater treatment systems. Some of the economic challenges are:

    - The high capital and operating costs of the wastewater treatment systems, especially for the advanced and innovative systems, which can affect the affordability and return on investment of the systems.
    - The low or negative value and demand of the wastewater treatment products, especially for the by-products and residues, which can affect the profitability and marketability of the systems.
    - The lack of adequate and appropriate incentives, subsidies, and policies, especially for the environmentally friendly and socially beneficial systems, which can affect the competitiveness and attractiveness of the systems.

- The lack of adequate and appropriate regulations, standards, and certifications, especially for the emerging and innovative systems, which can affect the compliance and acceptance of the systems.

c) **Social challenges:** Social challenges are the difficulties or limitations of the human and social aspects of the wastewater treatment systems, such as the perception, attitude, and behavior of the stakeholders. Social challenges can affect the awareness, participation, and support of the wastewater treatment systems. Some of the social challenges are:

- The low or negative awareness and knowledge of the wastewater treatment systems, especially for the benefits and risks of the systems, which can affect the information and education of the stakeholders.
- The low or negative attitude and preference of the wastewater treatment systems, especially for the aesthetics and convenience of the systems, which can affect the satisfaction and loyalty of the stakeholders.
- The low or negative behavior and practice of the wastewater treatment systems, especially for the adoption and use of the systems, which can affect the involvement and contribution of the stakeholders.
- The lack of adequate and appropriate communication, collaboration, and coordination among the stakeholders, especially for the planning, implementation, and evaluation of the systems, which can affect the cooperation and integration of the stakeholders.

d) **Environmental challenges:** Environmental challenges are the difficulties or limitations of the environmental and ecological aspects of the wastewater treatment systems, such as the impact, risk, and resilience of the systems. Environmental challenges can affect the compatibility, suitability, and durability of the wastewater treatment systems. Some of the environmental challenges are:

- The high or negative environmental impact of the wastewater treatment systems, especially for the energy consumption, greenhouse gas emission, and resource depletion of the systems, which can affect the environmental footprint and performance of the systems.
- The high or negative environmental risk of the wastewater treatment systems, especially for the pollution, contamination, and disturbance of the systems, which can affect the environmental quality and safety of the systems.
- The low or negative environmental resilience of the wastewater treatment systems, especially for the adaptation, mitigation, and restoration of the systems, which can affect the environmental stability and recovery of the systems.

10 Advances and Strategic Management of Feedlot Wastewater

The advances and strategic management of feedlot wastewater are the improvements and innovations of the wastewater treatment systems, as well as the integration and optimization of the wastewater management cycle to achieve the environmental, economic, and social goals of the livestock industry. The advances and strategic management of feedlot wastewater can involve the development and application of new or modified technologies, methods, and practices, as well as the coordination and collaboration of the stakeholders (de Oliveira Demarco, 2023). Some of the examples of the advances and strategic management of feedlot wastewater are:

- New or modified technologies: New or modified technologies are the devices, systems, or processes that are designed or adapted to enhance the performance, efficiency, and reliability of the wastewater treatment systems, as well as to reduce the cost, energy, and environmental impact of the systems. Some of the new or modified technologies are:
- Membrane bioreactors: Membrane bioreactors are a combination of the biological and physical methods of treating the liquid wastes from ranch and cattle lot operations by aerobic decomposition of the organic matter and membrane filtration of the suspended solids. Membrane bioreactors can produce a high-quality effluent that can be reused or discharged, as well as a concentrated sludge that can be further treated or disposed. Membrane bioreactors can reduce the footprint, energy, and sludge production of the wastewater treatment systems, as well as improve the removal of nutrients, pathogens, and contaminants from the wastewater (Du et al., 2023).
- Anaerobic membrane bioreactors: Anaerobic membrane bioreactors are a combination of the biological and physical methods of treating the liquid wastes from ranch and cattle lot operations by anaerobic decomposition of the organic matter and membrane filtration of the suspended solids. Anaerobic membrane bioreactors can produce a high-quality effluent that can be reused or discharged, as well as a biogas that can be used as a renewable energy source. Anaerobic membrane bioreactors can reduce the footprint, energy, and sludge production of the wastewater treatment systems, as well as improve the removal of nutrients, pathogens, and contaminants from the wastewater (Lou et al., 2020).
- Electrocoagulation: Electrocoagulation is a physical method of treating the liquid wastes from ranch and cattle lot operations by applying an electric current to the wastewater, which causes the coagulation and precipitation of the suspended solids, colloids, and dissolved metals. Electrocoagulation can produce a clarified effluent that can be reused or discharged, as well as a sludge that can be further treated or disposed. Electrocoagulation can reduce the chemical, energy, and sludge production of the wastewater treatment systems, as well as improve the removal of turbidity, color, odor, and contaminants from the wastewater (Khanaum & Borhan, n.d.).

- Ultraviolet disinfection: Ultraviolet disinfection is a physical method of treating the liquid wastes from ranch and cattle lot operations by exposing the wastewater to ultraviolet light, which inactivates the pathogens and degrades the organic compounds. Ultraviolet disinfection can produce a disinfected effluent that can be reused or discharged, as well as a by-product that can be further treated or disposed. Ultraviolet disinfection can reduce the chemical, energy, and sludge production of the wastewater treatment systems, as well as improve the removal of pathogens, antibiotics, hormones, and contaminants from the wastewater (Pascagaza Rubio, 2023).
- New or modified methods: New or modified methods are the techniques, procedures, or protocols that are developed or adapted to improve the operation, maintenance, and monitoring of the wastewater treatment systems, as well as to enhance the quality, safety, and compliance of the wastewater treatment processes and products. Some of the new or modified methods are:
- Process control and optimization: Process control and optimization are the techniques of adjusting and regulating the parameters and variables of the wastewater treatment systems, such as the pH, temperature, oxygen, retention time, and organic loading rate, to achieve the optimal performance, efficiency, and reliability of the systems. Process control and optimization can be done using sensors, controllers, and models, which can measure, analyze, and manipulate the wastewater treatment processes and products. Process control and optimization can reduce the variability, uncertainty, and risk of wastewater treatment systems, as well as improve the quality, safety, and compliance of wastewater treatment processes and products (Shomar et al., 2020).
- Sludge management and utilization: Sludge management and utilization are the procedures of handling, treating, and disposing or reusing the sludge that is produced by the wastewater treatment systems, such as the solid or liquid wastes, the by-products, or the residues. Sludge management and utilization can be done by using various methods, such as dewatering, drying, composting, anaerobic digestion, incineration, or land application, which can transform the physical, chemical, or biological characteristics of the sludge (Vaishnav et al., 2023). Sludge management and utilization can reduce the volume, weight, and odor of the sludge, as well as produce valuable or useful products, such as fertilizer, soil amendment, biogas, or heat.
- Quality assurance and quality control: Quality assurance and quality control are the protocols of verifying, validating, and regulating the quality and safety of the wastewater treatment processes and products, such as the effluent, the sludge, or the by-products. Quality assurance and quality control can be done by using various methods, such as sampling, testing, and auditing, which can measure, evaluate, and certify the wastewater treatment processes and products (Emert et al., 2023). Quality assurance and quality control can reduce the errors, defects, and failures of the wastewater treatment systems, as well as improve the compliance and acceptance of the wastewater treatment processes and products.

- New or modified practices: New or modified practices are the techniques, procedures, or protocols that are developed or adapted to improve the integration and optimization of the wastewater management cycle, such as the generation, collection, storage, treatment, and disposal or reuse of the wastewater. New or modified practices can involve the coordination and collaboration of the stakeholders, such as the livestock producers, the waste managers, the regulators, the researchers, and the consumers. These include:
- Integrated waste management: Integrated waste management is a practice of managing the wastewater as a resource rather than a waste, by applying the principles of prevention, reduction, reuse, recycling, recovery, and disposal of the wastewater, in a holistic and systematic manner. Integrated waste management can improve the environmental, economic, and social benefits of the wastewater management cycle, by reducing the environmental impact, increasing the value and demand, and enhancing the awareness and participation of the wastewater.
- Nutrient management: Nutrient management is a practice of managing the nutrient content and balance of the wastewater, by applying the principles of source reduction, nutrient recovery, and nutrient reuse of the wastewater, in a site-specific and crop-specific manner. Nutrient management can improve the environmental, economic, and social benefits of the wastewater management cycle, by reducing the nutrient loss, increasing the nutrient efficiency, and enhancing the crop yield and quality of the wastewater.
- Biogas management: Biogas management is a practice of managing the biogas production and utilization of the wastewater, by applying the principles of biogas capture, biogas purification, and biogas utilization of the wastewater, in a energy-efficient and cost-effective manner. Biogas management can improve the environmental, economic, and social benefits of the wastewater management cycle, by reducing the greenhouse gas emission, increasing the renewable energy generation, and enhancing the energy security and independence of the wastewater.

11 Improving Waste Management Strategies

Small-scale farm operations face different challenges and opportunities than large-scale ranch and cattle lot operations when it comes to waste management. Small-scale farm operations typically have less land, water, and financial resources, as well as less access to technologies, equipment, and infrastructure. However, small-scale farm operations also have more flexibility, diversity, and innovation potential, as well as more direct contact and interaction with the consumers and the community (Chen et al., 2023). Therefore, improving waste management strategies for small-scale farm operations requires a different approach than for large-scale operations. Some of the possible strategies for improving waste management for small-scale farm operations are:

a) Adopting integrated and holistic waste management practices, such as using the waste hierarchy principles of prevention, reduction, reuse, recycling, recovery, and disposal, and considering the environmental, economic, and social aspects of the waste management cycle.
b) Implementing appropriate and affordable waste treatment technologies, such as using simple and low-cost systems, such as composting, aerobic digestion, anaerobic digestion, or anaerobic lagoons, and utilizing the waste products, such as fertilizer, soil amendment, biogas, or heat.
c) Enhancing the waste management knowledge and skills, such as providing training, education, and extension services, as well as sharing information, experience, and best practices among the small-scale farm operators and the waste managers.
d) Developing the waste management market and policy, such as creating incentives, subsidies, and regulations, as well as establishing standards and certifications, for the waste management products and services, and promoting the value and demand of the waste management products and services among the consumers and the community.

12 Wastewater Valorization and Production of Value-Added Products

Cattle ranch lot wastewater represents a significant loss of valuable resources, such as water, energy, and nutrients, which could be recovered and reused for various purposes. Therefore, there is a need to develop and implement sustainable and efficient technologies for cattle ranch lot wastewater treatment and valorization, which can reduce the environmental and economic impacts of cattle ranching, while enhancing its productivity and profitability. Valorization refers to the process of converting waste into value-added products, such as bioplastics, biofuels, biohydrogen, and biogas, which can be used for various applications, such as packaging, transportation, electricity, and heating (Ghimire et al., 2019; Kumar et al., 2019). Valorization can also create new markets and opportunities for cattle ranchers, as well as reduce their dependence on fossil fuels and external inputs.

13 Potential Value-Added Products from Wastewater

Cattle ranch lot wastewater valorization is the process of converting cattle ranch lot wastewater into value-added products, such as bioplastics, biofuels, biohydrogen, and biogas, which can be used for various applications, such as packaging, transportation, electricity, and heating. Cattle ranch lot wastewater valorization can enhance the sustainability and profitability of cattle ranching operations, by creating new markets and opportunities, reducing the dependence on fossil fuels and external inputs, and mitigating the environmental and economic impacts of cattle ranch lot wastewater. Cattle ranch lot wastewater valorization can be achieved by using different biological, chemical, and physical processes, which can be integrated with the existing cattle ranch lot wastewater treatment systems, or implemented as standalone processes. Some of the potential value-added products from cattle ranch lot wastewater are (Ghimire et al., 2019; Kumar et al., 2019):

- **Bioplastics**: These are plastics that are derived from renewable biomass sources, such as cattle ranch lot wastewater, rather than from petroleum or natural gas. Bioplastics can have similar or superior properties and performance as conventional plastics, such as strength, flexibility, durability, and biodegradability, but with lower environmental impacts and carbon footprints. Bioplastics can be produced from cattle ranch lot wastewater by using different processes, such as fermentation, polymerization, and extrusion. Some of the common types of bioplastics that can be derived from cattle ranch lot wastewater are polyhydroxyalkanoates (PHAs), polylactic acid (PLA), and polybutylene succinate (PBS), which can be used for various applications, such as packaging, agriculture, medical, and textile (Ghimire et al., 2019; Kumar et al., 2019).
- **Biofuels**: These are fuels that are derived from renewable biomass sources, such as cattle ranch lot wastewater, rather than from petroleum or natural gas. Biofuels can have similar or superior properties and performance as conventional fuels, such as energy density, octane number, and combustion efficiency, but with lower environmental impacts and carbon footprints. Biofuels can be produced from cattle ranch lot wastewater by using different processes, such as transesterification, hydrothermal liquefaction, and pyrolysis. Some of the common types of biofuels that can be derived from cattle ranch lot wastewater are biodiesel, bio-oil, and biochar, which can be used for various applications, such as transportation, electricity, and heating
- **Biohydrogen**: This is hydrogen that is derived from renewable biomass sources, such as cattle ranch lot wastewater, rather than from fossil fuels or water electrolysis. Biohydrogen can have similar or superior properties and performance as conventional hydrogen, such as energy density, purity, and storage capacity, but with lower environmental impacts and carbon footprints. Biohydrogen can be produced from cattle ranch lot wastewater by using different processes, such as dark fermentation, photofermentation, and microbial electrolysis. Biohydrogen can be used for various applications, such as transportation, electricity, and heating, either directly or in combination with other fuels, such as biogas, natural gas, or syngas.
- **Biogas**: This is a mixture of methane and carbon dioxide, with traces of other gases, such as hydrogen sulfide, ammonia, and nitrogen, that is derived from renewable biomass sources, such as cattle ranch lot wastewater, rather than from fossil fuels or landfills. Biogas can have similar or superior properties and performance as conventional natural gas, such as energy density, calorific value, and combustion efficiency, but with lower environmental impacts and carbon footprints. Biogas can be produced from cattle ranch lot wastewater by using different processes,

such as anaerobic digestion, anaerobic fermentation, and anaerobic co-digestion. Biogas can be used for various applications, such as transportation, electricity, and heating, either directly or after upgrading to biomethane, which has higher methane content and lower impurities.

14 Conclusion

Cattle lot and ranch wastewater is a complex and variable mixture of organic matter, nutrients, pathogens, antibiotics, hormones, metals, and other contaminants, that can have significant negative impacts on the water, soil, and air quality, as well as the climate change and human health, if not properly managed. Feedlot wastewater treatment is the process of removing or reducing the pollutants in the livestock wastewater to meet the desired quality standards for disposal or reuse. The treatment can involve physical, chemical, biological, or thermal methods, or a combination of them, depending on the characteristics, quantity, and quality of the wastewater, as well as the availability, feasibility, and cost of the technology. The cattle lot wastewater treatment faces various challenges, such as technical, economic, social, and environmental, that require improvements and innovations of the wastewater treatment systems, as well as the integration and optimization of the wastewater management cycle. The feedlot wastewater treatment can provide various benefits, such as environmental, economic, and social benefits that can enhance the sustainability and profitability of the cattle lot and cattle ranch industry.

References

Adelagun, R. O. A. (2016). Technological options for phosphate removal and recovery from aqua system: A review. *Chemical Sciences Review and Letters., 5*(18), 19–34.

Akinnawo, S. O., Ayadi, P. O., & Oluwalope, M. T. (2023). Chemical coagulation and biological techniques for wastewater treatment. *Ovidius University Annals of Chemistry, 34*(1), 14–21.

Al-Hazmi, H. E., Mohammadi, A., Hejna, A., Majtacz, J., Esmaeili, A., Habibzadeh, S., Saeb, M. R., Badawi, M., Lima, E. C., & Mąkinia, J. (2023). Wastewater treatment for reuse in agriculture: Prospects and challenges. *Environmental Research,* 116711.

Al Ali, A., Ouda, M., Naddeo, V., Puig, S., & Hasan, S. W. (2021). Integrated electrochemical-adsorption process for the removal of trace heavy metals from wastewater. *Case Studies in Chemical and Environmental Engineering, 4,* Article 100147.

An, B.-H., Xu, D.-M., Geng, R., Cheng, Y., Qian, R.-B., Tang, X.-C., Fan, Z.-Q., & Chen, H.-B. (2023). The pretreatment effects of various target pollutant in real coal gasification gray water by coupling pulse electrocoagulation with chemical precipitation methods. *Chemosphere, 311,* Article 136898.

Björklund, J. et al. (2001). Emergy analysis of municipal wastewater treatment and generation of electricity by digestion of sewage sludge Resour. Conserv. Recyc.

Bolan, N., Sarkar, B., Vithanage, M., Singh, G., Tsang, D. C. W., Mukhopadhyay, R., Ramadass, K., Vinu, A., Sun, Y., & Ramanayaka, S. (2021). Distribution, behaviour, bioavailability and

remediation of poly-and per-fluoroalkyl substances (PFAS) in solid biowastes and biowaste-treated soil. *Environment International, 155*, Article 106600.
Chandrasasi, D., Haribowo, R., Wardana, B. (2021). Design for wastewater treatment plants (WWTP) for cattle in a single house model with a bio filter process. In: IOP Conference Series: Earth and Environmental Science, p. 012016.
Chen, X., Qi, Y., Wang, K., & Xing, L. (2023). Does farmer entrepreneurship promote willingness to participate in domestic waste management? An empirical analysis from China. *Journal of Material Cycles and Waste Management,* 1–12.
Cooprider, K., Mitloehner, F., Famula, T., Kebreab, E., Zhao, Y., & Van Eenennaam, A. (2011). Feedlot efficiency implications on greenhouse gas emissions and sustainability. *Journal of Animal Science, 89*, 2643–2656.
Costantini, M., Vázquez-Rowe, I., Manzardo, A., & Bacenetti, J. J. S. P. (2021). Consumption. *Environmental Impact Assessment of Beef Cattle Production in Semi-Intensive Systems in Paraguay., 27*, 269–281.
Dagnaisser, L. S., dos Santos, M. G. B., Rita, A. V. S., Chaves Cardoso, J., de Carvalho, D. F., & de Mendonça, H. V. (2022). Microalgae as bio-fertilizer: A new strategy for advancing modern agriculture, wastewater bioremediation, and atmospheric carbon mitigation. *Water, Air, & Soil Pollution, 233*(11), 477.
de Oliveira Demarco, J. (2023). Sustainable management of feedlot and agricultural runoff through bio-inspired bioreactor-constructed wetland configuration for high-quality reuse.
Du, X., Ma, R., Liang, Z., Kuang, K., Chu, L., Liu, L., Song, W., & Tian, J. (2023). Flux improvement and biofilm performance of an oblique gravity-driven ceramic membrane bioreactor coupled with electrocoagulation for roofing rainwater purification. *ACS ES&T Engineering, 3*(11), 1813–1825.
Elbana, T. A., Bakr, N., & Elbana, M. (2019). Reuse of treated wastewater in Egypt: Challenges and opportunities. *Unconventional Water Resources and Agriculture in Egypt*, 429–453.
Emert, A. D., Griffis-Kyle, K., Green, F. B., & Smith, P. N. (2023). Atmospheric transport of particulate matter and particulate-bound agrochemicals from beef cattle feedlots: human health implications for downwind agricultural communities. *Science of The Total Environment,* 164678.
Fito, J., & Van Hulle, S. W. H. (2021). Wastewater reclamation and reuse potentials in agriculture: Towards environmental sustainability. *Environment, Development and Sustainability, 23*, 2949–2972.
Gaur, V. K., Sharma, P., Sirohi, R., Awasthi, M. K., Dussap, C.-G., & Pandey, A. (2020). Assessing the impact of industrial waste on environment and mitigation strategies: A comprehensive review. *Journal of Hazardous Materials, 398*, Article 123019.
Gerber, P. J., Steinfeld, H., Henderson, B., Mottet, A., Opio, C., Dijkman, J., & Tempio, G. (2013). *Tackling climate change through livestock: a global assessment of emissions and mitigation opportunities*. Food and Agriculture Organization of the United Nations (FAO).
Ghimire, P. R., Agho, K. E., Renzaho, A. M., Nisha, M. K., Dibley, M., & Raynes-Greenow, C. (2019). Factors associated with perinatal mortality in Nepal: Evidence from Nepal demographic and health survey 2001–2016. *BMC Pregnancy and Childbirth, 19*, 1–12.
Gu, Y., Shen, S., Han, B., Tian, X., Yang, F., & Zhang, K. (2020). Family livestock waste: An ignored pollutant resource of antibiotic resistance genes. *Ecotoxicology and Environmental Safety, 197*, Article 110567.
Hossain, R., Islam, M. T., Ghose, A., & Sahajwalla, V. (2022). Full circle: Challenges and prospects for plastic waste management in Australia to achieve circular economy. *Journal of Cleaner Production,* 133127.
Hu, Y., Cheng, H., Ji, J., & Li, Y.-Y. (2020). A review of anaerobic membrane bioreactors for municipal wastewater treatment with a focus on multicomponent biogas and membrane fouling control. *Environmental Science: Water Research and Technology, 6*, 2641–2663. https://doi.org/10.1039/D0EW00528B

Hung, Y.T.; Wang, L.K.; Shammas, N.K. (2012). *Handbook of Environment and Waste Management: Air and Water Pollution Control*. World Scientific Publishing Company, Illustrated edition (May 30, 2012).

Ibekwe, A. M., Bhattacharjee, A. S., Phan, D., Ashworth, D., Schmidt, M. P., Murinda, S. E., Obayiuwana, A., Murry, M. A., Schwartz, G., & Lundquist, T. (2023). Potential reservoirs of antimicrobial resistance in livestock waste and treated wastewater that can be disseminated to agricultural land. *Science of the Total Environment, 872*, Article 162194.

Iorhemen, O. T., Hamza, R. A., Tay, J. H. (2016). Membrane bioreactor (MBR) technology for wastewater treatment and reclamation: membrane fouling. *Membranes, 6*, https://doi.org/10.3390/membranes6020033

Jin, L., Yuanqin, L., & Ping, Y. (2011). Study on treatment of wastewater from livestock and poultry farms. *Sichuan Environment, 30*, 139–143.

Kariyama, I. D., Li, W., Yu, S., Chen, L., Qi, R., Zhang, H., Li, X., Deng, X., Lin, J., & Wu, B. (2023). Simplified modeling of high-solids anaerobic digestion of dairy manure in a pilot-scale stirred tank anaerobic digester. *Journal of the ASABE*, 0.

Kesari, K. K., Soni, R., Jamal, Q. M. S., Tripathi, P., Lal, J. A., Jha, N. K., Siddiqui, M. H., Kumar, P., Tripathi, V., & Ruokolainen, J. (2021). Wastewater treatment and reuse: A review of its applications and health implications. *Water, Air, & Soil Pollution, 232*, 1–28.

Knight, R. L., Payne, V. W., Jr., Borer, R. E., Clarke, R. A., Jr., & Pries, J. H. (2000). Constructed wetlands for livestock wastewater management. *Ecological Engineering, 15*, 41–55.

Kumar, V., Parihar, R. D., Sharma, A., Bakshi, P., Sidhu, G. P. S., Bali, A. S., & Rodrigo-Comino, J. (2019). Global evaluation of heavy metal content in surface water bodies: A meta-analysis using heavy metal pollution indices and multivariate statistical analyses. *Chemosphere, 236*, Article 124364.

Li, R., Han, Z., Shen, H., Qi, F., & Sun, D. (2021). Volatile sulfur compound emissions and health risk assessment from an A^2/O wastewater treatment plant. *Science of the Total Environment, 794*, Article 148741.

Long, N. V., Le, T. D. N., Nguyen, H., Khanh, D. V., The, N. T. M., Do, D. T., & Cheng, Y. (2022). From a hard to soft approach for flood management in the Vietnamese Mekong delta: Integrating ecological engineering for Urban sustainability in my Tho city. *Water, 14*(7), 1079.

Lou, E. G., Harb, M., Smith, A. L., & Stadler, L. B. (2020). Livestock manure improved antibiotic resistance gene removal during co-treatment of domestic wastewater in an anaerobic membrane bioreactor. *Environmental Science: Water Research & Technology, 6*(10), 2832–2842.

Meng, F., Chae, S.-R., Drews, A., Kraume, M., Shin, H.-S., & Yang, F. (2009). Recent advances in membrane bioreactors (MBRs): Membrane fouling and membrane material. *Water Research, 43*, 1489–1512. https://doi.org/10.1016/j.watres.2008.12.044

Ming, T. T., Hyun, K. T., & Myun, J. (2007). Characterization of livestock wastewater at various stages of wastewater treatment plant. *Malaysian Journal of Analytical Sciences, 11*, 23–28.

Nam, K., Heo, S., Kim, S., & Yoo, C. (2023). A multi-agent AI reinforcement-based digital multi-solution for optimal operation of a full-scale wastewater treatment plant under various influent conditions. *Journal of Water Process Engineering, 52*, Article 103533.

Novikova, A., & Startiene, G. (2018). Analysis of farming system outputs and methods of their evaluation. *Research for Rural Development, 2*, 138–145.

Oladoja, N. A., Adelagun, R. O. A., Ahmad, A. L., & Ololade, I. A. (2015). Phosphate recovery from aqua culture wastewater using thermally treated gastropod shells. *Process Safety and Environmental Protection, 38*, 296–308.

Park, M.-J., Cho, M. C., Kim, K., Lee, Y. W., Jeong, B. H., Lee, B. (2013). Characteristics of livestock wastewater treatment technology using high temperature aeration process. In. Pastoor, J., Loy, D., Trenkle, A., Lawrence, J., 2012. Comparing fed cattle performance in open lot and bedded confinement feedlot facilities. *The Professional Animal Scientist, 28*, 410–416

Pascagaza Rubio, H. D. (2023). Development of strategies to reduce pathogens in water troughs in cow-calf farms.

Rane, N., Choudhary, S., & Rane, J. (2023). Leading-edge artificial intelligence (AI), Machine Learning (ML), Blockchain, and Internet of Things (IoT) technologies for enhanced wastewater treatment systems. (October 31, 2023).

Shomar, B., Al-Darwish, K., & Vincent, A. (2020). Optimization of wastewater treatment processes using molecular bacteriology. *Journal of Water Process Engineering, 33*, Article 101030.

Schlink, A., Nguyen, M., & Viljoen, G. (2010). Water requirements for livestock production: A global perspective. *Revue Scientifique Et Technique, 29*, 603–619.

Sweeten, J. M., & Wolfe, M. L. (1994). Manure and wastewater management systems for open lot dairy operations. *Transactions of the ASABE, 37*, 1145–1154.

Sweeten, S. (2014). Manure and wastewater management for Cattle feed lots. *Res. Envi. Contam. Toxicol., 167*, 125–153.

Thao, N. T. P., Nga, N. T. T., Kim, H. T. T., Kien, T. T., Hieu, T. T., Thang, N. V., Tri, N. L. M., & Le, T. H. (2023). Combination of biochar filtration and ozonation processes in livestock wastewater treatment and application for soil cultivation. *Case Studies in Chemical and Environmental Engineering, 7*, Article 100286.

Ungureanu, N., Vlăduț, V., Istrate, I.-A., Zăbavă, B.Ș., Tociu, C., Ferdeș, M., Dincă, M. (2019). Advanced electrochemical treatment of the wastewater from cattle farm.

USEPA. (2004). Guidelines for water reuse. U.S. Environmental Protection Agency, Washington, DC, EPA/625/R-04/108 (NTIS PB2005 106542), 2004. Retrieved from: https://cfpub.epa.gov/si/si_public_record_report.cfm?Lab=NRMRL&dirEntryId=129543

Vaishnav, S., Saini, T., Chauhan, A., Gaur, G. K., Tiwari, R., Dutt, T., & Tarafdar, A. (2023). Livestock and poultry farm wastewater treatment and its valorization for generating value-added products: Recent updates and way forward. *Bioresource Technology,* 129170.

Vanderholm, D. H., Day, D. L., Muehling, A. J., Wang, L. K., Hung, Y.-T., Butler, E., Wang, M.-H. S., Yehya, H. (2022). Management and treatment of livestock wastes, In: Waste treatment in the biotechnology, agricultural and food industries: 1. Springer, pp 1–77.

Visvanathan, C., Ben Aim, R., & Parameshwaran, K. (2000). Membrane separation bioreactors for wastewater treatment. *Critical Reviews in Environmental Science and Technology, 30*, 1–48.

Wang, J., Song, A., Huang, Y., Liao, Q., Xia, A., Zhu, X., & Zhu, X. (2021). Domesticating Chlorella vulgaris with gradually increased the concentration of digested piggery wastewater to bio-remove ammonia nitrogen. *Algal Research, 60*, Article 102526.

Winchell, L. J., Wells, M. J. M., Ross, J. J., Fonoll, X., Norton, J. W., Jr., Kuplicki, S., Khan, M., & Bell, K. Y. (2022). Per-and polyfluoroalkyl substances presence, pathways, and cycling through drinking water and wastewater treatment. *Journal of Environmental Engineering, 148*(1), 3121003.

Ziara, R. M. M. (2015). Water and energy use and wastewater production in a beef packing plant. In Singh, N. K., Yadav, M., Singh, V., Padhiyar, H., Kumar, V., Bhatia, S. K., & Show, P.-L. (2023). Artificial intelligence and machine learning-based monitoring and design of biological wastewater treatment systems. *Bioresource Technology, 369*, 128486.

Management of Wastewater from Sea Food Processing Industries: A Review

Samson O. Akpotu, Simphiwe M. Nelana, and Fanyana M. Mtunzi

Abstract The management of partially treated or untreated wastewater from seafood processing industries presents considerable challenges due to the negative impacts it poses on the environment and public health. This review article is aimed at providing a comprehensive analysis of the management practices associated with seafood wastewater. As seafood processing plays a vital role in global nutrition and food security, the generation of substantial amounts of wastewater containing various pollutants such as organic matter, nutrients, and suspended solids poses a serious concern. This article explores the challenges faced in seafood wastewater management and discusses the pollution potential, characteristics, technologies, strategies, and best practices that can effectively address these challenges. Adopting a holistic approach, this review covers topics such as wastewater production, environmental impacts, regulatory frameworks, remediation methods, and valuable resource recovery in sustainable management of seafood processing wastewater. The objective is to promote the development of sustainable, efficient, and best practices for wastewater management in the seafood processing industry, ensuring the protection of the environment and public health.

Keywords Seafood processing · Wastewater management · Sustainable wastewater management · Environment impacts · Regulatory framework

S. O. Akpotu (✉) · S. M. Nelana · F. M. Mtunzi
Department of Biotechnology and Chemistry, Vaal University of Technology, Gauteng, South Africa
e-mail: akpotusamson@gmail.com

S. M. Nelana
e-mail: simphiwen@vut.ac.za

F. M. Mtunzi
e-mail: fanyana@vut.ac.za

N. A. Oladoja and E. I. Unuabonah (eds.), *Strategic Management of Wastewater from Intensive Rural Industries*, Springer Water,
https://doi.org/10.1007/978-3-031-90314-4_11

1 Introduction

Global population is on the rise with its attendant increase for food, water, and energy. Food sustainability and safety is crucial, and freshwater availability plays a significant role in achieving these goals. However, modern food production systems face two prominent challenges: excessive freshwater use and associated environmental hazards (Bhagwat, 2019). The seafood processing industry plays a vital role in meeting the global demand for fish and seafood products. However, the processing operations involved in transforming raw seafood into consumer-ready products generate significant amounts of wastewater containing various organic and inorganic pollutants. Processing operations in the seafood industry include washing, degutting, filleting, salting, icing, freezing, and canning which generate substantial quantities of organic-rich wastewater (Tay et al., 2004). Improper management of seafood wastewater can result in adverse environmental impacts, including water pollution, ecological disruption, and threat to public health. Therefore, it is imperative to implement effective management practices for wastewater from seafood processing industries to ensure sustainable and responsible practices in this sector.

Over the years, there has been a tremendous rise in global seafood production with an annual growth rate of 5.8% recorded from 2001–2006 (Jolly et al., 2023). Estimated seafood production in 2018 according to the FAO was approximately 180 million tons (Action, 2020). A substantial amount of this production was processed into value-added products such as canned goods, frozen food and fillets. Wastewater from seafood processing is characterised by its complex composition with high organic load (Venugopal & Sasidharan, 2021). It contains organic matter, nutrients, salts, chemicals, heavy metals, and suspended solids. However, the volume of wastewater produced depends on the size of the processor, specific processing operations, and water management practices. Typical wastewater generation rates range from 1 to 10 cubic meters per ton of seafood processed, with an average usage of approximately 5 cubic meters/ton (Tay et al., 2004). With global seafood production projected to continue rising to meet food demand, the environmental impacts of seafood processing wastewater are also becoming a growing concern. If not effectively managed, these pollutants can have detrimental effects on water bodies, ecosystems, and public health.

This chapter provides a comprehensive analysis of wastewater management from seafood processing industry, employing a multidimensional approach. The chapter further explores the application of an integrated framework for seafood wastewater management, wastewater generation, environmental impacts of wastewater, adherence to regulatory frameworks, treatment methodologies, and resource recovery. To attain these objectives, the exploration and implementation of innovative and sustainable approaches for seafood wastewater treatment and management are deemed vital. By conducting a critical evaluation of the current state of knowledge and identifying existing gaps, this chapter significantly contributes to the advancement of efficient and sustainable wastewater management practices within the seafood processing industry.

1.1 Objectives of Seafood Wastewater Management

The management objectives of wastewater from seafood processing industry may be perceived through three (3) distinct standpoints: environmental/human health, economic and social.

The primary objective of seafood wastewater management is to mitigate the environmental impact resulting from the discharge of wastewater from seafood processing industries. This includes various measures, such as reducing the pollution load in wastewater, preventing water contamination, preserving aquatic ecosystems, and upholding water quality standards (Show, 2008). The goal is to ensure that the released wastewater poses no environmental risks, including the receiving water bodies, marine life, and sensitive habitats. Furthermore, this is to safeguard public health by appropriately managing seafood processing wastewater. This involves minimizing the release of pathogens, antimicrobial compounds, and other contaminants that may potentially enter the food chain or pose harm to human health (Venugopal, 2021).

Two economic objectives are central to seafood wastewater management: (a) resource recovery (b) operational efficiency and (c) cost-effectiveness. The first objective primarily focuses on maximising the extraction of valuable resources from seafood processing wastewater. This involves the reclamation of nutrients, bioactive compounds, and organic matter for reuse in various applications. Resource recovery reduces waste and its environmental impacts and creates economic opportunities, such as producing biogas, biofertilizers, or bio-based products (Choudhury et al., 2022). The second and third objectives are focused on enhancing operational efficiency, optimizing energy usage, and achieving cost-effectiveness in seafood wastewater management. By implementing innovative treatment technologies, employing process optimization strategies, and adopting water conservation measures, the overall efficiency of wastewater management practices can be significantly improved (Tom et al., 2021).

The social objectives of seafood wastewater management encompass two key aspects: (a) social responsibility and sustainability and (b) adherence to regulatory compliance. The first social objective, which pertains to promoting sustainability and social responsibility, entails considering the broader social and environmental implications of wastewater management within the seafood processing industries. This involves actively engaging with stakeholders, cultivating positive community relations, and supporting initiatives that foster social well-being and contribute to the sustainable development of local communities (Tom et al., 2021). The second crucial social objective in the management of seafood processing wastewater is regulatory compliance. Complying with both national and international regulations, adhering to effluent standards, and securing necessary environmental permits are vital to meet legal requirements and prevent potential penalties or sanctions (Venugopal & Sasidharan, 2021). By strictly adhering to these regulatory frameworks, seafood-processing facilities operate within acceptable environmental limits, thereby promoting sustainable development.

Addressing these objectives ensures that the management of wastewater from seafood processing industries plays a pivotal role in the industry's long-term viability and environmental integrity. Moreover, it safeguards ecosystems, public health, and the welfare of local communities.

1.2 Characteristics of Seafood Industry Processing Wastewater

Seafood processing industries produce significant volumes of wastewater as a by-product of processing activities. The wastewater is often a complex matrix, and its composition may vary depending on the type of processing, water usage, management practices, the type of seafood being processed, and the technology used in processing. An understanding of the sources and factors influencing seafood wastewater generation is critical in the development of effective management strategies in seafood processing industries.

Wastewater derived from the seafood industry can be categorized into two distinct groups: (a) high volume and low strength, and (b) low volume and high strength. The former category comprises processing effluents from aquaculture farms, process water used for unloading, washdown, and fluming at various stages of seafood processing, including washing, cutting, peeling, filleting, and packaging (Kyllönen et al., 2021). The latter category encompasses wastewater resulting from different processing operations.

Effluents generated from fish production operations, such as blood or slime removal, generally have a saline nature, containing chlorides, sulphates, and sodium from soap, detergents, and disinfectants, all of which contribute to the seafood wastewater (Laly et al., 2019). Process effluent also contains a significant amount of organic matter in the form of total suspended solids (TSS), as well as nutrients like phosphorus and nitrogen, fats, oils, grease, and proteins (Ahmad et al., 2021). Additionally, the cleaning water employed in sanitation practices within seafood processing facilities potentially contains detergents, sanitisers, residual seafood waste, and microbial contaminants (Huss, 1994). Moreover, product rinsing is carried out to eliminate excess chemicals, salts, and other contaminants from the seafood products. The rinsing water may contain residual seafood waste, residual chemicals, and traces of microorganisms (Venugopal & Sasidharan, 2021).

The proteins present in the wastewater include myofibrillar proteins, collagen, enzymes, soluble peptides, gelatin, and volatile amines, which occur in colloidal, soluble, or particulate forms. Over time, the TSS in the effluent sediments can lead to the clogging of pumps and other equipment during handling. The fats, oils, and grease constitute the most objectionable components of the effluents, whose decomposition gives rise to unpleasant odours (Guerrero et al., 1998).

Processing waters in the herring industry are rich in prominent amino acids such as leucine, glutamic acid, and glutamine, as well as calcium and magnesium as the primary minerals (Forghani et al., 2023; Sar et al., 2021). Approximately, 15% of herring protein can potentially seep into cooking waters as waste, which presents environmental concerns. Studies have detected diverse microorganisms, including potentially harmful pathogens, in the effluents of the fish processing industry (Søtoft et al., 2015). The cooking of snow crab and processing of fish species such as pollock, cod, and salmon lead to seafood effluents containing high levels of total suspended solids (TSS) (Tremblay et al., 2020). In the process of marinating herring in brine, the herring absorbs approximately 33% salt and 60% acetic acid. Subsequently, the post-production brine contains about 6.4% to 7.3% dry matter, 1.4% to 2.6% protein, and 0.5% to 2.7% fat (Sar et al., 2021; Søtoft et al., 2015). Fish (tuna) processing wastewater contains BOD, TSS and FOGs in concentrations of 7–20 mg/L, 4–17 mg/L, and 2–13 mg/L, respectively. By-products like condensate and juices from tuna canning are rich in nitrogenous compounds, fats, oil, pigments, and minerals (Uttamangkabovorn et al., 2005).

Effluents from canning and marination processes contain 2–5% (w/v) salt, which may result to stress in microorganisms and affect treatment efficacy. Phosphorus and nitrogen are common causes of water quality impairment, resulting in high BOD and COD values, an indication that these substances negatively affect the oxygen balance (Søtoft et al., 2015). BOD values measure the oxygen required for microbial flora to oxidize organic matter, while COD measures the chemically oxidizable compounds, with COD levels generally higher than BOD_5, as they include a broader range of compounds that can be chemically oxidized. During summer, pathogenic microorganisms in effluents can be more prevalent, and improperly stored effluent with high nutrient content can lead to rapid bacterial growth and unpleasant odours. There is also concern about the potential presence of antibiotic-resistant organisms, posing risks to public health safety.

Carawan et al. (1986) carried out studies on the characterisation of effluent seafood processing plant and reported values of 100–800 mg/L for TSS, 40–300 mg/L for fat, oil and grease, 400–2000 mg/L for COD, 200–1000 for BOD. Fish meal plants were reported to have values of 70–20,000 mg/L for TSS, 150–42,000 mg/L for COD, and 20–5000 mg/L for fat, oil and grease (Islam et al., 2004). Jamieson et al. (2017) characterised shellfish and finfish effluent and found the concentration of BOD_5 (179–296 mg/L), COD (458–1717 mg/L), turbidity (28.8–88.3 NTU), total suspended solid (27.2–120.1 mg/L) and NH_3-N (1.5–12.9 mg/L).

1.3 Wastewater Generation in Seafood Processing Industry

Seafood processing industries generate significant volumes of wastewater as byproducts of various processes. The composition and characteristics of wastewater vary depending on the specific processes involved, the types of processed seafood, and the level of technological advancement in the facilities. Understanding the sources and factors influencing wastewater generation is crucial for developing effective management strategies for the seafood processing industries (Hayes & Gallagher, 2019).

1.3.1 Types and Sources of Wastewater

Wastewater in seafood processing industries can be categorized into several types based on its source and composition. The main sources of wastewater include process effluent, cleaning water, and product rinse water.

Process effluents are generated during different stages of seafood processing, including washing, sorting, cutting, peeling, filleting, and packaging. Process effluents contain organic matter, suspended solids, oils, fats, proteins, and various contaminants, such as bacteria, enzymes, and residual chemicals used in processing operations (Murali et al., 2021). Cleaning water is derived from sanitation practices in seafood-processing facilities. This requires the use of water for equipment and facility cleaning. Cleaning water can contain detergents, sanitizers, residual seafood waste, and microbial contaminants (Gall et al., 2004). Product rinsing is carried out to remove excess chemicals, salts, and other contaminants from seafood products. The water used in rinsing water may contain residual chemicals and seafood waste, and low concentration of microorganisms (Ghaly et al., 2013).

1.4 Factors Influencing Composition and Wastewater Volume of Seafood Effluent

Numerous factors affect the composition and volume of wastewater generated in the seafood processing industry. These factors include processing methods, seafood type, processing facility size and capacity, cleaning practices and technology.

2 Environmental and Health Impacts of Seafood Processing Wastewater

2.1 Water Pollution

Indiscriminate discharge of untreated or inadequately treated seafood processing wastewater can severely impact water bodies. This leads to water pollution in both surface and groundwater resources (Ching & Redzwan, 2017). The main 4 main aspects of water pollution due to seafood processing wastewater which are oxygen depletion due to nutrient load, toxic substances, pathogens and microorganisms, and reproductive and developmental disruptions.

Elevated organic loads in seafood processing wastewater can reduce oxygen levels in water bodies, causing hypoxic or anoxic conditions (Anh et al., 2021a). This brings harm to fish and aquatic organisms, impacting their survival and reproduction. Fish processing wastewater contains significant fish parts and blood, leading to high levels of dissolved and suspended organic matter. The fish processing sector also contributes substantial organic pollution and increased salinity to receiving waters (Ching & Redzwan, 2017). The excessive water consumption and discharge of high-strength post-processing wastewater in the fish processing industry are of global concern (Venugopal & Sasidharan, 2021). Additionally, soluble solids, though often overlooked, can affect aquatic life by reducing light penetration in water. Seafood processing wastewater contains high concentrations of organic matter (BOD/COD), suspended solids (SS), and nutrients such. These pollutants lead in water to eutrophication, causing excessive algal growth, oxygen depletion, and disruption of aquatic ecosystems.

Seafood processing effluent may contain toxic substances such as heavy metals, pesticides, and residual chemicals used in processing operations (Bashir et al., 2020). If these pollutants are present in elevated concentrations, they can have detrimental effects on aquatic life, impairing their growth, reproduction, and overall health (Austin, 1998). For example, the larval stages of some fish species and benthic invertebrates may experience high mortality rates or impaired growth and development in the presence of pollutants (Bashir et al., 2020).

Wastewater from seafood processing facilities can harbour microorganisms and pathogens including bacteria, viruses, and parasites (Show, 2008). If seafood effluent is discharged into water bodies without proper treatment, these pathogens can contaminate water sources, posing risks to human health and recreational activities (Ching & Redzwan, 2017). Certain elements within seafood processing wastewater, such as residual chemicals or endocrine-disrupting compounds, have the potential to disrupt the reproductive and developmental processes of aquatic organisms. These disruptions can lead to reproductive abnormalities, reduced fertility, and diminished survival rates among affected species (Ching & Redzwan, 2017).

2.2 *Ecological Consequences*

The discharge of untreated seafood processing wastewater can have a range of ecological consequences, significantly impacting the health and biodiversity of aquatic ecosystems. Biodiversity loss, disruption of aquatic habitats, bioaccumulation in food chain are some key ecological impacts associated with seafood processing wastewater (Ahmad et al., 2022).

Water pollution caused by seafood processing wastewater may have detrimental effects on aquatic flora and fauna. This causes a decline in biodiversity. Increased nutrient levels, oxygen depletion, and the presence of toxic substances can disrupt the ecological equilibrium, reducing the abundance and diversity of aquatic species (Ching & Redzwan, 2017). Eutrophication and oxygen depletion caused by seafood processing wastewater can result in the degradation and loss of crucial aquatic habitats, such as wetlands, coastal zones, and estuaries. These habitats serve as vital breeding grounds, feeding areas, and shelters for various aquatic species, and their deterioration can have severe ecological consequences (Iber & Kasan, 2021).

An increase in nutrient concentrations of seafood effluent has been observed due to increased farming period. Seafood effluent have been known to contain elevated amounts of chloride from brine solutions and processing wastewater and organic nitrogen (0–300 mg/L) (Chowdhury et al., 2010). Seafood effluent have been known to have BOD, suspended solids, alkaline and acid concentrations of 200,000 mg/L, 120,000 mg/L, pH 11 and pH 3.5, respectively (Carawan, 1991{Park, 2001 #171). Such concentrations have been known to have profound effects on microbes present in water. Casé et al. (2008) noticed that Copepoda and diatom were replaced by rotifers, cyanobacteria and protozoans. This change was attributed to the increase in nutrient concentrations of effluent from a Pacific white shrimp farm over a period which greatly affected the type and density of plankton.

Pollutants present in seafood processing wastewater can enter the food chain through uptake and accumulation in aquatic organisms. This bioaccumulation poses risks to higher trophic levels, including humans consuming contaminated seafood products, leading to potential health concerns-Effective management of seafood processing wastewater is imperative for mitigating the environmental and health impacts of seafood effluents. The implementation of efficient treatment methods and strict adherence to rigorous regulatory standards are essential in minimizing pollution, preserving water quality, and safeguarding aquatic ecosystems.

2.3 Public Health Risks and Human Exposure

There are health risks associated when humans consume or exposed to water contaminated with seafood effluent. This includes the risk of contracting waterborne diseases and the contaminated water may be unsafe for recreational purposes. Furthermore, workers in seafood processing plants may encounter occupational health risks due to direct exposure to wastewater. Contact with contaminated water, aerosols, or bioaerosols during processing operations may increase the likelihood of skin and respiratory infections, allergic reactions, and other health issues (Liney et al., 2006).

Inadequate wastewater management practices can result in the infiltration of pollutants into groundwater sources. This may lead to the contamination of drinking water supplies, and subsequently exposing individuals to dangerous chemical substances (Ahmed, 1991). Furthermore, there is a risk of chemical contaminants bioaccumulating from seafood processing wastewater bioaccumulating in aquatic organisms, including commercially harvested seafood. The consumption of such contaminated seafood products can result in the ingestion of harmful substances, leading to adverse health effects in humans (Wahiduzzaman et al., 2021). Occupational exposure in workers involved in seafood processing may be through direct contact or inhalation which may be detrimental to health. The use of cleaning agents, disinfectants, and processing chemicals may cause a health risk for workers if adequate safety measures and personal protective equipment are not used (Mobo et al., 2010).

Addressing these contamination concerns is of utmost importance in a bid to protect public health, ensure worker safety, and safeguard the environment from potential adverse impacts of seafood effluent. The implementation of efficient treatment methods and adherence to regulatory standards play pivotal roles in maintaining the safety and integrity of seafood products.

3 Regulatory Framework and Compliance

Wastewater management from seafood processing industries, pertaining to environmental pollution, falls under the purview of multiple national and international regulations and standards. The primary objective of these regulatory frameworks is to ensure the sustainable development of the seafood processing sector, while safeguarding environmental and public health. The implementation of these regulatory frameworks occurs at both national and international levels. Different nations have formulated specific guidelines and regulations to govern the management of wastewater from seafood processing industries. These regulations encompass provisions such as seafood wastewater discharge permits, monitoring prerequisites, and enforcement mechanisms.

Several nations have instituted specific effluent standards for seafood processing industries to safeguard the quality of the discharged wastewater. These standards generally delineate limits for various parameters, encompassing biochemical oxygen demand (BOD), total suspended solids (TSS), nitrogen, phosphorus, and pH values. Seafood processing facilities are obligated to adhere to these specified standards. For instance, in the United States, the Environmental Protection Agency (USEPA) exercises authority over seafood effluent regulation via several legislative instruments. These include the Clean Water Act (CWA), the National Pollutant Discharge Elimination Systems (NPDES) permit program, and the Seafood Processing Effluent Guidelines and Standards established in 1974 and 1975. The CWA necessitates the complete cessation of pollutant discharge into U.S. waters and has set quality standards for surface water. The NPDES permit outline limits on pollutant discharges, mandates monitoring and reporting requisites, and ensures adherence to effluent standards (https://www.epa.gov/npdes). In Canada, Fisheries and Oceans Canada serves as the federal authority responsible for the protection of aquatic environments and the management of fisheries, as well as oceanic and freshwater resources (Ayles et al., 2016). This organization plays a critical role in fostering the health and sustainability of these ecosystems. Within the European Union, the Industrial Emissions Directive (IED) and the Water Framework Directive (WFD) delineate standards and stipulations for wastewater discharge emanating from seafood processing industries. These directives are designed with the objective of ensuring the ecological health of water bodies and managing pollutant emissions originating from industrial activities, inclusive of seafood processing by the European Commission (Directive, 2000).

In Australia, the national water quality guidelines, inclusive of those specifically pertinent to seafood processing wastewater, have been established by the Australian and New Zealand Environment and Conservation Council (ANZECC) in collaboration with the Agriculture and Resource Management Council of Australia and New Zealand (ARMCANZ) (ANZECC & ARMCANZ) (https://www.waterquality.gov.au/anz-guidelines/resources/previous-guidelines/anzecc-armcanz-2000).

International organisations have formulated standards to foster sustainable wastewater management within seafood processing industries. These standards act as referential frameworks for national regulations and facilitate international cooperation to tackle shared environmental issues. The Food and Agriculture Organization (FAO) of the United Nations, for instance, has conceived the "Code of Conduct for Responsible Fisheries". This directive offers insights into responsible fisheries practices, which includes the management of seafood processing wastewater to reduce environmental impacts and assure food safety (www.fao.org/3/v9878e/v9878e.pdf). Additionally, the International Organization for Standardization (ISO) has promulgated the ISO 14001 standard, presenting a framework for environmental management systems (EMS). Enterprises within the seafood processing sector can utilize ISO 14001 to institute and uphold efficacious environmental management practices, encompassing wastewater management (https://www.iso.org/iso-14001-environmental-management.html). In 2018, the Global Agriculture Alliance Best Aquaculture Practices Seafood Processing Plant Standard set the threshold for seafood effluent with the following parameters: pH (6.0–9.5), TSS (200 mg/L), soluble phosphorus (< 10 mg/L), total ammonia nitrogen (20 mg/L), 5-day BOD (< 500 mg/L), oil and gas (< 30 mg/L).

In Africa, countries are faced with numerous challenges which hampers effective resource management, policy formulation, enforcement of regulatory framework and compliance on seafood effluent discharge. This could be attributed to weak institutional capacity, limited human and financial resources, inadequate governance oversight, fragmented policies and limited stakeholder participation, incoherent policies and overlapping mandates across government agencies which causes conflicting regulations on effluent discharge. However, there are some regional efforts aimed at aquaculture policy framework regulation. The Pan-African Fisheries and Aquaculture Policy Framework is aimed at strengthening institutional capacity, promoting good governance and the harmonisation of policies across regional economic zone and regional fishery bodies for sustainable aquaculture development which includes the management of seafood effluent. For example, the East African zone has a Code of Hygienic Practice responsible for setting standards for processing, handling, and storage of fish/aquaculture related products, including setting the guidelines for wastewater treatment and discharge (https://law.resource.org/pub/eac/ibr/eas.62.2.2000.html). Furthermore, there are specific country with regulatory framework. The three largest countries engaged in aquaculture in Africa is South Africa, Nigeria, and Egypt. South Africa has recently developed aquaculture legislation led by the Department of Environment, Forestry and Fisheries (DAFF) in a bid to streamline the fragmented regulatory framework (from 13 government agencies) and provide enabling environment for seafood wastewater management (Adeleke et al., 2020). In

Egypt, the ministry of Agriculture and Land Reclamation is saddled with the responsibility of managing aquaculture and fisheries sector, whereas General Authority for Fish Resources Development (GAFRD) is responsible for planning, controlling and implementing policies on aquaculture and fish industries (Adeleke et al., 2020; Macfadyen et al., 2012). However, in Nigeria there is no specific legislation on aquaculture; however, the Inland Fisheries Decree of 1992 task the Minister in charge of fisheries matter to regulating aquaculture (Adeleke et al., 2020).

Adherence to both national and international regulations and standards is a prerequisite for seafood processing facilities to attenuate the environmental and health ramifications of wastewater and guarantee sustainable operations. This compliance not only underscores their commitment to responsible environmental stewardship but also aligns with societal expectations.

4 Conventional Wastewater Treatment Methods

As a result of standards set by regulatory agency, seafood effluents are subjected to treatment before being discharged into receiving water bodies. The selection of appropriate water treatment depends on effluent type, properties, and volume (Gärde, 2011). Effluents are subjected to primary pretreatment, which is mainly a physical process, other treatment method include chemical, biological, and advanced treatment technology (Fig. 1).

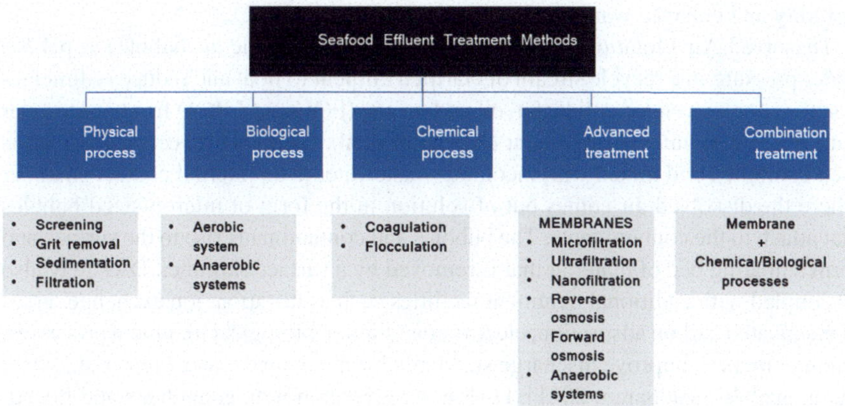

Fig. 1 Seafood effluent treatment methods

4.1 Physical Processes

Conventional wastewater treatment methods employ various physical processes to remove contaminants from wastewater generated by seafood processing industries. These physical processes aim to separate the solid and liquid components, reduce suspended solids, and remove certain pollutants through physical means. The following are the important physical processes used in wastewater treatment (Razzak et al., 2022).

The primary stage of wastewater treatment involves the process of screening, during which wastewater is allowed to pass through specialized screens designed to eliminate sizable particles and solid entities, such as shells and bones, greater than 0.7 mm. This process helps to avert clogging and prevent damage to downstream equipment (Cheremisinoff, 2019). For instance, fat and grease may form hydrophobic sludge with lipids which clog pipes and hinder the sedimentation process. This necessitates the use of a decanter and a sludge blanket to rapidly remove settled sludge (Abomohra et al., 2020).

Immediately after the primary screening is the phase that involves eliminating grit and segregating inorganic particles from wastewater (Chahal et al., 2016; Englande Jr et al., 2015). This is carried out through centrifugation or sedimentation, enabling denser particles to settle before the grit is disposed or additional treatment (Yan et al., 2014). Sedimentation employs gravity to settle suspended solids in settling tanks or clarifiers (Mays, 2013). After sedimentation, filtration is applied to eliminate finer suspended solids that might have eluded prior treatment stages. Porous filtration media, including membranes, activated carbon, and sand, are used to decrease turbidity and enhance water quality (Gholikandi et al., 2012).

Dissolved Air Flotation (DAF) systems uses microscopic air bubbles at pH 5.0 under pressure in a recycle stream of clarified effluent to float and reduce sedimentation time for suspended solids, fat, oil and grease (FOG), and BOD from wastewater within a few minutes of treatment (Abuhasel et al., 2021). This recycled stream is then combined and mixed with incoming wastewater in an internal contact chamber where the dissolved air comes out of solution in the form of micron-sized bubbles that attach to the contaminants. The bubbles and contaminants rise to the surface and form a floating bed of material that is removed by a surface skimmer. DAF may also be coupled with additional treatment facilities such as filtration, ion exchange, granular activated carbon adsorption, electro-coagulation, biological treatment processes, among others to improve discharge standards for the treated water (Teh et al., 2016; Venugopal & Sasidharan, 2021). DAF in combination with coagulants and flocculants can remove up to 50% of TSS and 80% of FOG. DAF system can be started and shut down easily to accommodate fluctuations in seafood processing (Venugopal & Sasidharan, 2021).

4.2 Biological Processes

Biological processes utilise anaerobic, aerobic and/or microorganisms to degrade organic pollutants in seafood effluent, transforming the pollutants to less harmful substances (Chowdhury et al., 2010).

Aerobic systems are dependent on oxygen to support microorganism (bacteria and fungi) growth. This method involves the use of aerobic biofilter lagoons, trickling filters, activated sludge processes, rotating discs, and sequential batch reactors (SBRs) to degrade organic pollutants. Microbial communities drive these processes to biochemically degrade organic pollutants. In activated sludge systems, an optimized mix of microorganisms reduces organic matter in seafood wastewater using dissolved oxygen and nitrogen compounds, substantially lowering nutrients, FOG, and soluble BOD_5 concentration (Chowdhury et al., 2010). However, this technique is energy intensive which may increase cost (Parvathy et al., 2017).

Anaerobic treatment is a sustainable method for seafood effluent treatment without oxygen (Rashidi & Moghaddam, 2021). This process utilises anaerobic microbes to break down organic matter into biogas and sludge. It includes hydrolysis, liquefaction, fermentation, acetic acid and hydrogen formation, and eventually methane production through acetate and CO_2 reduction by hydrogen. Treatment efficacy depends on factors like pH, pretreatment, organic loading, temperature, mixing, nitrogen concentration, and carbon-to-nitrogen ratio (Chowdhury et al., 2010; Sillapacharoenkul & Sinbuathong, 2020). In a study by Ching and Redzwan (2017), lab-scale aerobic bioreactors treated saline fish processing wastewater. Three wastewater concentrations with 0–7% NaCl content were tested, with optimal conditions at 8–10 day HRTs. Lower HRTs led to decreased biomass growth, substrate and pollutant removal. Above 2.5% NaCl, inhibition occurred, reducing biomass growth, TSS, and pollutant removal, notably COD. At 1% NaCl, up to 86% COD and 73% NH_3-N removal were achieved at 10-day HRT with 30 times diluted wastewater. Aerobic bioreactors effectively treated wastewater up to 2.5% salinity, but performance declined at higher salt levels.

Anaerobic systems, including fluidized bed reactor (FBR), anaerobic filter (AF), up-flow anaerobic sludge blanket reactor (UASBR), anaerobic biochemical reactor (ABR), and more, can be tailored as needed (Rashidi & Moghaddam, 2021). These systems offer benefits like minimal sludge generation, high organic loading capacity, methane-rich biogas production, process stability, reduced nutrient requirements, and cost-effective operation. Notably, processes like Ultrasonic-assisted Membrane Anaerobic System (UASBR), AF, and AFBR can achieve 80–90% organic compound removal while generating gas. This aids in lowering organic load and generating renewable energy (Chowdhury et al., 2010).

In a study by Najafpour et al. (2006), a 3-stage rotating biological contractor (RBC) was used to treat wastewater from food canning industries. A 5-day batch test resulted in 96.4% BOD removal and increasing hydraulic retention time from 24 to 48 h improved COD removal efficiency from 85.3% to 97.4%. Enhanced disc submergence from 31 to 36% increased COD removal efficiency from 74.9% to

87.5%. Lower submergence of 23.7% improved removal efficiency due to better flow. A rise in rotational speed from 3 to 11 rpm enhanced COD removal from 62.7 to 93.7% by boosting oxygen transfer.

Jemli et al. (2015) used a membrane bioreactor (MBR) and a continuous stirred tank reactor (CSTR) for treating saline wastewater from a fish canning factory. MBR achieved 97.9–98.6% TOC removal efficiency and over 99.9% BOD removal at organic loading rate (OLRs) of 0.5 to 1.4 kg/m^3/day. MBR outperformed CSTR in organic matter removal and biomass production. Increasing OLR in CSTR led to poor sludge settleability. MBR showed diverse microbial communities, including halotolerant taxa. High OLRs increased membrane fouling, necessitating more frequent cleaning. MBR effectively treated high salinity wastewater, surpassing CSTR but facing membrane fouling challenges.

These biological methods can be applied individually or combined based on seafood processing wastewater attributes and desired effluent standards. They provide efficient and eco-friendly approaches to diminish organic pollutants, eliminate nutrients, and enhance overall discharged effluent quality.

4.3 Chemical Processes

Chemical processes are extensively utilised in the treatment of seafood effluent to eliminate pollutants and enhance water quality of seafood processing industry (Venugopal & Sasidharan, 2021). These methods utilise chemicals to facilitate separation, precipitation, or chemical conversion of contaminants. Key chemical processes include:

Coagulation and flocculation are widely used in wastewater treatment (Teh et al., 2016). Coagulants like aluminium sulphate or ferric chloride when added to seafood effluent neutralize particle charges on suspended particles, as well as promoting aggregation. Subsequent addition of flocculants, such as polyacrylamides, forms larger flocs for easier settling or filtration (Lee et al., 2014). pH adjustment optimizes treatment processes and removes specific pollutants. Altering pH enhances metal removal via precipitation or co-precipitation (Prokkola et al., 2020). The pH of seafood effluent is adjusted by adding acid or base (Venugopal & Sasidharan, 2021). In the precipitation process, insoluble solid precipitates are formed from soluble contaminants. Precipitants like calcium hydroxide or sodium hydroxide when added to seafood effluent forms precipitates with pollutants such as heavy metals or nutrient. Thereafter, the precipitates are separated through sedimentation or filtration (Chi et al., 2018). Oxidation and reduction are utilised to transform pollutants into less harmful forms. Chemical oxidants, like chlorine or hydrogen peroxide, degrade organic pollutants or applied in disinfection of seafood wastewater. Sulphites or ferrous salts can be used to eliminate certain pollutants from wastewater and convert those contaminants to insoluble forms (Matei et al., 2022). Chlorination has been applied in the disinfection of wastewater using sodium hypochlorite or chlorine gas to kill pathogens or microbes in wastewater (Kesar & Bhatti, 2022). Thereafter, dichlorination with sodium bisulphite or activated carbon is used to remove the residual chlorine to reduce the environmental impacts on the receiving water body. These chemical methods are utilised in secondary or tertiary treatment stages in wastewater plants. They enhance contaminant removal and improve effluent quality before discharge or reuse (Venugopal & Sasidharan, 2021).

4.4 Advanced Treatment Techniques

Membrane filtration represents an advanced treatment technique employing semi-permeable membranes to selectively separate and eliminate organic contaminants. This method is particularly valuable in the treatment of wastewater derived from seafood effluent (Islam et al., 2004). The are four primary membrane-based separation processes encompass microfiltration (MF), ultrafiltration (UF), nanofiltration (NF), and reverse osmosis (RO). The main difference exhibited by these processes, apart from their pressure requirements, is their membrane pore sizes (Obotey Ezugbe and Rathilal, 2020). These processes rely on hydraulic pressure to achieve separation. Membrane filtration have been effectively applied for the removal of pathogens, bacteria, endotoxins, and viruses (Ahmed et al., 2023). Membrane filtration technology offers the distinct advantage of affordability, featuring low costs and minimal energy consumption (Venugopal & Sasidharan, 2021).

MF utilises a process that uses pore size between 0.1 and 10 microns, pressure of 1–3 bar, molecular weight cut-off (MWCO) of 100–500 kilo Dalton and an average permeability of 500 (L/m^2h bar) (Obotey Ezugbe and Rathilal, 2020). It is efficient in the removal of bacteria, fat, oil and grease, organics, large colloidal particles and suspended solid from wastewater. The MF membrane is porous, asymmetric, or symmetric. MF membranes are used as a pretreatment step to protect downstream processes or for the production of reuse water (Alzahrani & Mohammad, 2014). Anh et al. (2021b) applied MF for the treatment of saline wastewater from a fish processing plant. The results showed that MF can effectively remove bacteria, suspended solids, and other particles from the wastewater. This makes it a promising option for the treatment of saline wastewater from fish processing. A study by Kuca and Szaniawska (2009), investigated the efficacy of microfiltration techniques using a 150 kDa ceramic membrane with BSA model solutions, NaCl, and salted fish industry effluent. They explored parameters for brine regeneration. pH, NaCl concentration, and pressure difference effects on BSA rejection and permeate flux were assessed at 20 °C and 4 m/s cross-flow velocity. Industrial fish brine, pre-treated with 100 and 1 mm bag filters, achieved high protein (81%), BOD (72%), and COD (60%) rejection without permeate recycling.

In ultrafiltration (UF), membrane with smaller pore sizes ranging from 0.001 to 0.1 microns are employed. Pressure is typically between 2 and 5 bar, MWCO of 20–150 kilo Dalton and an average permeability of 150 (L/m^2h bar) (Obotey Ezugbe and Rathilal, 2020). UF membranes are microporous and asymmetric (Muro et al., 2012).

UF efficient in the removal of organics, pigments, sugar and recovering significant quantities of protein. UF provides efficient separation, while retaining compounds, making it suitable for seafood effluent treatment and water reuse . Benhabiles et al. (2013) utilised UF to recover protein from shrimp shell. Analysis of the result revealed that up to 96% of the proteins was removed from shrimp shells through incubation in NaOH, and a solute rejection coefficient of 97% was obtained in the UF process to recover proteins from deproteinized shell wastewater. Furthermore, chemical oxygen demand of the permeate was reduced by 87%. Gringer et al. (2015) investigated the efficiency of ceramic ultrafiltration in recovering high-value biomolecules such as proteins, fatty acids, minerals, and phenolic compounds from two herring spice brines. The retentates contained 75–82% of the protein and 75–100% of the fatty acids compared to the level in the initial brines. The ceramic membranes retained up to 42% COD, > 95% TSS, and > 85% iron, and the permeates generated were fat-free and contained approx. 2% of the proteins compared to the unfiltered brines. In another study by Mota et al. (2022) the performance of a capillary polyethersulfone ultrafiltration (UF) membrane was evaluated for the removal of two benchmark waterborne fish pathogens, the infectious pancreatic necrosis virus (IPNV) and the bacterium Aeromonas salmonicida, from aquaculture water. The UF membrane bench-scale unit was tested at two temperatures according to salmonid aquaculture: low (4–7 °C) and high (16–19 °C). Both pathogen solution and permeate were evaluated using a cell culture method to estimate the colony-forming units (CFU/ml) for bacteria presence. The membrane permeate was negative for both virus and bacteria for all analysis and for both low and high temperatures. The results from this bench-scale study were positive and encouraged the application of UF membrane technology in aquaculture water treatment to prevent virus and bacteria outbreaks.

In nanofiltration (NF), the membrane has an even smaller pore size between 0.001–0.01 microns. The pressure is between 5 and 15 bar, MWCO of 2–20 kilo Dalton and an average permeability of 10–20 (L/m^2h bar) (Obotey Ezugbe and Rathilal, 2020). NF membranes are made up of tight porous, thin composite film and are asymmetric in nature (Mota et al., 2022). NF membranes effectively remove divalent anions and cations such as magnesium, sulphate, and calcium, as well as lactose, sucrose, and pigments. Additionally, it allows the passage of smaller organic compounds and monovalent ions (Mota et al., 2022). NF have potential application in water softening, wastewater treatment, vegetable oil and other industries (Venugopal & Sasidharan, 2021). Yeong et al. (2002) investigated the potential application of NF membranes in the treatment of wastewater from fish and surimi industries. Additionally, the study explored the possibility of recovering protein from wastewater, which could be concentrated by utilising. A polyamide NF membrane of 500 Da was used to treat fish and surimi washing wastewater generated in the laboratory and subjected to pre-treatment. Permeation experiments showed that NF was capable of reducing COD and TSS up to 93% and 87%, respectively. The study also found that the polyamide NF membrane fouled much more slowly than other membranes. A study carried out by (Ali et al., 2020) utilised NF to treat wastewater from anodised bath. The water was treated for colour rinse, alkaline pickling rinse, acidic pickling rinse, and anodizing rinse, as well as a mixture of all the wastewater streams. A commercial membrane (NF99HF) with water permeability of 10 L/(m^2·h·bar) was used. The nanofiltration of the colour rinse offered high-quality permeate with heavy metals below the detection limit and high permeability, whereas the nanofiltration of the alkaline pickling rinse exhibited no permeability. The NF of the neutralized mixed wastewater, after the removal of precipitate, produced high-quality permeate with a stable permeability of 1 L/(m^2·h·bar).

In reverse osmosis (OS), the membrane is made up of tiny pores of less than 0.001 microns. The pressure is between 15 and 75 bar, MWCO of 0.2–2 kilo Dalton and an average permeability of 5–10 (L/m^2h bar) (Obotey Ezugbe and Rathilal, 2020). NF membranes are made up of semi porous asymmetric thin film composite (Mota et al., 2022). RO membranes can remove almost all contaminants including organic compounds, microorganisms, and dissolved salts from wastewater. This technology produces high-quality water that is suitable for various applications, including water reuse or discharge in sensitive environments. Liu et al. (2007) used wind-driven RO system for aquaculture wastewater treatment. This technology ensured that a fish tank becomes a closed aquaculture production system with zero waste discharge. The freshwater generated from the RO system can be utilised as freshwater for fish culture, also, the brine was further processed into fish feed by a duckweed-covered reactor.

Forward osmosis (FO) is a membrane-based separation process that is becoming more popular for wastewater treatment due to its ability to operate without external pressure, utilize osmotic gradients, cost effectiveness and be less vulnerable to fouling (Lutchmiah et al., 2014). Recent studies have evaluated the use of FO in wastewater treatment and highlighted its potential in the removal of suspended solids, organic matter, and other contaminants (Korenak et al., 2017). FO has also been identified as a promising technology for high water recovery of proteins from seafood processing wastewaters. However, optimization of parameters is essential to increase FO performance and prompt its feasibility in wastewater treatment. The focus lies with new membrane developments, draw solutions to enhance wastewater treatment and energy recovery, and operating conditions. Fouling mitigation, decreased concentration polarization, and increased FO performance are essential to prompt FO-wastewater feasibility (Goh et al., 2019).

4.5 Combination Treatment Process

Traditional single-step treatment methods often struggle to meet stringent environmental regulations, and in most situations, they may not be effective for successful treatment of a particular effluent while achieving resource recovery goals. For this reason, combination treatment approaches, integrating multiple technologies to enhance treatment efficiency, reduce cost, produce high-quality effluents suitable for reuse and minimize environmental impact.

In a study by Achour et al. (2000), an integrated system consisting of a decanter (to remove fats and oil), anaerobic digester, and aerated sludge bio-reactor effectively reduced tuna processing effluent COD by 95%. This was achieved this with low energy use and minimal sludge production, whereas individual processes had limited COD reduction success. Mannacharaju et al. (2020) utilised a series of reactors, including a baffled moving bed biofilm reactor (BMBBR), up-flow anaerobic sludge blanket (UASB) reactor, fluidized immobilized cell carbon oxidation (FICCO) reactor, and chemoautotrophic activated carbon oxidation (CAACO) reactor in the treatment of fish processing wastewater. Exceptional removal efficiency of COD at 99 ± 0.1%, protein at 99 ± 0.2%, lipids at 100%, and oil and grease at 100% was obtained.

Recent development in integrated seafood wastewater treatment involves clean energy generation as a by-product of seafood wastewater treatment. A study by Jamal et al. (2020) used an air cathode microbial fuel cell (ACMFC) reactor to in the treatment of halophiles at different organic load (OL) concentrations. The results showed that the halophiles in ACMFC effectively removed total chemical oxygen demand (TCOD) and total suspended solids (TSS) from seafood processing wastewater and produce electricity. Nookwam et al. (2022) used microbial fuel cells (MFCs) fed with anaerobic digester effluent from the rubber industry. Power density at 55.43 ± 1.08 W/m^3 was recorded, while simultaneously removing COD, nitrogen, and phosphorus by 72.4 ± 0.9%, 40.5 ± 0.8%, and 24.4 ± 1.5%, respectively. Optimizing the organic loading rates and introducing photosynthetic oleaginous microalgae in the cathodic chamber, the MFCs achieved a power density of 116.9 ± 15.5 W/m^3, sequentially treated wastewater, and produced microalgal biomass (465 ± 10 g/m^3) with a high lipid content of 38.17 ± 0.01%.

Membrane processes can be integrated with conventional methods for better efficiency in the treatment of seafood effluent. The treatment may combine either biological or chemical methods. Sridang et al. (2008) evaluated the performance and microbial community structure of a submerged membrane bioreactor (MBR) for seafood processing wastewater treatment and reuse. The MBR showed good efficiency in treating high-strength wastewater with organic loading variation characteristic of real seafood processing wastewater. Analysis of the results showed the importance of ecological microbial compositions in each stage of development that were different in MBR in comparison with conventional activated sludge. The MBR sludge was stress tolerant to salinity and variation of COD and suspended solid. Furthermore, high removal efficiencies in terms of BOD and COD were recorded. Thongsai et al. (2022) investigated the use of anaerobic membrane bioreactor (AnMBR) as a secondary treatment unit for anaerobically treated wastewater from a full-scale digester of a fish canning factory. An analysis of the results showed reactors achieved COD removal greater than 84%, while nitrogen removals values of 31.5–41.5% were achieved without biological nitrification/denitrification. The overall membrane fouling rate was effectively controlled in higher intermittent liquid circulation conditions. A sizable energy saving with the anaerobic membrane process was recorded when compared to conventional activated sludge. A study by Cristóvão et al. (2015) assessed the reuse of fish canning industrial wastewater by combining conventional treatments followed by a polishing step by reverse osmosis (RO) and ultraviolet (UV) disinfection. The overall removal efficiencies obtained for oil and grease, suspended solid, dissolved organic carbon, anions and bacteria were high. The final clarified effluent was found to have the quality requirements to be recycled or reused in an industrial plant which allows for the reduction in effluent quantity to be discharged. Chung et al. (2023) investigated the use of microbubbles as alternative cleaning agents for MF membranes used for food oily wastewater treatment. The incorporation of microbubbles into NaOH solution for cleaning increased the flux recovery by 235%, and MB-assisted cleaning proved a promising technology for integration into current clean-in-place (CIP) operations to improve their cleaning efficiency with reduced water and chemical usage.

Overall, membrane filtration is a promising technology for the treatment of seafood processing wastewater due to its compact system design, high removal efficiency, and potential for water reuse. However, there are numerous challenges associated with membrane, such as fouling, scaling, and operational costs. Membrane fouling can reduce their effectiveness and increase operational costs; hence, it is essential to optimise operating conditions and cleaning protocols to mitigate fouling. Despite these challenges, membrane filtration has is a viable option for the treatment of seafood processing wastewater.

4.6 Resource Recovery from Seafood

Seafood effluents have been known to contain valuable resources with potential applications in allied, pharmaceutical and food industries (Venugopal & Sasidharan, 2021). The "circular economy biorefinery concept" aims to promote sustainable production of bioenergy, biochemicals, biofuels, and other bioproducts by reusing and recycling waste streams (Fig. 2). This concept reduces an industry's carbon footprint and promotes economic growth (Choudhury et al., 2022). The sustainability of seafood processing depends on the recognition of solid seafood waste and wastewater as valuable resources containing different ingredients, energy, and water. Seafood effluent treatment options is primarily designed to mitigate environmental risks, as well as offer prospects for recovering valuable bio-compounds with potential applications in nutrition, healthcare, pharmaceuticals, and related sectors. While both valorisation processes and waste treatments yield comparable effects, the substantial advantages lie in the valorisation of fish by-products (Venugopal & Sasidharan, 2021).

Solid seafood waste is known to contain high value resources such as minerals, fish oil, chitin, peptides, proteins, lipids, polysaccharides, gelatin and collagen. Nitrogen and phosphorus are the major dissolved nutrients which are present in seafood effluents. Several techniques have been developed to extract these compounds such as the use of membranes, enzymes, fermentation, microwave, ultrasound and supercritical fluid assisted extraction (Goh et al., 2022).

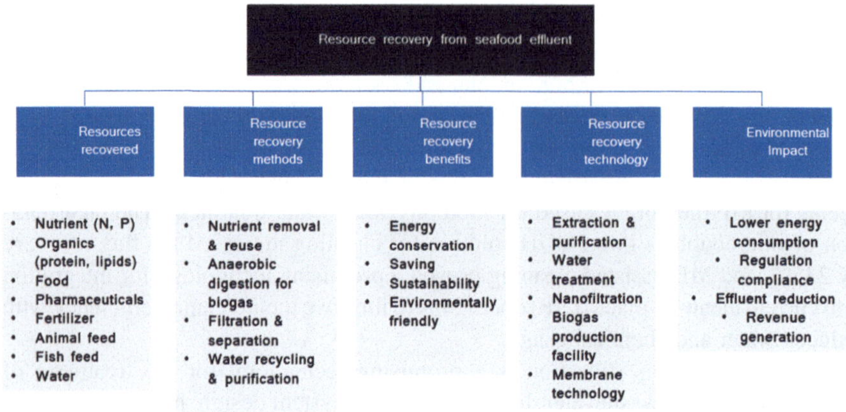

Fig. 2 Life cycle approach to resource recovery from seafood effluent

4.6.1 Lipids

Lipids recovery from seafood wastewater has significant economic and environmental implications. Lipids, including fats and oils are valuable products that can be used in various industries such as food, pharmaceuticals, and cosmetics. Fish oils are rich in omega-3 PUFA and vitamins A and D. Certain fish species can have up to 35% omega-3 fatty acids, with EPA and DHA at around 10% of the oil (Venugopal, 2008). Fish wastes, specifically the livers of albacore, cod, salmon, shark, haddock, and tuna, are good sources of oil. Lipids, can be extracted from seafood effluents by centrifugation, solvent extraction, biological methods, electrocoagulation, freezing and thawing, low temperature crystallisation, enzymatic methods, urea complexation, alkaline hydrolysis, supercritical fluid extraction, microwave-assisted extraction and molecular distillation (Venugopal, 2021). The choice of extraction depends on the characteristics of the seafood effluent, recovered lipids purity, cost, and environmental issues.

4.6.2 Protein

Protein is an important nutrient source, and it is in high demand to meet nutritional needs. For this reason, protein recovered from seafood effluent must be safe for consumption and must meet the required environmental regulation (Henchion et al., 2017). However, based on the complex matrix of seafood effluent, protein recovery is tedious and little quantities are isolated during recovery. Numerous techniques can be used to recover protein from seafood wastewater, nonetheless, membrane technology (MF and UF) coupled with flocculation methods are mostly used. Other methods for protein recovery from seafood wastewater includes chemical precipitation, enzymatic hydrolysis, electro-flocculation, microbial conversion, acid and alkaline hydrolysis (Forghani et al., 2020).

4.6.3 Carotenoids

Carotenoid recovery from seafood wastewater is an environmentally friendly and economically valuable process. Carotenoids, like astaxanthin, can be extracted from seafood wastewater using fish scales waste as a natural adsorbent (Stepnowski et al., 2004). Other methods such as solvent extraction, enzymatic hydrolysis, and membrane separation, have been employed to extract carotenoids from seafood processing effluents. It protects cells from oxidative stress (Venugopal & Sasidharan, 2021). Additionally, UF membrane processes have been explored as an eco-friendly and technically efficient to isolate, concentrate and recover carotenoids from seafood effluent (Pérez-Santín et al., 2013).

4.6.4 Flavours Compound Recovery

Seafood aromas consist of mixtures of volatile, low-molecular-weight organic compounds (Bourseau et al., 2014). These aromas are primarily composed of alcohols, aldehydes, ketones, and compounds containing nitrogen and sulfur. Some of these aromas can contribute to undesirable tastes and odours (Martínez et al., 2014). Lipid autoxidation and the thermal degradation of fatty acids results primarily into aldehydes. These compounds play a significant role in food aroma, which may result sometimes lead into unpleasant oxidized odours. In contrast, alcohol have a lesser impact on flavour, unless they are present in substantial amounts. Ketones represents a smaller portion of aroma and result from the thermal degradation and oxidation of polyunsaturated fatty acids. They are thought to contribute to the floral and fruity aroma of seafood products (Tremblay et al., 2020).

In the industrial processing of seafood, aromatic compounds undergo various transformations, including oxidation processes through chemical or microbial pathways. These reactions produce sulfur-containing volatile compounds that can shift the overall seafood odour from pleasant to undesirable (Varlet & Fernandez, 2010).

Numerous techniques have been applied for the recovery of aromas from seafood wastewater (Cristóvão et al., 2015). This can be classed into membrane processes (UF, RO, NF), biological and physicochemical methods (centrifugation, coagulation/flocculation, dissolved air flotation and activated carbon adsorption) (Murali et al., 2021). Recently, integrated techniques which involves the combination of membrane processes and conventional separation technologies have been applied in obtaining high quality flavours and aroma (Venugopal & Sasidharan, 2021). In a study carried out by Vandanjon et al. (2002), RO was efficient in the recovery of all aromatic compounds from tuna, shrimps, and buckies wastewater as compared to NF. However, UF pretreatment was required to maintain acceptable RO permeation fluxes. In another study, Cros et al. (2006) used a membrane process of RO and electrodialysis in the recovery of aromatic compounds from shrimp cooking wastewater. The process was deemed to be economically. Flavour and aroma were recovered from tuna cooking wastewater with a 2-step process of MF pre-treatment before NF concentration allows factor of 5 with good permeation fluxes (Walha et al., 2011).

4.7 *Agricultural Industry Applications*

4.7.1 Bioenergy Production from Seafood Effluent

There is a shortage of energy globally, due to depletion in fossil fuel reserves and this has necessitated the need for cleaner and renewable energy sources. Recent studies have shown that fish waste and algae can be easily processed to obtain bio-oil or syngas as precursors in synthetic fuel production (Alvarado-Ramírez et al., 2023). Fishery waste is a great potential source for good quality fish oil, which can be used for the production of biofuels and bio-derivatives (Yuvaraj et al., 2019). Biofuel or biogas is a derivative of fish waste or oil, and it is produced majorly through anaerobic digestion (Arvanitoyannis & Tserkezou, 2014). The composition of biogas is CO_2 (25–50%), CH_4 (50–75%), H_2S, H_2 and NH_3 (1–2% each) and negligible quantities of oxygen and nitrogen (R). Methane is the major component of biogas, and it has a calorific value of 39,800 kJ/m^3. Treated seafood effluent sludge can be a potential source of biogas. Biogas system for seafood effluents processing plants consists of digester which is connected to a sedimentation column with an aerobic filter and a zeolite column. Total nitrogen and wastewater contaminants are reduced by the zeolite column, while the digester reduced the volatile, soluble, suspended, and total solids (Jayasinghe & Hawboldt, 2013).

Kato et al. (2004) investigated fish waste oil potential for the generation of biodiesel fuel. Fish oil was pretreated by filtration and subjected to ozonation (8000 ppm, 5 g/h, 16 g/m^3) and mixed in the presence of calcium phosphate and iron oxide for 1 h at room temperature. The oil obtained had higher density and heating values, pour and lower flash points when compared to the properties of commercial oils. They further observed that there was low soot, no SO_4 and CO_2 emissions. Jayasinghe and Hawboldt (2013) in a study analysed the feasibility of using oil derived from fish processing waste as a biofuel for on-site application in lower grade fuel. The authors used effluents from farmed Atlantic salmon viscera, and a modified fishmeal process recovered 17.8 wt.% oil from the effluent, with high monounsaturated fatty acids in lipids. The viscosity of the oil was (29 cP), specific heat capacity (1.72 kJ/kg °C) and cold flow properties similar to No. 6 fuel oil and marine vessel fuels, and low sulfur content. In a study by Paone et al. (2021), biobased limonene was used to extract fish oil rich in omega-3, vitamin D3, and zeaxanthin from anchovy fillet leftovers and used the solid residue (anchovy sludge) as a starting substrate for the production of biogas by anaerobic digestion. The anchovy sludge showed a good methane yield, proving to be an ideal substrate for co-digestion along with other carbon-rich wastes and residues. Analysis of the results demonstrate that anchovy leftovers after the fish oil extraction process can be efficiently used as a starting co-substrate to produce biogas in a modern biorefinery.

4.7.2 Fertilizer from Seafood Effluent

Seafood effluent, rich in nutrients and organic matter, offers a valuable resource for liquid fertilizer and soil enhancement in agriculture (Ahuja et al., 2020). Fish waste can be transformed into fertilizer using a green, simple, and effective process that comprehensively utilizes shrimp shell waste. Hu et al. (2020) developed a green and efficient process that effectively transforms fish waste into fertilizer using shrimp shell waste. These fish-based fertilizers facilitate nutrient recycling from the sea to terrestrial ecosystems (Ahuja et al., 2020). However, caution should be exercised in the application of these fertilizers due to the potential presence of disease pathogens (Guimarães et al., 2018). Treated wastewater can also be used for irrigation to provide essential nutrients to crops and enhance soil fertility, reducing the need for synthetic fertilizers and closing the nutrient cycle.

Udoh and Iren (2016) assessed the suitability of fishpond wastewater as a fertilizer in the cultivation of a longer season crop. An analysis of the results revealed that soil treated with fishpond effluent had higher levels of soil organic matter, nitrogen, phosphorus, calcium, and bulk density. It also showed that the performance of the fertilizers was comparable to pig and poultry manure. In a study by Kim (2011), a scaled-up conversion process of fish waste to liquid fertilizer was carried out using a 5 L ribbon-type reactor with mixed microorganisms, resulting in a degradation rate of 3.61 g/h performed using mixed microorganisms that degraded approximately 430 g of fish waste. The biodegraded broth of fish waste possessed comparable fertilizing ability to commercial fertilizers in hydroponic culture. The scaled-up production achieved a more satisfactory fish waste degradation rate than the flask-scale production, and the biodegraded broth did not undergo putrefaction for 6 months due to the addition of 1% lactate.

The valorisation of byproducts and bioactive compounds from seafood processing wastewater promotes the efficient use of resources, reduces waste generation, and creates economic opportunities. It aligns with the principles of circular economy by transforming waste into valuable products, fostering sustainability, and minimizing environmental impacts.

4.7.3 Feed from Seafood Effluent

Proteins have been recovered from seafood wastewater sludge through sludge disintegration, followed by ultra-sonication, precipitation and drying of soluble proteins (Hwang et al., 2008). Gao et al. (2020) compared six methods for extracting proteins from excess activated sludge: acid-thermal, alkaline-thermal, single enzyme, composite enzyme, ultrasonic assisted enzyme, and thermal assisted enzyme. The chemical methods had a higher protein extraction rate of over 75% compared to the enzymatic methods. Ultrasonic assisted enzyme had a significantly higher RP compared to enzymatic methods, with an average increase of over 39%. The proteins extracted by ultrasonic assisted enzyme had a high content of amino acids, making them suitable for animal feed. The sludge dewatering performance was significantly improved by all six hydrolysis methods. Hui et al. (2022) extracted proteins from waste activated sludge (WAS) using thermal alkaline treatment (TAH) and the extraction rate of protein was optimised. Analysis of the result obtained showed that the extracted protein from the WAS had a relative molecular weight of approximately 6.5–16 kDa and have potential as animal feed.

4.8 Conclusions

Seafood processing industries produce significant wastewater with high organic content, varying physicochemical properties, and nutrient levels. Discharging untreated or improperly treated wastewater poses environmental and health risks, including water pollution, ecological impacts, and public health threats. To ensure responsible wastewater management, regulatory frameworks and compliance measures are vital, encompassing effluent standards, discharge limits, and environmental impact assessments. While conventional treatment methods are common, they may not meet stringent water quality standards. Thus, advanced treatment technologies such as membrane filtration, advanced oxidation processes, and electrochemical treatment show promise for pollutant removal and wastewater treatment. Embracing resource recovery and circular economy approaches, including nutrient reuse, bioenergy generation, and by-product valorisation, promotes sustainability and value-added product creation. Industry collaboration and stakeholder engagement through partnerships, knowledge-sharing platforms, and joint research efforts are critical for advancing sustainable practices, fostering innovation, and addressing shared challenges in wastewater management.

Acknowledgements The authors acknowledge Vaal University of Technology Research Directorate and the South Africa National Research Foundation (NRF) for Provision of Resources and Funding (Grant number: 129632), respectively.

References

Abomohra, A.E.-F., Elsayed, M., Esakkimuthu, S., El-Sheekh, M., & Hanelt, D. (2020). Potential of fat, oil and grease (FOG) for biodiesel production: A critical review on the recent progress and future perspectives. *Progress in Energy and Combustion Science, 81*, Article 100868.

Abuhasel, K., Kchaou, M., Alquraish, M., Munusamy, Y., & Jeng, Y. T. (2021). Oily wastewater treatment: Overview of conventional and modern methods, challenges, and future opportunities. *Water, 13*, 980.

Achour, M., Khelifi, O., Bouazizi, I., & Hamdi, M. (2000). Design of an integrated bioprocess for the treatment of tuna processing liquid effluents. *Process Biochemistry, 35*, 1013–1017.

Action, S. I. (2020). World fisheries and aquaculture. *Food and Agriculture Organization*, 1–244.

Adeleke, B., Robertson-Andersson, D., Moodley, G., & Taylor, S. (2020). Aquaculture in Africa: A comparative review of Egypt, Nigeria, and Uganda vis-a-vis South Africa. *Reviews in Fisheries Science & Aquaculture, 29*, 167–197.

Ahmad, A., Abdullah, S. R. S., Hasan, H. A., Othman, A. R., & Ismail, N. I. (2021). Aquaculture industry: Supply and demand, best practices, effluent and its current issues and treatment technology. *Journal of Environmental Management, 287*, Article 112271.

Ahmad, A. L., Chin, J. Y., Harun, M. H. Z. M., & Low, S. C. (2022). Environmental impacts and imperative technologies towards sustainable treatment of aquaculture wastewater: A review. *Journal of Water Process Engineering, 46*, Article 102553.

AHMED, F. E. (1991). Occurrence of chemical contaminants in seafood and variability of contaminant levels. *Seafood Safety*. National Academies Press (US).

Ahmed, W., Smith, W. J., Sirikanchana, K., Kitajima, M., Bivins, A., & Simpson, S. L. (2023). Influence of membrane pore-size on the recovery of endogenous viruses from wastewater using an adsorption-extraction method. *Journal of Virological Methods, 317*, Article 114732.

Ahuja, I., Dauksas, E., Remme, J. F., Richardsen, R., & Løes, A.-K. (2020). Fish and fish waste-based fertilizers in organic farming–With status in Norway: A review. *Waste Management, 115*, 95–112.

Ali, A., Nymann, M. C., Christensen, M. L., & Quist-Jensen, C. A. (2020). Industrial wastewater treatment by nanofiltration—a case study on the anodizing industry. *Membranes, 10*, 85.

Alvarado-Ramírez, L., Santiesteban-Romero, B., Poss, G., Sosa-Hernández, J. E., Iqbal, H., Parra-Saldívar, R., Bonaccorso, A. D., & Melchor-Martínez, E. M. (2023). Sustainable production of biofuels and bioderivatives from aquaculture and marine waste. *Frontiers in Chemical Engineering, 4*, 1072761.

Alzahrani, S., & Mohammad, A. W. (2014). Challenges and trends in membrane technology implementation for produced water treatment: A review. *Journal of Water Process Engineering, 4*, 107–133.

Anh, H. T. H., Shahsavari, E., Bott, N. J., & Ball, A. S. (2021a). Bioaugmentation of seafood processing wastewater enhances the removal of inorganic nitrogen and chemical oxygen demand. *Aquaculture, 542*, Article 736818.

Anh, H. T. H., Shahsavari, E., Bott, N. J., & Ball, A. S. (2021b). Options for improved treatment of saline wastewater from fish and shellfish processing. *Frontiers in Environmental Science, 9*, Article 689580.

Arvanitoyannis, I. S. & Tserkezou, P. (2014). Fish waste management. *Seafood processing: Technology, Quality and Safety*, 263–309.

Austin, B. (1998). The effects of pollution on fish health. *Journal of applied microbiology, 85*, 234S–242S.

Ayles, B., Porta, L., & Clarke, R. M. (2016). Development of an integrated fisheries co-management framework for new and emerging commercial fisheries in the Canadian Beaufort Sea. *Marine Policy, 72*, 246–254.

Bashir, I., Lone, F. A., Bhat, R. A., Mir, S. A., Dar, Z. A. & Dar, S. A. (2020). Concerns and threats of contamination on aquatic ecosystems. *Bioremediation and Biotechnology: Sustainable Approaches to Pollution Degradation*, 1–26.

Benhabiles, M., Abdi, N., Drouiche, N., Lounici, H., Pauss, A., Goosen, M. F., & Mameri, N. (2013). Protein recovery by ultrafiltration during isolation of chitin from shrimp shells Parapenaeus longirostris. *Food Hydrocolloids, 32*, 28–34.

Bhagwat, V. R. (2019). Safety of water used in food production. *Food safety and human health*. Elsevier.

Bourseau, P., Massé, A., Cros, S., Vandanjon, L., & Jaouen, P. (2014). Recovery of aroma compounds from seafood cooking juices by membrane processes. *Journal of Food Engineering, 128*, 157–166.

Carawan, R., Green, D., Thomas, F., & Thomas, S. (1986). *Reduction in waste load from a seafood processing plant* (p. 27695). North Carolina State University, Raleigh, NC.
Carawan, R. E. (1991). *Processing Plant Waste Management Guidelines: Aquatic Fishery Products*, NC Pollution Prevention Program.
Casé, M., Leça, E. E., Leitão, S. N., Sant, E. E., Schwamborn, R. & De Moraes Junior, A. T. (2008). Plankton community as an indicator of water quality in tropical shrimp culture ponds. *Marine Pollution Bulletin, 56*, 1343–1352.
Chahal, C., van den Akker, B., Young, F., Franco, C., Blackbeard, J., & Monis, P. (2016). Pathogen and particle associations in wastewater: Significance and implications for treatment and disinfection processes. *Advances in Applied Microbiology, 97*, 63–119.
Cheremisinoff, P. N. (2019). *Handbook of water and wastewater treatment technology*, Routledge.
Chi, Y.-L., Guo, L.-F., Xu, Y., Liu, J.-W., Xu, W., & Zhao, H.-Z. (2018). Rapid removal of bound water from dredged sediments using novel hybrid coagulants. *Separation and Purification Technology, 205*, 169–175.
Ching, Y. C., & Redzwan, G. (2017). Biological treatment of fish processing saline wastewater for reuse as liquid fertilizer. *Sustainability, 9*, 1062.
Choudhury, A., Lepine, C., Witarsa, F., & Good, C. (2022). Anaerobic digestion challenges and resource recovery opportunities from land-based aquaculture waste and seafood processing byproducts: A review. *Bioresource Technology, 354*, Article 127144.
Chowdhury, P., Viraraghavan, T., & Srinivasan, A. (2010). Biological treatment processes for fish processing wastewater—A review. *Bioresource Technology, 101*, 439–449.
Chung, M. M. S., Bao, Y., Domingo, J. A. V., & Huang, J.-Y. (2023). Enhancing cleaning of micro-filtration membranes fouled by food oily wastewater using microbubbles. *Food and Bioproducts Processing, 138*, 53–59.
Cristóvão, R. O., Botelho, C. M., Martins, R. J., Loureiro, J. M., & Boaventura, R. A. (2015). Fish canning industry wastewater treatment for water reuse—A case study. *Journal of Cleaner Production, 87*, 603–612.
Cros, S., Lignot, B., Jaouen, P., & Bourseau, P. (2006). Technical and economical evaluation of an integrated membrane process capable both to produce an aroma concentrate and to reject clean water from shrimp cooking juices. *Journal of Food Engineering, 77*, 697–707.
Directive, W. F. (2000). EU water framework directive. *EC Directive, 60.*
Englande, J. R., A., Krenkel, P. & Shamas, J. (2015). Wastewater treatment & water reclamation. *Reference Module in Earth Systems and Environmental Sciences.*
Forghani, B., Bordes, R., Ström, A., & Undeland, I. (2020). Recovery of a protein-rich biomass from shrimp (Pandalus borealis) boiling water: A colloidal study. *Food Chemistry, 302*, Article 125299.
Forghani, B., Sørensen, A.-D.M., Sloth, J. J., & Undeland, I. (2023). Liquid side streams from mussel and herring processing as sources of potential income. *ACS Omega, 8*, 8355–8365.
Gall, K., Scott, V. N., Collette, R., Jahncke, M. L., Hicks, D. & Wiedmann, M. (2004). Implementing targeted good manufacturing practices and sanitation procedures to minimize Listeria contamination of smoked seafood products.
Gao, J., Weng, W., Yan, Y., Wang, Y., & Wang, Q. (2020). Comparison of protein extraction methods from excess activated sludge. *Chemosphere, 249*, Article 126107.
Gärde, J. (2011). Treatment of waste water from a fish conservation industry: Flocculation experiment and analysis.
Ghaly, A., Ramakrishnan, V., Brooks, M., Budge, S., & Dave, D. (2013). Fish processing wastes as a potential source of proteins. Amino acids and oils: A critical review. *J. Microb. Biochem. Technol, 5*, 107–129.
Gholikandi, G. B., Dehghanifard, E., Sepehr, M. N., Torabian, A., Moalej, S., Dehnavi, A., Yari, A., & Asgari, A. (2012). Performance evaluation of different filter media in turbidity removal from water by application of modified qualitative indices. *Iranian Journal of Public Health, 41*, 87.

Goh, P., Wong, K., & Ismail, A. (2022). Membrane technology: A versatile tool for saline wastewater treatment and resource recovery. *Desalination, 521*, Article 115377.

Goh, P. S., Ismail, A. F., Ng, B. C., & Abdullah, M. S. (2019). Recent progresses of forward osmosis membranes formulation and design for wastewater treatment. *Water, 11*, 2043.

Gringer, N., Hosseini, S. V., Svendsen, T., Undeland, I., Christensen, M. L., & Baron, C. P. (2015). Recovery of biomolecules from marinated herring (Clupea harengus) brine using ultrafiltration through ceramic membranes. *LWT-Food Science and Technology, 63*, 423–429.

Guerrero, L., Omil, F., Mendez, R., & Lema, J. (1998). Protein recovery during the overall treatment of wastewaters from fish-meal factories. *Bioresource Technology, 63*, 221–229.

Guimarães, J. T., Souza, A. L., Brígida, A. I. S., Furtado, A. A., Chicrala, P. C., Santos, V. R., Alves, R. R., Luiz, D. B., & Mesquita, E. F. (2018). Quantification and characterization of effluents from the seafood processing industry aiming at water reuse: A pilot study. *Journal of Water Process Engineering, 26*, 138–145.

Hayes, M., & Gallagher, M. (2019). Processing and recovery of valuable components from pelagic blood-water waste streams: A review and recommendations. *Journal of Cleaner Production, 215*, 410–422.

Henchion, M., Hayes, M., Mullen, A. M., Fenelon, M., & Tiwari, B. (2017). Future protein supply and demand: Strategies and factors influencing a sustainable equilibrium. *Foods, 6*, 53.

Hu, X., Tian, Z., Li, X., Wang, S., Pei, H., Sun, H., & Zhang, Z. (2020). Green, simple, and effective process for the comprehensive utilization of shrimp shell waste. *ACS Omega, 5*, 19227–19235.

Hui, W., Zhou, J., & Jin, R. (2022). Proteins recovery from waste activated sludge by thermal alkaline treatment. *Journal of Environmental Chemical Engineering, 10*, Article 107311.

HUSS, H. (1994). Cleaning and sanitation in seafood processing. *Assurance of seafood quality*. The Food and Agriculture Organization, United Nations Rome.

Hwang, J., Zhang, L., Seo, S., Lee, Y.-W., & Jahng, D. (2008). Protein recovery from excess sludge for its use as animal feed. *Bioresource Technology, 99*, 8949–8954.

Iber, B. T. & Kasan, N. A. (2021). Recent advances in Shrimp aquaculture wastewater management. *Heliyon, 7*.

Islam, M. S., Khan, S., & Tanaka, M. (2004). Waste loading in shrimp and fish processing effluents: Potential source of hazards to the coastal and nearshore environments. *Marine Pollution Bulletin, 49*, 103–110.

Jamal, M. T., Pugazhendi, A., & Jeyakumar, R. B. (2020). Application of halophiles in air cathode MFC for seafood industrial wastewater treatment and energy production under high saline condition. *Environmental Technology & Innovation, 20*, Article 101119.

Jamieson, B. L., Gagnon, G. A., & Gonçalves, A. A. (2017). Physicochemical characterization of Atlantic Canadian seafood processing plant effluent. *Marine Pollution Bulletin, 116*, 137–142.

Jayasinghe, P., & Hawboldt, K. (2013). Biofuels from fish processing plant effluents–waste characterization and oil extraction and quality. *Sustainable Energy Technologies and Assessments, 4*, 36–44.

Jemli, M., Karray, F., Feki, F., Loukil, S., Mhiri, N., Aloui, F., & Sayadi, S. (2015). Biological treatment of fish processing wastewater: A case study from Sfax City (Southeastern Tunisia). *Journal of Environmental Sciences, 30*, 102–112.

Jolly, C. M., Nyandat, B., Yang, Z., Ridler, N., Matias, F., Zhang, Z., Murekezi, P. & Menezes, A. (2023). Dynamics of aquaculture governance. *Journal of the World Aquaculture Society*.

Kato, S., Kunisawa, K., Kojima, T., & Murakami, S. (2004). Evaluation of ozone treated fish waste oil as a fuel for transportation. *Journal of Chemical Engineering of Japan, 37*, 863–870.

Kesar, S. & Bhatti, M. S. (2022). Chlorination of secondary treated wastewater with sodium hypochlorite (NaOCl): An effective single alternate to other disinfectants. *Heliyon, 8*.

Kim, J. K. (2011). Scaled-up bioconversion of fish waste to liquid fertilizer using a 5 L ribbon-type reactor. *Journal of environmental management, 92*, 2441–2446

Korenak, J., Basu, S., Balakrishnan, M., Hélix-Nielsen, C., & Petrinic, I. (2017). Forward osmosis in wastewater treatment processes. *Acta Chimica Slovenica, 64*, 83–94.

Kuca, M., & Szaniawska, D. (2009). Application of microfiltration and ceramic membranes for treatment of salted aqueous effluents from fish processing. *Desalination, 241*, 227–235.

Kyllönen, H., Heikkinen, J., Järvelä, E., Sorsamäki, L., Siipola, V., & Grönroos, A. (2021). Wastewater purification with nutrient and carbon recovery in a mobile resource container. *Membranes, 11*, 975.

Laly, S., Murthy, L. & Asohk Kumar, K. (2019). Fish industry waste: Treatments and environmental impacts. Mumbai Research Centre of ICAR: Central Institute of Fisheries Technology.

Lee, C. S., Robinson, J., & Chong, M. F. (2014). A review on application of flocculants in wastewater treatment. *Process Safety and Environmental Protection, 92*, 489–508.

Liney, K. E., Hagger, J. A., Tyler, C. R., Depledge, M. H., Galloway, T. S., & Jobling, S. (2006). Health effects in fish of long-term exposure to effluents from wastewater treatment works. *Environmental Health Perspectives, 114*, 81–89.

Liu, C., Xia, W., & Park, J. W. (2007). A wind-driven reverse osmosis system for aquaculture wastewater reuse and nutrient recovery. *Desalination, 202*, 24–30.

Lutchmiah, K., Verliefde, A., Roest, K., Rietveld, L. C., & Cornelissen, E. (2014). Forward osmosis for application in wastewater treatment: A review. *Water Research, 58*, 179–197.

Macfadyen, G., Nasr-Alla, A. M., Al-Kenawy, D., Fathi, M., Hebicha, H., Diab, A. M., Hussein, S. M., Abou-Zeid, R. M., & El-Naggar, G. (2012). Value-chain analysis—An assessment methodology to estimate Egyptian aquaculture sector performance. *Aquaculture, 362*, 18–27.

Mannacharaju, M., Kannan Villalan, A., Shenbagam, B., Karmegam, P. M., Natarajan, P., Somasundaram, S., Arumugam, G. & Ganesan, S. (2020). Towards sustainable system configuration for the treatment of fish processing wastewater using bioreactors. *Environmental Science and Pollution Research, 27*, 353–365.

Martínez, R., Sanz, M. T., & Beltrán, S. (2014). Pervaporation investigation of recovery of volatile compounds from brown crab boiling juice. *Food Science and Technology International, 20*, 511–526.

Matei, E., Predescu, A. M., Şăulean, A. A., Râpă, M., Sohaciu, M. G., Coman, G., Berbecaru, A.-C., Predescu, C., Vâju, D., & Vlad, G. (2022). Ferrous industrial wastes—Valuable resources for water and wastewater decontamination. *International Journal of Environmental Research and Public Health, 19*, 13951.

MAYS, L. W. (2013). A brief history of water filtration/sedimentation. *Water Science and Technology: Water Supply,* 13**,** 735–742

Mobo, B. H. P., Rabinowitz, P. M., Conti, L. A. & Taiwo, O. A. (2010). Occupational health of animal workers. *Human-Animal Medicine, 343.*

Mota, V. C., Brenne, H., Kojen, M., Marhaug, K. R., & Jakobsen, M. E. (2022). Evaluation of an ultrafiltration membrane for the removal of fish viruses and bacteria in aquaculture water. *Frontiers in Marine Science, 9*, 1037017.

Murali, S., Krishnan, V. S., Amulya, P., Alfiya, P., Delfiya, D. A., & Samuel, M. P. (2021). Energy and water consumption pattern in seafood processing industries and its optimization methodologies. *Cleaner Engineering and Technology, 4*, Article 100242.

Muro, C., Riera, F. & Del Carmen Díaz, M. (2012). Membrane separation process in wastewater treatment of food industry. *Food Industrial Processes–Methods and Equipment***,** 253–280.

Najafpour, G., Zinatizadeh, A., & Lee, L. (2006). Performance of a three-stage aerobic RBC reactor in food canning wastewater treatment. *Biochemical Engineering jJournal, 30*, 297–302.

Nookwam, K., Cheirsilp, B., Maneechote, W., Boonsawang, P., & Sukkasem, C. (2022). Microbial fuel cells with photosynthetic-cathodic chamber in vertical cascade for integrated bioelectricity, biodiesel feedstock production and wastewater treatment. *Bioresource Technology, 346*, Article 126559.

Obotey Ezugbe, E. & Rathilal, S. (2020). Membrane technologies in wastewater treatment: A review. *Membranes, 10***,** 89.

Paone, E., Fazzino, F., Pizzone, D. M., Scurria, A., Pagliaro, M., Ciriminna, R., & Calabrò, P. S. (2021). Towards the anchovy biorefinery: Biogas production from anchovy processing waste after fish oil extraction with biobased limonene. *Sustainability, 13*, 2428.

Parvathy, U., Jeyakumari, K. R. A. & Zynudheen, A. (2017). Biological treatment systems for fish processing wastewater–A Review.

Pérez-Santín, E., Calvo, M., López-Caballero, M., Montero, P., & Gómez-Guillén, M. (2013). Compositional properties and bioactive potential of waste material from shrimp cooking juice. *LWT-Food Science and Technology, 54*, 87–94.

Prokkola, H., Nurmesniemi, E.-T., & Lassi, U. (2020). Removal of metals by sulphide precipitation using na2s and hs−-solution. *ChemEngineering, 4*, 51.

Rashidi, S., & Moghaddam, A. H. (2021). Investigation and optimization of anaerobic system for treatment of seafood processing wastewater. *Chemical Papers, 75*, 4649–4660.

Razzak, S. A., Faruque, M. O., Alsheikh, Z., Alsheikhmohamad, L., Alkuroud, D., Alfayez, A., Hossain, S. Z., & Hossain, M. M. (2022). A comprehensive review on conventional and biological-driven heavy metals removal from industrial wastewater. *Environmental Advances, 7*, Article 100168.

Sar, T., Ferreira, J. A., & Taherzadeh, M. J. (2021). Conversion of fish processing wastewater into fish feed ingredients through submerged cultivation of Aspergillus oryzae. *Systems Microbiology and Biomanufacturing, 1*, 100–110.

Show, K.-Y. (2008). Seafood wastewater treatment. *Handbook of water and energy management in food processing*. Elsevier.

Sillapacharoenkul, B., & Sinbuathong, N. (2020). Anaerobic biological treatment of frozen seafood wastewater. *Environmental Progress & Sustainable Energy, 39*, Article e13418.

Søtoft, L. F., Lizarazu, J. M., Parjikolaei, B. R., Karring, H., & Christensen, K. V. (2015). Membrane fractionation of herring marinade for separation and recovery of fats, proteins, amino acids, salt, acetic acid and water. *Journal of Food Engineering, 158*, 39–47.

Sridang, P. C., Pottier, A., Wisniewski, C., & Grasmick, A. (2008). Performance and microbial surveying in submerged membrane bioreactor for seafood processing wastewater treatment. *Journal of Membrane Science, 317*, 43–49.

Stepnowski, P., Olafsson, G., Helgason, H., & Jastorff, B. (2004). Recovery of astaxanthin from seafood wastewater utilizing fish scales waste. *Chemosphere, 54*, 413–417.

Tay, J.-H., Show, K.-Y., & Hung, Y.-T. (2004). *Seafood processing wastewater treatment*. CRC Press.

Teh, C. Y., Budiman, P. M., Shak, K. P. Y., & Wu, T. Y. (2016). Recent advancement of coagulation–flocculation and its application in wastewater treatment. *Industrial & Engineering Chemistry Research, 55*, 4363–4389.

Thongsai, A., Phuttaro, C., Saritpongteeraka, K., Charnnok, B., Bae, J., Noophan, P. L., & Chaiprapat, S. (2022). Efficacy of anaerobic membrane bioreactor under intermittent liquid circulation and its potential energy saving against a conventional activated sludge for industrial wastewater treatment. *Energy, 244*, Article 122556.

Tom, A. P., Jayakumar, J. S., Biju, M., Somarajan, J., & Ibrahim, M. A. (2021). Aquaculture wastewater treatment technologies and their sustainability: A review. *Energy Nexus, 4*, Article 100022.

Tremblay, A., Corcuff, R., Goulet, C., Godefroy, S. B., Doyen, A., & Beaulieu, L. (2020). Valorization of snow crab (Chionoecetes opilio) cooking effluents for food applications. *Journal of the Science of Food and Agriculture, 100*, 384–393.

Udoh, D. J. & Iren, O. B. (2016). Evaluation of fish pond effluent as an organic fertilizer in comparison with poultry and pig manures in the cultivation of Pineapple (Ananas comosus). *International Organization of Scientific Research-Journal of Agriculture and Veterinary Science (IOSR-JAVS), 9*, 5–10.

Uttamangkabovorn, M., Prasertsan, P., & Kittikun, A. H. (2005). Water conservation in canned tuna (pet food) plant in Thailand. *Journal of Cleaner Production, 13*, 547–555.

Vandanjon, L., Cros, S., Jaouen, P., Quéméneur, F., & Bourseau, P. (2002). Recovery by nanofiltration and reverse osmosis of marine flavours from seafood cooking waters. *Desalination, 144*, 379–385.

Varlet, V., & Fernandez, X. (2010). Sulfur-containing volatile compounds in seafood: Occurrence, odorant properties and mechanisms of formation. *Food Science and Technology International, 16*, 463–503.

Venugopal, V. (2008). *Marine products for healthcare: functional and bioactive nutraceutical compounds from the ocean*, CRC press.

Venugopal, V. (2021). Valorization of seafood processing discards: Bioconversion and bio-refinery approaches. *Frontiers in Sustainable Food Systems, 5*, 611835.

Venugopal, V., & Sasidharan, A. (2021). Seafood industry effluents: Environmental hazards, treatment and resource recovery. *Journal of Environmental Chemical Engineering, 9*, Article 104758.

Wahiduzzaman, M., Islam, M. M., Sikder, A. H. F. & Parveen, Z. 2021. Bioaccumulation and heavy metal contamination in fish species of the Dhaleswari River of Bangladesh and related human health implications. *Biological Trace Element Research*, 1–13.

Walha, K., Amar, R. B., Massé, A., Bourseau, P., Cardinal, M., Cornet, J., Prost, C., & Jaouen, P. (2011). Aromas potentiality of tuna cooking juice concentrated by nanofiltration. *LWT-Food Science and Technology, 44*, 153–157.

Yan, P., Ji, F., Fan, J., & Zong, S.-A. (2014). Grit separation module performance and influencing factors for grit removal efficiency from activated sludge. *Separation Science and Technology, 49*, 121–129.

Yeong, W. T., Mohammad, A. W., Anuar, N. & Rahman, R. A. (2002). Potential use of nanofiltration membrane in treatment of wastewater from fish and surimi industries. *Songklanakarin Journal of Science and Technology, 24*.

Yuvaraj, D., Bharathiraja, B., Rithika, J., Dhanasree, S., Ezhilarasi, V., Lavanya, A., & Praveenkumar, R. (2019). Production of biofuels from fish wastes: An overview. *Biofuels, 10*, 301–307.

Environmental Laws, Regulations and Economic Policy Instruments for the Management of Wastewater from Intensive Rural Industries in Nigeria

Ngozi Chinwa Ole, Rasaki Stephen Daudi, Empire Nyekwere Hechime, and Emmanuel I. Unuabonah

Abstract Effective wastewater management in Nigeria's intensive rural industries—such as agricultural processing, livestock farming, aquaculture, and small-scale manufacturing—is vital for reducing environmental pollution and safeguarding public health. However, significant challenges hinder progress, including poor facility maintenance, unreliable electricity, inadequate policies, and climate change impacts such as irregular rainfall, flooding, and water scarcity. Financial constraints in the water and sanitation sectors and ineffective waste management systems exacerbate water contamination and ecosystem degradation. Urbanisation, industrialisation, and population growth further contribute to untreated wastewater discharge, while poverty and corruption undermine resources and infrastructure investments. The Nigerian government has established institutions such as the Federal Ministry of Water Resources (FMWR), the National Water Resources Institute (NWRI), and the River Basin Development Authorities (RBDAs) to oversee policy formation, training, and water supply. Additionally, agencies like the Federal Ministry of Environment

Rasaki Stephen Daudi, Empire Nyekwere Hechime, Emmanuel I. Unuabonah contributed equally to the chapter.

N. C. Ole (✉)
Christopher University, Mowe, Nigeria
e-mail: olengozi@gmail.com

N. C. Ole · R. S. Daudi · E. I. Unuabonah (Deceased)
African Centre of Excellence for Water and Environment Research (ACEWATER), Redeemer's University, Osun State, Ede, Nigeria

R. S. Daudi · E. I. Unuabonah (Deceased)
Redeemer's University, Osun State, Ede, Nigeria

E. N. Hechime
Faculty of Law, Bowen University, Osun State, Bowen, Australia

E. I. Unuabonah (Deceased)
Department of Chemical Sciences, Redeemer's University, Osun State, Ede, Nigeria

N. A. Oladoja and E. I. Unuabonah (eds.), *Strategic Management of Wastewater from Intensive Rural Industries*, Springer Water,
https://doi.org/10.1007/978-3-031-90314-4_12

(FMENV) and the National Environmental Standards and Regulations Enforcement Agency (NESREA) enforce environmental laws and wastewater discharge regulations. Despite these efforts, enforcement remains weak, and compliance with national standards is limited. This chapter highlights the need for more robust environmental policies, effective enforcement mechanisms, and improved resource allocation to address wastewater management issues. Reshaping existing laws and integrating multi-sectoral approaches are essential to overcoming challenges, ensuring sustainable wastewater management, and protecting Nigeria's ecosystems and public health.

Keywords Wastewater management · Rural industries · Environmental laws · Nigeria · Pollution control · Infrastructure · Public health

1 Introduction

Industrial wastewater, which has a complex composition and many fatal toxins with strong destructive power to the environment, is one of the major sources of environmental pollution (especially, the water environment) globally. The rapid industrial transformation over the past three decades has led to significant economic development and also brought corresponding increase in industrial wastewater dicharges.[1] In particular, the prevalence of intensive rural industries that often adopt outdated and inefficient technologies leads to the generation of large amounts of industrial wastewater, which are mostly disposed off without proper treatment.[2] From food to fashion, oil to chemicals, and mining to pharmaceuticals, industries are failing to stem the flow of wastewater into our natural environments.[3]

Water, covering about two-thirds of the earth's surface (Amaefule, et al., 2023) is among the planet's most abundant resources. Yet, access to clean, safe water remains a global challenge, especially in communities housing intensive rural industries. Water pollution and scarcity have become critical threats due to population growth, industrialisation, and inadequate wastewater management. Despite advancements, untreated wastewater continues to be discharged into the environment in many developing countries, with municipal, agricultural, and industrial sources contributing heavily to pollution. In intensive industrial communities, water quality is projected to deteriorate further unless urgent measures are taken to manage wastewater sustainably (Zahoor & Mushtaq, 2023).

The increasing global demand for safe water highlights the need for a circular economy approach to wastewater management to replace disposal-based systems. Sustainable wastewater management is essential for environmental health and public well-being. Some nations, like the U.S. and Australia, have enacted regulations to improve wastewater management and protect water resources. Although Nigeria has environmental laws that could address wastewater management issues, studies have shown that little has been achieved due to the lack of robust wastewater production and management frameworks. Studies reveal challenges like cost-intensive methods and limited public awareness about water reuse. Raising awareness, fostering investments, and developing innovative policies could ensure sustainable regional wastewater management.

Advanced countries have built strong systems to clean water and wastewater for many years. According to the WHO, these systems help protect people's health and keep the environment safe by preventing problems like dirty water, the spread of diseases, and damage to plants and animals (WHO, Retrieved January 11, 2025). This WHO study looks at how water has been cleaned in the past and the newest ways to make it safe. These methods include simple techniques like boiling and more advanced ones like using special filters, ultraviolet light, or reverse osmosis to clean water.

Meanwhile, the challenges that make it hard to treat water include poor electricity supply, lack of suitable facilities, lousy waste management, poverty, weak laws, and corruption in Nigeria (Omohwovo, 2024). Thanks to new technology, there are better ways to clean water, like using bacteria to naturally remove harmful substances (Ritchie & Roser, 2020). Although it can be expensive to set up proper water treatment systems, they are essential for keeping people healthy and protecting the environment (World Bank, 2021). It's also vital for everyone—families, businesses, and industries—to learn about waste and wastewater and how to handle it properly. Local groups and community efforts can help teach people and encourage better waste disposal (Global Water Partnership, 2019). In addition to wastewater management issues, climate change affects water, causing floods, droughts, and rising sea levels, making it harder to get clean water. Leveraging the instrumentality of the law, the Nigerian government is working on solutions, like the Climate Change Act of 2021, to deal with these problems (Climate Action Tracker, 2023).

Undoubtedly, effective laws and policies lie at the height of finding solutions to wastewater management issues and lack of access to clean water. The Nigerian government have made attempts at this by establishing institutions such as the Federal Ministry of Water Resources (FMWR), the National Water Resources Institute (NWRI), and the River Basin Development Authorities (RBDAs) to oversee policy formation, training, and water supply. Additionally, agencies like the Federal Ministry of Environment (FMENV) and the National Environmental Standards and Regulations Enforcement Agency (NESREA) enforce environmental laws and wastewater discharge regulations. Despite these efforts, enforcement remains weak, and compliance with national standards is limited. This exemplifies that scientific approaches need to be corroborated with legal frameworks to achieve desired results in this respect. To better appreciate the roles environmental laws, regulations, and economic policies could play in wastewater management in communities housing intensive rural industries, subsequent paragraphs are dedicated to clarifying key concepts and examining the potential impacts of relevant laws and policies, extant government agencies dealing with wastewater management issues.

2 Contextualising Wastewater

The U.S. Environmental Protection Agency defined wastewater as "waste material that includes industrial liquid waste and sewage waste that is collected in towns and urban areas" (Environmental Protection Agency, 2021). That is, wastewater refers to liquid substances that have been used or contaminated owing to organic and inorganic pollutants, including by-products of production such as chemicals, animal waste, and residues from agricultural processing. Consequently, wastewater has significant environmental and social impacts, including water pollution that affects drinking water quality, soil contamination that degrades soil health, public health risks from exposure to untreated wastewater, and ecosystem damage from pollutants harming aquatic life.

3 Wastewater in Intensive Rural Industries

Intensive Rural Industries refer to industries in rural areas with high activity and production (Seedat-Khan, Ogunsola, Okocha, Owoyomi, & Ramnund-Mansingh, 2023). Typical examples are livestock farming in the agricultural sector, palm oil processing and production in the food processing industry, using chemicals and dyes in the textile industry, etc.

In Nigeria, wastewater abounds, especially in communities where intensive rural industries operate. The lack of proper waste treatment facilities and reliance on outdated practices exacerbates the situation. Characterised by manure, chemical dyes, decayed food, and other agricultural, chemical, or organic waste, wastewater from these industries is frequently discharged into rivers, open fields, or drainage systems without treatment, causing environmental and health problems (Echioma, 2024).

4 The Legal Framework for Managing Wastewater from Intensive Rural Industries in Nigeria

The Organization for Economic Cooperation and Development (OECD) developed principles to improve water governance through efficient, inclusive, and effective policies. These OECD Principles on WaterGovernance, endorsed in 2015 (Ahopelto et al., 2024), focus on addressing systemic governance gaps in water management. The OECD framework recognises that many water crises stem from governance failures and fosters stakeholder dialogue and collaborative problem-solving. This initiative has demonstrated positive outcomes, such as enhanced access to clean water, improved sanitation, and strengthened water security. However, broader impacts, including food, energy, and environmental sustainability, remain insufficiently addressed, necessitating innovative frameworks.

In Nigeria, government agencies manage natural resources, including water. Over time, deregulation sought to make water access a fundamental right to meet the needs of a growing population. Unfortunately, this shift lacked an integrated water policy to ensure effective governance. The absence of such a policy has led to reliance on unregulated practices, such as artisanal drilling of boreholes, often ignoring ecological risks. Furthermore, poor stakeholder engagement in water supply planning has undermined water security and sanitation efforts.

This is not to say the Nigerian government has not made efforts to curb this issue, as there are extant laws focussed on waste and wastewater management. Notably, topical laws governing wastewater management in Nigeria include the National Environmental (Surface and Ground Water Quality Control) Regulations, 2011, regulating water quality standards and defining penalties for pollution; the Harmful Waste (Special Criminal Provisions) Act (Cap H1 LFN 2004) prohibiting hazardous waste disposal in waterways; the National Guidelines and Standards for Water Quality in Nigeria which establishes the benchmarks for potable and natural mineral water. On the authority of these laws, facilities must obtain permits to handle toxic waste, discharge effluents, or manage pollution sources. Violations result in fines, environmental restoration obligations, and even imprisonment for repeat offenders.

Furthermore, the Nigerian government has, over the years, developed institutions to ensure environmental preservation and conservation. The Federal Ministry of Water Resources (FMWR) was established in 1977, following the creation of the National Water Resources Institute (NWRI) and River Basin Development Authorities (RBDA) in 1976 (O., A., & V., 2024). The NWRI focuses on research and workforce training, while the FMWR formulates policies and provides guidance. Conversely, RBDAs supply water for domestic, industrial, and agricultural uses (O., A., & V., 2024). In 1988, Decree 58 led to the formation of the Federal Environmental Protection Agency (FEPA), tasked with setting environmental standards for water quality, effluent discharge, and air pollution (N. M. A., 2023). By 1999, FEPA transitioned into the Federal Ministry of Environment (FMENV), although enforcement of environmental laws remained weak. The National Environmental Standards and Regulations Enforcement Agency (NESREA) was established in 2007 to address this gap, focusing on biodiversity preservation, sustainable resource use, and enforcing environmental laws and treaties (Gadzama & Okeokwo, 2024).

Given its statutory obligations, the NESREA plays a pivotal role in wastewater management and regulation. And true to its mandate, the NESREA employs a "command and control" approach to regulate industrial pollution. This includes Mandating that facilities treat wastewater before discharge, enforcing effluent discharge standards based on available best technologies, and requiring facilities to install pollution control equipment and take responsibility for wastewater management until final discharge—non-compliance results in penalties such as fines, imprisonment, or facility closures. However, many facilities struggle to meet Nigeria's high wastewater treatment standards, which are stringent even by international comparisons.

At the state level, the FMWR oversees the National Council on Water Resources (NCWR), which includes representatives from the FMENV, state commissioners, and other stakeholders (Omokaro, Idama, Airueghian, & Micheal, 2024). State Ministries of Water Resources regulate water policies locally, but significant disparities exist in water management and service delivery across states. For example, the Lagos State Environmental Protection Agency (LASEPA) actively enforces laws against discharging untreated waste into public drains or water bodies, with penalties for violators (Babaniyi, Akor, & Adebomi, 2024).

In addition, the Nigerian government's climate change laws and policies are further attempts at curbing the impact of wastewater management issues on the environment. As earlier stated, climate change exacerbates water-related challenges such as erratic rainfall, rising sea levels, and severe droughts. Nigeria's National Climate Change Policy (2015) provides a strategic framework for mitigating these impacts and promoting climate resilience. The Climate Change Act of 2021 further solidifies Nigeria's commitment to net-zero carbon emissions by 2060, as announced at COP26 (Tracker, 2025). The Act emphasises inclusive green growth, sustainable development, and comprehensive legal frameworks to address climate change risks and vulnerabilities. More regional and national efforts have equally been channelled towards climate change issues, including the establishment of the National Council on Climate Change (Raimi & Lukman, 2023), which coordinates Nigeria's climate action plans; the Niger Delta Climate Conference in 2019, which mobilised stakeholders to create shared solutions for climate management; and the issuance of Nigeria's first sovereign green bond to fund environmental initiatives.

5 Conclusion

Nigeria's regulatory landscape for water and wastewater treatment is robust on paper but faces challenges in enforcement and infrastructure. Climate change further complicates water governance, underscoring the need for more muscular institutional coordination, stakeholder engagement, and innovative solutions. Through consistent policy reforms, effective enforcement, and alignment with global best practices, Nigeria can secure sustainable water resources and build resilience against climate impacts.

Acknowledgements The authors acknowledge the Redeemer's University, Nigeria, for her support.

References

Ahopelto, L., Sojamo, S., Belinskij, A., Soininen, N., & Keskinen, M. (2024). Water governance for water security: Analysing institutional strengths and challenges in Finland. *International Journal of Water Resources Development, 40*(2), 153–173.

Amaefule, E. O., Amarachi, A. S., Abuka, C. O., Nwaogazie, F. O., Nwabuko, C. N., & Divine, C. C. (2023). Water and wastewater treatment in Nigeria: Advancements, challenges, climate change and socioeconomic impacts. *Path of Science, 9*(8), 2010–2031. https://doi.org/10.22178/pos.95-4

Babaniyi, G. G., Akor, U. J., & Adebomi, J. I. (2024). Baban legislation and programs supporting envronmental protection. In B. R. Sesan Abiodun Aransiola, *Prospects for Soil Regeneration and Its Impact on Environmental Protection*. Switzerland: Springer.

Climate action tracker. (2023). *Home*. Retrieved January 17th, 2024, from https://climateactiontracker.org/

Echioma, I. A. (2024). Evaluation of water pollution in Nigeria. *Omanarp International Journal Of Law, 1*(1).

Environmental protection agency. (2021). *Wastewater*. Retrieved November 12, 2024, from Environmental Protection Agency. (nhttps://www.epa.ie/our-services/compliance--enforcement/wastewater/

Gadzama, C. L., & Okeokwo, T. (2024). Analysis of some statutory framework on environmental pollution in Nigeria. *American Journal of Public Law and Political Education, 1*(2), 37–50.

Global water partnership. (2019). *Homepage*. Retrieved November 2024, 11th, from https://www.gwp.org/

Adione, N. M. A. (2023). An environmental conservation approach to handling oil and gas in Nigeria: An assessment of legal framework. *Igwebuike Journal: An African Journal of Arts & Humanities, 9*(3), 2–6.

Okorafor, O. O., Nwaobi, A. F., & Ugwah, V. N. (2024). Optimization of water policy, regulation and technology for sustainable food production in Nigeria: A review. *Open Access Library Journal, 11*(6), 1–12.

Omohwovo, E. J. (2024). Wastewater management in Africa: challenges and recommendations. *Environmental Health Insights, 18*, 11786302241289680.

Omokaro, G. O., Idama, V., Airueghian, E. O., & Micheal, I. (2024). Water resources, pollution, integrated management and practices in Nigeria–An overview. *3*(1), 5–18.

Raimi, L., & Lukman, F. M. (2023). Rethinking sustainable development under climate change in Nigeria: A strategic analysis. In L. Raimi & M. L. Fatima (Eds.), *Corporate Resilience: Risk, Sustainability and Future Crises* (pp. 73–91). Emerald Publishing Limited.

Ritchie, H., & Roser, M. (2020). Clean water: Our World in data. Retrieved December 26, 2024, from https://www.ncbi.nlm.nih.gov/pmc/articles/PMC7098060/

Seedat-Khan, M., Ogunsola, V. I., Okocha, R. O., Owoyomi, A. V., & Ramnund-Mansingh, A. (2023). Sustainable wastewater management: An imperative for urban development in Lagos, Nigeria. *Seedat-Khan, M., Ogunsola, V. I., Okocha, R. O., Owoyomi, A. V., & Ramnund-Mansingh, A. (2023). Sustainable wastewaSustainable Development, 31*(5), Seedat-Khan, M., Ogunsola, V. I., Okocha, R. O., Owoyomi, A. V., & Ramnund-Mansingh, A. (2023). Sustainable waste3291–3302.

Tracker, C. A. (2025). *Climate Action Tracker*. Retrieved January 10th, 2025, from https://climateactiontracker.org/

WHO. (2025). *Drinking Water*. Retrieved January 10, 2025, from https://www.who.int/news-room/fact-sheets/detail/drinking-water

World Bank. (2021). *Water Overview*. Retrieved January 15, 2025, from https://www.worldbank.org/en/topic/water/overview

Zahoor, I., & Mushtaq, A. (2023). Water pollution from agricultural activities: A critical global review. *Int. J. Chem. Biochem. Sci*, 164–176.

The manufacturer's authorised representative in the EU is Springer Nature Customer Service Centre GmbH, Europaplatz 3, 69115 Heidelberg, Germany. If you have any concerns regarding our products, please contact ProductSafety@springernature.com

Printed and bound by CPI Group (UK) Ltd, Croydon, CR0 4YY
05/10/2026
02228130-0009